COMPUTERIZED ENGINE CONTROLS

SECOND EDITION

Dick H. King

Glendale Community College

With an introduction by

Chris Johanson

&

Tony Owens

 Delmar Publishers Inc.®

NOTICE TO THE READER

Introduction reproduced with permission from Johanson and Owens; ELECTRONIC ENGINE CONTROLS, copyright © 1987 by Delmar Publishers Inc.

Delmar Staff
Associate Editor: Joan Gill
Editing Manager: Barbara A. Christie
Project Editor: Christopher Chien
Design Coordinator: Susan Mathews

For information, address Delmar Publishers Inc.
2 Computer Drive West, Box 15-015
Albany, New York 12212-5015

Cover Photo
Manfred Kage/PETER ARNOLD, INC.

Book Packager
Scharff Associates, Ltd.
R D 1 Box 276
New Ringgold, PA 17960-9637

COPYRIGHT © 1989
BY DELMAR PUBLISHERS INC.

Printed in the United States of America
Published simultaneously in Canada
by Nelson Canada
A Division of International Thomson Limited

10 9 8 7 6 5 4 3 2 1

Library of Congress Cataloging in Publication Data

King, Dick H.
  Computerized engine controls/Dick H. King; with an introduction
by Chris Johanson & Tony Owens.—2nd ed.
      p.   cm.
  Includes index.
  ISBN 0-8273-3472-9 (pbk.). ISBN 0-8273-3473-7 (instructor's guide)
    1. Automobiles—Motors—Control systems. 2. Automobiles—
Electronic equipment. 3. Microprocessors. I. Title
TL214.C64K56 1989   88-30022
629.2'58—dc 19   CIP

CONTENTS

Contents

Chapter 5   General Motors' Port Fuel Injection

Chapter 6   Cadillac's Digital Fuel Injection

Chapter 7   Ford's Electronic Engine Control IV

Chapter 8   Ford's EEC I, EEC II, and EEC III

Chapter 9   Ford's Microprocessor Control Unit

Chapter 10   Chrysler's Single-point and Multipoint Fuel Injection

Chapter 11   Chrysler's Oxygen Feedback System

Chapter 12   Robert Bosch Corporation's Motronic

Appendix: 1989 Automobiles/343

Index/349

PREFACE

The application of the microprocessor with its related components and circuits has made automotive technology both exciting and fast paced. Recent technological developments and those that follow will require entry level automotive service technicians to be well trained in the principles of automotive technology and to be career-long students. Those who respond positively to this requirement will find the task challenging but achievable and rewarding.

This text was written in response to a widely recognized need within the industry: to help student-technicians get a commanding grasp of how computerized engine control systems work and how to diagnose problems within them. The student-technician who studies this text will soon come to realize that no single component or circuit within any given computerized engine control system, other than the computer itself, is complicated.

Computerized Engine Controls is written with the assumption that the reader is familiar with the principles of traditional engine, electrical system, and fuel system operation.

While it is recognized that a computerized engine control system does in fact become an integral part of an engine's electrical and fuel system, it is much too significant and complex to be taught as just a unit in an engine performance textbook or class. For purposes of instruction, this topic should be taken out of context and examined as a stand-alone system. Once the student-technician fully understands the system's purpose, operation, and diagnostic approach, the diagnostic procedures within the service manual will put the system back in its proper perspective as an integral part of the engine's support system.

Computerized Engine Controls presents each popular, multifunction, computer control system in a separate chapter. Each system is fully covered, with enough specific information and detail to enable the reader to get a complete and clear picture of how the system works. This text is written with the premise that understanding how the system works and what it should be doing not only makes the diagnosis process easier, but also makes the diagnostic literature much easier to understand. Correctly interpreting diagnostic procedure directions is often a problem if the technician is not aware of what the procedure is trying to measure, what normal readings or responses should be, and what conditions will cause abnormal readings or responses.

Frequently, information contained in a manufacturer's service manual is presented in this text to help make a specific concept more clear and to help acquaint the reader with information found in the service manual. Other references are made to the service manual to impress upon the reader the importance of using the service manual in diagnostic procedures. This text is dedicated to helping the student-technician acquire the necessary knowledge to be able to diagnose and repair driveability problems with a computerized engine control system.

Objectives are provided at the beginning of each chapter to help the reader identify the major concepts to be presented. Review questions are provided at the end of each chapter, which help the reader assess his or her recall and comprehension concerning the most important concepts. The review questions are also written to reinforce the objectives.

Personal safety concerns peculiar to specific computerized engine control systems are highlighted where applicable. Terms peculiar to computerized engine control systems are highlighted when they are first introduced and their meaning is explained in the related text. A glossary of such terms is also provided at the end of each chapter. For those chapters where a significant number of abbreviations are used, a list of abbreviations is also provided at the end. Background information that is not of specific concern to understanding a given concept but might be of general interest to the reader is inserted in selected locations throughout the text.

SUGGESTIONS ON HOW TO USE TEXT

The Introduction should be read first. This chapter provides background on the operation of many components that are common to most systems. The remaining chapters, which are specific to individual systems, can be read in any order. Many students may not wish to study each specific chapter or their instructors may not choose to assign the study of each chapter. It is suggested, however, that at least three of the specific system chapters be selected for study following the completion of the Introduction. Remaining chapters can then be skimmed or serve as a reference for future use.

ACKNOWLEDGMENTS

I am grateful to several people who have helped with the many activities involved in the creation of this book: to Therese Murphy who generously gave of her personal time to type the proposal chapter; to Margery Rothschild for her hours of work and consistently quick turnaround time in processing illustrations; to Tim Turpin for the photographs used in the book; to Renault Catalano for his consultation concerning electronics, and to the following for their critical reviews and/or answers to questions:

Dave Clark
Chrysler Motors Zone Office in Phoenix, Arizona

Richard Fleming
General Motors Training Center in Tempe, Arizona

Douglas Gossett
Jefferson State Vocational Technical School & Manpower Skill Center

Roger Megorden
Golden West College

Ed Ralbovsky
State University of New York at Canton

David Reyes and Richard Wilson
Ford Motor Co. Service School in Phoenix, Arizona

INTRODUCTION TO ELECTRICITY AND ELECTRONICS

The electrical system is a critical part of every automobile engine. To understand the electrical components in an engine's control system, you must have a basic understanding of electrical theory. This introduction provides a brief overview of the basic principles of electricity. It also gives a brief explanation of the uses of electricity in an automobile.

Semiconductor electronics is concerned with the flow of current in solid-state devices. This type of current flow provides the basis for the modern engine control systems used in automobiles. For this reason, the automotive technician should have a basic understanding of electronic theory and devices. In electricity, the flow of free electrons from atom to atom, in a conductor such as copper or aluminum, produces current. In semiconductor electronics, electrons do not flow to other atoms but move into different arrangements within materials called semiconductors.

You may wish to use this introduction as a reference and review it when you are studying the later, detailed chapters on troubleshooting control systems.

ATOMIC THEORY

To understand how electricity works, you must first understand the structure of the atom. *Atoms* are the building blocks of all materials. Every substance, including man, is made up of atoms. Atoms of the same type may be grouped together. These groups of atoms are called *elements.* There are over 100 elements, each one representing a single type of atom. Different types of atoms may also be grouped together. There are millions of these groups, called *compounds,* representing all the possible combinations between different types of atoms.

All atoms are constructed according to the same basic design:

- In the center of every atom is a nucleus.
- This nucleus is made up of positively charged substances called *protons.*
- This nucleus also contains substances with no charge, called *neutrons.* These particles are not important for our discussion of basic electricity.
- Orbiting around every nucleus are negatively charged substances called *electrons.*
- Every different type of atom has a different number of protons and electrons.
- In every atom, the number of protons equals the number of electrons. This means that the total electrical charge of the atom is zero.

Figure I-1 shows a simplified drawing of an atom. This atom of lithium has three protons and three electrons.

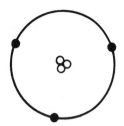

FIGURE I-1   Simple Atom

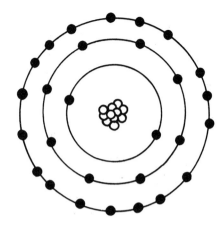

FIGURE I-3   Tightly Connected Nickel Atom

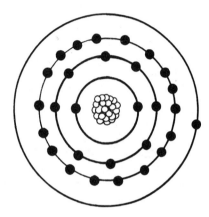

FIGURE I-2   Electron Arrangement of Copper Atom

In all atoms, the electrons are arranged in different orbits, called *shells.* Each shell contains a specific number of electrons. Figure I-2 shows a copper atom. This atom has 29 electrons and 29 protons. The 29 electrons are arranged in shells.

Notice in Figure I-2 that the outer shell has only one electron. This shell needs 32 electrons to be completely full. This means that the one electron in the outer shell is loosely tied to the atom and can be easily removed.

Figure I-3 shows a simplified drawing of a nickel atom. This atom contains 28 protons in the nucleus and 28 electrons arranged in shells.

The third shell of any atom can contain only 18 electrons. In the nickel atom, the outer shell is completely filled. Therefore, its elec-

trons are closely tied to the atom and cannot be removed easily.

The looseness or tightness of electrons is important in electrical theory. Electricity is caused by the *flow of electrons* from one atom to another. The ability to give up electrons, either easily or with difficulty, determines whether an atom is electrically active. The copper atom, with its one electron in the outer shell, is very *active* electrically. This means that it gives up electrons easily and is very useful as an electrical conductor. The nickel atom, whose electrons are tightly held, is *not active* electrically.

Note: It is also useful to know that an element that is electrically active is also chemically active. Copper tarnishes very easily (combines chemically with the oxygen in the air); nickel resists tarnishing (does not combine easily with oxygen) and is often used for plating other metals.

FLOW OF ELECTRICITY _____

Electricity is the movement of electrons from one place to another. This is why we speak of the "flow" of electricity; its flow is similar to the flow of water. However, to create this flow, there must be a source of energy. This source must be strong enough to force electrons out of their shell.

Several kinds of energy are strong enough to force electrons out of their shell and cause electrical flow. Some of these energy sources are

- Heat
- Friction
- Light
- Pressure
- Chemical reaction
- Magnetism

In an automobile, electricity is produced by the last two sources of energy, chemical reaction and magnetism.

- *Chemical Reaction.* The battery in an automobile is a source of chemical energy. The chemical reaction in the battery provides a source of electrons. As these electrons are drawn from the battery, the chemical reaction continues. This process goes on providing electrons until the chemicals in the battery are exhausted.
- *Magnetism.* Electricity and magnetism are interrelated; one can be used to produce the other. The negative charge of an electron produces a magnetic field, similar to the pole of a bar or horseshoe magnet. When there is no current flow in a conductor (such as a wire), the magnetic fields of the electrons tend to cancel each other out. However, when current flows in the wire, the electrons are all moving in the same direction, and their magnetic fields begin to work together. This creates a strong magnetic field around the wire. This field is called an *electromagnetic field,* because it is produced by the movement of electrons.

Moving a wire (a conductor) through an already existing magnetic field (such as a permanent magnet) can produce current flow (electricity). This happens because the magnetic fields of the electrons in the wire react to the already existing magnetism. This process of producing electricity through magnetism is called *induction.* In a generator, the coil of wire is moved through the magnetic field. In an alternator, the magnetic field is moved through the coil of wire. In both cases, electrical flow is produced.

For electrical flow to occur (and continue), three things must be present. These are:

1. An *excess* of electrons in one place
2. A *lack* of electrons in another place
3. A *path* between the two places

A simple example of this process is shown in the battery and light arrangement in Figure I-4.

The chemical reaction in the battery causes a lack of electrons at the positive (+) terminal and an excess of electrons at the negative (–) terminal. This creates an electrical imbalance, causing the electrons to flow through the path provided by the wire.

The path in this case is copper wire. The electron produced by the battery attaches itself to the first copper atom. Remember that the copper atom has one loosely held electron. The new electron pushes this loose electron off the other side of the copper atom, where it attaches itself to the next atom. Again the new electron pushes *that* atom's original electron off, and so on, until an electron is released at the other end of the path. Figure I-5 shows a simplified drawing of this process.

It takes much longer to describe this process than for it to occur, because electrons travel at the *speed of light.* In addition, electrons are also extremely small. Therefore, in the flow of electricity, millions of electrons are moving past a given point on a wire.

3

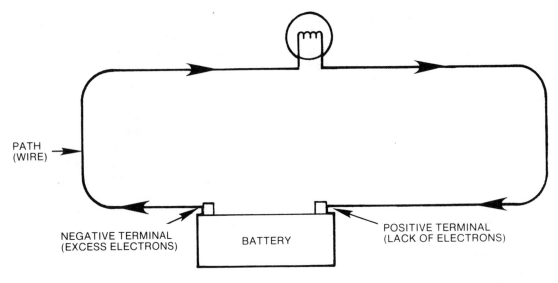

FIGURE I-4   Electrical Flow

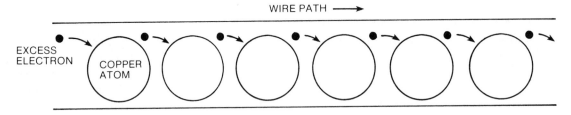

FIGURE I-5   Electron Movement

ELECTRICAL MEASUREMENT

To understand electricity more fully, you should know how it is measured. Remember that electricity is actually the flow of electrons. This flow is called *current*. This current, like the flow of water or any other substance, can be measured. When a substance flows, it meets *resistance*. The resistance to electrical flow can be measured. When a flowing substance meets resistance, *pressure* is built up. This pressure, which helps the flow to overcome resistance, can also be measured. The following section will explain the measurement of

- Flow (current)
- Resistance
- Pressure

Electrical Flow (Current)

Any measurement of flow includes a unit of time. Flow is measured as an *amount* flowing past a certain place (point) in a given time. For example, we can measure the flow of water as 100 gallons per minute.

Electrical flow, or current, is measured in the same way. Current is the number of electrons flowing past a given point in a given amount of time. The unit for measuring electrical current is the *ampere,* usually called an

4

amp. The instrument used to measure electrical current is called an *ammeter.*

In the flow of electricity, millions of electrons are moving past any given point at the speed of light. Also, the electrical charge of any one electron is extremely small. It takes millions of electrons to make a charge that can be measured. For these reasons, one amp means that 6.25 million million electrons are flowing past a given point in one second (Figure I-6).

> 1 Amp = 6,250,000,000,000,000 Electrons Flowing Past Point A in One Second

There are two different types of electrical flow, or current. These are direct current (DC) and alternating current (AC). In direct current, the electrons flow in one direction only. The example of the battery and light shown earlier is based upon direct current. In alternating current, the electrons change direction at a fixed rate. Automobile electrical systems use direct current (DC). Only direct current will be explained in this chapter.

Resistance

In every atom, the electrons *resist* being moved out of their shell. The amount of resistance to flow depends upon the type of atom. As explained earlier, in some atoms (such as those in copper) there is very little resistance to electron flow. This is because the outer electron is loosely held. In other substances (such as nickel) there is more resistance to flow, because the outer electrons are tightly held.

The resistance to flowing current produces heat. This heat can be measured to indicate the amount of resistance. The unit of resistance is called an *ohm.* Resistance can be measured by an instrument called an *ohmmeter.* Differing amounts of resistance are measured by changing scales (such as × 1, × 10, × 100) on the ohmmeter.

Pressure

In electrical flow, some force is needed to move the electrons between atoms. This force is the pressure that exists between the positive and negative points (the electrical imbalance). This force, also called *electromotive force (EMF)* is measured in units called *volts.* One volt is the amount of pressure (force) required to move one amp of current through a resistance of one ohm. Voltage is measured by an instrument called a *voltmeter.*

In any electrical flow, current, resistance, and pressure work together in a mathematical relationship. This relationship is expressed in a basic law of electricity—Ohm's law. If you know two of these three factors, you can use Ohm's law to find the third. Ohm's law can be expressed as a triangle, as shown in Figure I-7. Cover the part of the triangle that you wish to

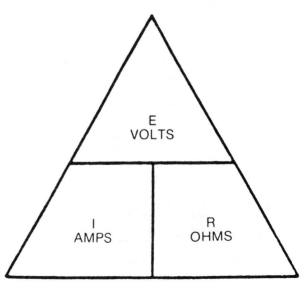

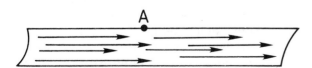

FIGURE I-6   Electron Flow

FIGURE I-7   Ohm's Law Triangle

find, and the remainder will give you the appropriate formula. For example:

$$R = \frac{E}{I}$$

The following formulas make up Ohm's law:

$$E = I \times R$$
$$I = E \div R$$
$$R = E \div I$$

Ohm's law is seldom used by the automotive technician. If you do need to use it, simply refer to the formulas above, put in the known quantities, and perform the mathematical calculations.

For example, a set of fog lights with six ohms of resistance is installed on a 12-volt automobile. How many amperes does it use to operate?

$$I \text{ (unknown)} = E \text{ (12 volts)} \div R \text{ (6 ohms)}$$
$$I = \frac{12}{6}$$
$$I = 2 \text{ amps}$$

If *ohms* are unknown:

$$R \text{ (unknown)} = E \text{ (12 volts)} \div I \text{ (2 amps)}$$
$$R = \frac{12}{2}$$
$$R = 6 \text{ ohms}$$

If *volts* are unknown:

$$E \text{ (unknown)} = I \text{ (2 amps)} \times R \text{ (6 ohms)}$$
$$E = 2 \times 6$$
$$E = 12 \text{ volts}$$

CONDUCTORS AND INSULATORS _____

To understand how electrical flow is used, you must be familiar with the principles and types of conductors and insulators.

Conductors

Conductors are materials with low resistance. This means that they are made of materials that offer very low resistance to the flow of electrons. These are materials, such as copper, whose atoms have *free electrons*—electrons held loosely in the outer shell. Because of their free electrons, these materials are excellent carriers (conductors) of electric current.

WARNING: The human body is a good conductor of electricity. Remember this when you are working with the electrical system of an automobile. Always observe *all* electrical safety rules.

Because of its low resistance to electron flow, a good conductor can also carry large amounts of current without producing excessive amounts of heat. A good conductor is also strong, lightweight, and easy to use. For all of these reasons, the materials that make the best electrical conductors are copper, aluminum, and copper-covered aluminum. Most conductors are wires made out of these materials.

- *Copper* is the most commonly used electrical conductor. It has low resistance and is lightweight. Copper wire does not break easily and does not weaken quickly.
- *Aluminum* is a common substitute for copper. It is often less expensive than copper and has low resistance, is lightweight, and is strong. However, because aluminum is not as good a conductor as copper, larger wires must be used to carry the same amount of current. Aluminum wire can also react chemically with other materials, such as brass. When this happens, resistance increases, heat is produced, connections expand and contract with the flow of current, and serious

overheating may occur. For these reasons, special instructions and precautions must be followed when aluminum wire is used as a conductor.

- *Copper-covered Aluminum* is about 10 percent copper and about 90 percent aluminum. It has the conductor properties of copper and the other advantages (cost, weight, handling) of aluminum.

Insulators

Insulators are materials with very high resistance. This means that they are made of materials that offer high resistance to the flow of electrons. These are materials, such as rubber or glass, whose atoms have tightly held electrons (whose outer shells are full or nearly full). Because these atoms have no free electrons, they do not carry electricity very well. For this reason, these materials are used as insulators.

A good insulator material should have high resistance to electron flow, should be able to withstand the heat caused by this resistance, should resist moisture and corrosion, and be able to withstand high or continued voltage.

Some good insulators with these properties include most plastics, rubber, cloth, glass, wood, and paper. In electrical applications, a good conductor should be covered by a good insulator.

Rubber is one of the most common insulator materials. It is easy to handle and resists moisture well. However, rubber will not withstand very high temperatures, so it is sometimes combined with other materials. A common type of plastic insulation is called *thermoplastic*.

Insulator Condition. Always be aware of the condition of the insulation material on any electrical wiring system. Broken, torn, or otherwise damaged insulation should be replaced. Insulation in poor condition can cause shorts and other problems. Damaged insulation can also be dangerous.

CIRCUITS

When electrons are able to flow along a path between two points, an electrical *circuit* is formed. In a complete circuit, the resistance is low enough to allow the pressure (voltage) to push the electrons (current) through the path between the two points. In a complete circuit, the flow of electricity can be controlled and applied to do useful work.

Circuit Parts

An electrical circuit may consist of several basic types of electrical hardware. A circuit includes four basic parts.

1. *Power Source* provides the energy to create electron flow (excess electrons), for example, a battery or generator.
2. *Conductors* provide a path on which the current can flow, for example, electric wiring.
3. *Load* is any device that uses the electrical flow to perform a working function, for example, a light bulb or heater blower motor.
4. *Control* is a device that directs the flow of electrons, for example, a switch.

Circuit Symbols

Symbols are used on electrical drawings to indicate circuit devices and connections. Figure I-8 shows some of these symbols.

Circuit Types

Three basic types of circuits are used in automobile wiring systems. These are:

1. Series circuits
2. Parallel circuits
3. Series-parallel circuits

Each type of circuit has different amperage, voltage, and resistance factors. These differences are explained in the following sections.

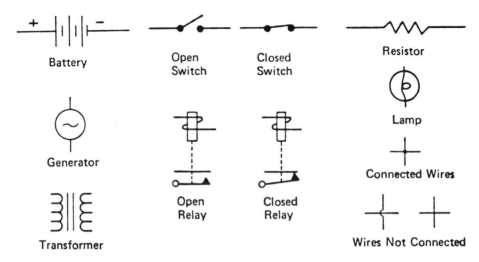

FIGURE I-8   Electrical Symbols

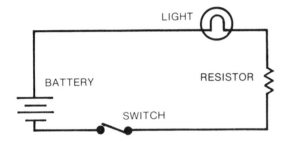

FIGURE I-9   Series Circuit

Series Circuits. In a series circuit, all the current flows through all the parts of the circuit. This means that the current is the same throughout the circuit. Current flows from the negative side of the source (the excess electrons), through each device or load, and to the positive side of the source (the lack of electrons). If the current does not flow through the load, the circuit is not complete. Figure I-9 shows a simple series circuit used to power a light.

In a complete circuit, when the current flows through the load, it encounters resistance. This causes a loss of electrical pressure, or voltage. This is called a *voltage drop.* In a complete circuit, the total voltage drop across all loads equals the total applied voltage.

It is important to remember two main characteristics of series circuits. These are:

1. The current flows (amps) are the same through each load in the circuit.
2. The total voltage drop equals the total applied voltage.

Parallel Circuits. In a parallel circuit, the loads (resistance units) are side by side and connected at the ends. This means that the current flow is split and flows through the parallel wiring. Figure I-10 shows an example.

In a parallel circuit, the current divides according to the load in each branch. Because the current is divided, the current in any one branch is always less than the total current of the circuit. Also, if the current stops in one branch, it can continue to flow in the other branch or branches.

The *voltage* in each branch of a parallel circuit is the same as the total applied voltage. The total *resistance* in a parallel circuit is always less than the lowest resistance in any one branch.

When more branches are added to a parallel circuit, the total resistance is decreased and the total current is increased.

Series-Parallel Circuits. Series circuits and parallel circuits can be combined to form series-parallel circuits. In this type of circuit, a load can be parallel with some loads and in series with others. Figure I-11 shows an example.

The current, voltage, and resistance in a series-parallel circuit depend upon the part of the circuit you are working with. The series part is figured as a series circuit, and the parallel part is figured as a parallel circuit.

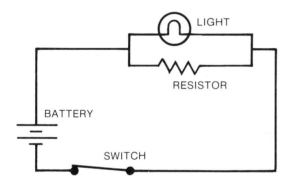

FIGURE I-10   Parallel Circuit

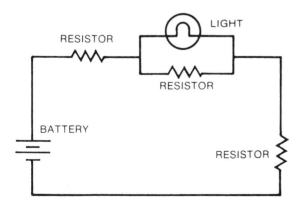

FIGURE I-11   Series-Parallel Circuit

Polarity

Current in a direct current system travels in one direction only. The direction of flow is called its *polarity*. The term comes from the two poles (terminals) of any device that produces electricity. One pole, as mentioned earlier, will have more electrons than the other pole. Electrons will flow from the pole with more electrons to the pole with less electrons. In any functioning system, the same pole will always have more electrons.

There are two theories of polarity:

1. *Conventional theory* states that current flows from the positive pole to the negative. This means that when a car's headlights are turned on, electrons leave the positive terminal of the battery and travel through the headlight switch and headlights on the way to the negative terminal of the battery. This theory was the basis of electrical design for over 100 years.

2. Later it was discovered that electrons actually travel from negative to positive. This is called the *electron theory*. According to this theory, electrons leave the battery negative terminal and travel through the headlights and headlight switch on the way to the battery positive terminal.

Most automotive diagrams continue to use the conventional theory. This does not usually cause problems since most automotive electrical equipment can be understood according to either theory. This is not the case when working with electronic components (such as diodes or transistors); they cannot operate on the wrong polarity. When working with diagrams of components that depend on proper polarity, it is important to determine which theory is being used.

Open Circuits

In an open circuit, the resistance is so high that the electrical pressure (voltage) cannot

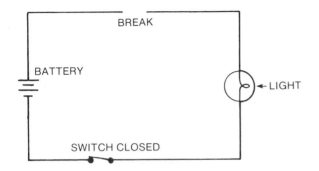

FIGURE I-12   Open Series Circuit

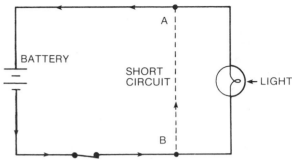

FIGURE I-13   Short Circuit

push the electrons through the circuit. An open circuit may be intentional, as when you turn off the switch to the headlights. This breaks the flow of electricity and creates an open circuit. A circuit breaker or fuse is simply a device that opens the circuit and prevents electrical flow when current drain becomes excessive.

An open circuit can also be accidental. This may be caused by a faulty electrical device, a bad connection, or bad wiring. Figure I-12 shows an open series circuit.

When a series circuit is broken as shown in Figure I-12, the current will stop flowing and the load unit, in this case a light, will not work. In a parallel or series-parallel circuit, the effect of the break depends upon where the break occurs. A break in one place may open the entire circuit. A break in one of the parallel sections may affect only one of the electrical units in the circuit.

Short Circuits

In a short circuit, the electrical current bypasses the device it is supposed to power and takes another path. This alternate path is actually a shortcut and this type of circuit is often called a *short*. A short happens because currents always follow the path of least resistance, the shortest path, to complete the circuit. A short circuit is completed, but in the wrong direction. Figure I-13 shows an example.

The purpose of insulation is to prevent short circuits. The insulation material contains the current flow along its proper path. A break in insulation can sometimes cause a short circuit. When this happens, the current flow is usually stopped by a fuse or circuit breaker. If the current were not stopped, a fire could occur or equipment could be damaged. This is the main purpose of the fuses in an automobile.

ELECTRICITY IN AUTOMOBILES

The electrical system is an important part of every automobile. It is especially important in the engine control system. The following list is an indication of some basic uses of electricity in automobiles.

1. *Lights.* The filament in an electric light bulb is a special type of high resistance wire. The filament glows from the heat caused by the resistance to current flow.

2. *Solenoids.* A solenoid is a winding (that is, many turns of wire around a central core) with a metal shaft inside of it. The shaft is free to move. When the winding has current applied to it, it creates an intense magnetic field. This field attracts the metal shaft, pulling it into the windings. The shaft

operates a mechanical device attached to it by a linkage. Examples of solenoid-operated devices in automobiles are vacuum control solenoids, starter solenoids, fuel injectors, and electric door locks.

3. *Ignition Coils.* This coil uses two windings. In the primary winding, the current flow produces a magnetic field. In the secondary winding, the collapse of the magnetic field induces current flow. The difference in the size of the windings produces a high voltage to the spark plugs.

4. *Motors.* Magnetic fields inside the motor cause its armature and output shaft to spin. Examples of electric motors in automobiles are starter motors, power seat and window motors, windshield-wiper motors, and carburetor idle speed control motors.

5. *Resistors.* Resistors offer a great deal of opposition (resistance) to the flow of current. They are used to reduce the amount of current flowing to other electrical units, protecting the units from overloading. This resistance to current produces a large amount of heat. This heat is used in turn signal flashers and other electromechanical devices. In other cases, it is unwanted and must be removed.

SUMMARY OF ELECTRICAL THEORY

Electrons flow through a conductor by jumping from atom to atom inside of the conductor material; this flow is called *current*. Electrons are pushed through a conductive circuit by electrical pressure (*voltage*), caused by the difference in the number of electrons on each side of the circuit. Direct current circuits always have a definite polarity, flowing from negative to positive. *Resistance* in the circuit is caused by the reluctance of the atoms to re-

lease their electrons. A circuit through which current can flow is a *complete circuit*. A circuit through which current cannot flow is an *open circuit*. A circuit that bypasses its intended path is a *short circuit*. The three types of circuits are *series, parallel,* and *series-parallel*.

SEMICONDUCTOR THEORY

A semiconductor is a material or device that can function as either a conductor or an insulator, depending upon how its structure is arranged. Semiconductor materials have less resistance than an insulator, but more resistance than a conductor. Some common semiconductor materials include silicon (Si) and germanium (Ge).

Note: Although electrons in a circuit actually flow from negative to positive (the electron theory of current flow), most automotive electrical diagrams continue to assume that current flow is from positive to negative (the conventional theory of current flow).

For the sake of a consistent presentation, all these electrical diagrams use the conventional theory with current flowing from positive to negative.

When examining an actual electrical diagram, always take care to determine which current flow theory is being used, as this can make a major difference in the operation of electronic components.

Atomic Structure

In semiconductor applications, materials such as silicon and germanium have a crystal structure. This means that their atoms do not lose and gain electrons as the atoms in conductors do. Instead, the atoms in these semi-conductor materials share outer electrons with each other. In this type of atomic structure, the electrons are tightly held and the element is stable.

Because the electrons are not free, crystals such as silicon and germanium cannot conduct current. These materials are called electrically inert materials. In order to function as semiconductors, a small amount of trace element must be added. The addition of these traces, called *impurities,* allows the material to function as a semiconductor. The type of impurity added determines what type of semiconductor will be produced.

N-Type Semiconductors

N-type semiconductors have loose, or excess, electrons and therefore have a negative charge. This enables them to carry current. N-type semiconductors are produced by adding an impurity with five electrons in the outer ring (called pentavalent atoms). Four of these electrons will fit into the crystal structure, but the fifth will be *free.* This excess of electrons produces the negative charge. Figure I-14 shows an example.

P-Type Semiconductors

P-type semiconductors are positively charged materials. This enables them to carry current. P-type semiconductors are produced by adding an impurity with three electrons in the outer ring (trivalent atoms). When this element is added to silicon or germanium, the three outer electrons fit into the pattern of the crystal, leaving a *hole* where a fourth electron would fit. This hole is actually a positively charged empty space. This hole carries the current in the P-type semiconductor. Figure I-15 shows an example of a P-type semiconductor.

Hole Flow. To understand how semiconductors carry current without losing electrons, you must understand the concept of *hole flow.*

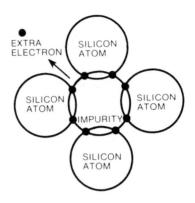

FIGURE I-14 Atomic Structure of N-Type Semi-conductor

The holes in a P-type semiconductor, being positively charged, attract electrons. Although the electrons cannot be freed from their atom, they can rearrange their pattern and fill a hole in a nearby atom. Whenever this happens, the electron leaves a hole. This hole in turn is filled by another electron, and the process continues. The electrons move toward the positive side of the structure, and the holes move to the negative side. This is the principle by which semiconductors carry current.

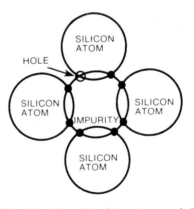

FIGURE I-15 Atomic Structure of P-Type Semi-conductor

Semiconductor Uses

Because semiconductors have no moving parts, they seldom wear out or need adjustment. Semiconductors, or solid-state devices, are also small, require very little power to operate, are very reliable, and generate very little heat. For all these reasons, semiconductors are being used in many applications, including automobiles. The next section will explain how semiconductor materials, both N-type and P-type, are used as electronic devices.

DIODES AND TRANSISTORS

Because a semiconductor can function as both a conductor *and* an insulator, it is very useful as a switching device. How a semiconductor functions depends upon the way current flows (or tries to flow) through it. Two common semiconductor devices are diodes and transistors.

Diodes

The diode is the simplest semiconductor device. It is formed by joining P-type semiconductor material with N-type semiconductor material. The point at which the two materials are joined is called the *junction.* Figure I-16 shows an example.

A diode allows current to flow in one direction, but not in the opposite direction. Therefore, it can function as a switch, acting as either conductor or insulator, depending upon the direction of current flow.

One automotive application of diodes is in the alternator, where they function as a one-way valve for current flow. All charging systems, whether alternators or generators, produce alternating current. In the now-vanished generator, current was rectified (changed from AC to DC) by a rotating commutator and a set of brushes. In the alternator, current is rectified by the use of diodes. The diodes are arranged

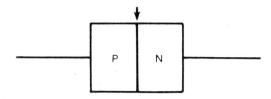

FIGURE I-16   Diode

so that current can leave the alternator in one direction only (as direct current).

When the electric current from the alternator is applied to the N side of the diode, the current (with its flowing electrons) pushes electrons in the N side toward the junction. These electrons cross the junction and fill holes in the P side. When this happens, electrons flow through the P side and out into the circuit. The diode is now functioning as a conductor. Figure I-17 shows an example.

However, if alternator current is applied to the P side of the diode, the incoming electrons tend to fill the holes in the P side atoms. This stabilizes the P side, making it an excellent insulator. At the same time, excess electrons in the N side are pulled out by the positive pull from the other side of the alternator circuit. This makes the N side into an excellent insulator. The entire diode is now functioning as an insulator, and current cannot flow through. Figure I-18 shows an example. When the diode is used in this way, current can flow out of the alternator and into the battery in one direction only.

A variation of the diode is the *zener diode.* This device functions like a standard diode until a certain voltage is reached. When the voltage level reaches this point, the zener diode will allow current to flow in the reverse direction. Zener diodes are often used in electronic voltage regulators.

Transistors

A transistor is an electronic device produced by joining three sections of semicon-

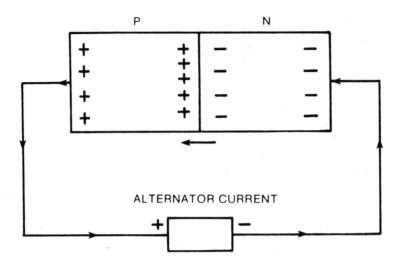

FIGURE I-17   Current Flow Through Diode (Diode Functioning as a Conductor)

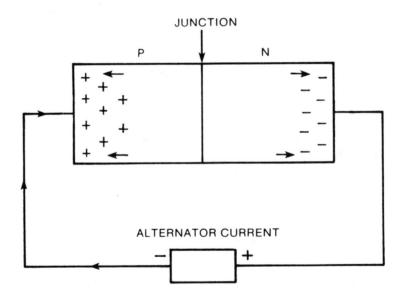

FIGURE I-18   Current Flow Blocked by Diode (Diode Functioning as an Insulator)

ductor materials. Like the diode, it is very useful as a switching device, functioning as either a conductor or as an insulator. Figure I-19 shows two common transistors, which come in many different sizes and types.

A transistor resembles a diode with an extra side. It may consist of two P-type materials and one N-type material, or it may consist of two N-type materials and one P-type material. These are called PNP and NPN types. In both types, junctions occur where the materials are joined. Figure I-20 shows a PNP junction transistor in a circuit. Notice that each of the three sections has a lead connected to it. This allows

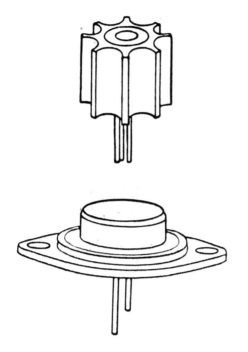

FIGURE I-19   Transistors

any of the three sections to be connected to the circuit. The center section is called the *base.*

Transistors can function as insulators, as shown in Figure I-21. As long as the switch is open, and no current is flowing into the center N section, the transistor is blocking current

flow in the circuit. The junctions are thus acting as insulators.

However, if the switch is closed and current is applied to the center N section, the transistor changes functions. The incoming electrons destabilize the center section and both outside sections; this makes all of them into conductors. Figure I-22 shows an example.

An NPN-type transistor will function in the same manner, except that current will flow out of the center P section, instead of into it. The basic operation remains the same.

A very small amount of current applied to the center section (base) of a transistor can control a much larger amount of current flowing through the entire transistor. This fact allows transistors to be used as signal amplifiers in radios, stereos, calculators, home computers, and electronic engine controls. An example of a transistor application in automobiles is the early type of electronic ignition system shown in Figure I-23.

Instead of running the entire coil primary voltage through the contact points, a small amount of voltage is bled off from the P section of an NPN transistor. As long as current flows out of the P section, the transistor is a conductor and current flows through the entire system. When the points open, current stops flowing out of the P section, the transistor becomes

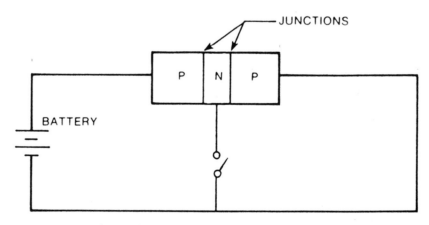

FIGURE I-20   PNP Junction Transistor

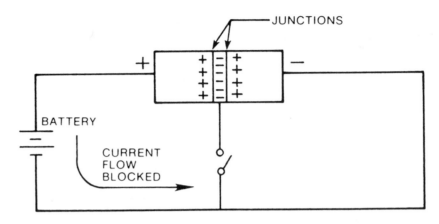

FIGURE I-21   Transistor as Insulator

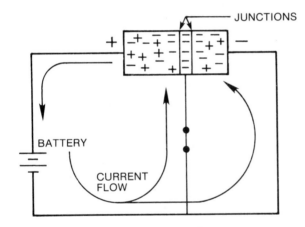

FIGURE I-22   Transistor as Conductor

an insulator, and current stops flowing in the system. This causes the coil's magnetic field to collapse, firing the coil secondary. The advantage of this system is longer point life since the points carry very little current from the transistor base.

SEMICONDUCTOR CIRCUITS

One transistor or diode is limited in its ability to do complex tasks. However, when many semiconductors are combined into a circuit, they can perform complex functions. An example of this is the electronic voltage regulator.

To control alternator voltage, the electronic regulator uses a zener diode, set to break down or let current pass, at the maximum voltage that the battery can tolerate.

The sequence of events is detailed below. Before beginning the sequence, look at Figure I-24. When the vehicle is first started, current flows from the battery's positive terminal (note that we are using the conventional current flow theory), through the field coil of the alternator, and through an NPN transistor before returning to the battery through the negative terminal. Current flow through the field coil creates a magnetic field around the alternator stator, producing current to charge the battery.

When the alternator begins charging the battery, current flow will be from the alternator stator to the battery's positive terminal. The field coil will receive input current from either alternator or battery, depending on whether or not the alternator is charging. An input voltage is also sent to the zener diode at all times.

Following is the sequence of events.

1. The charging voltage exceeds the zener diode limit.

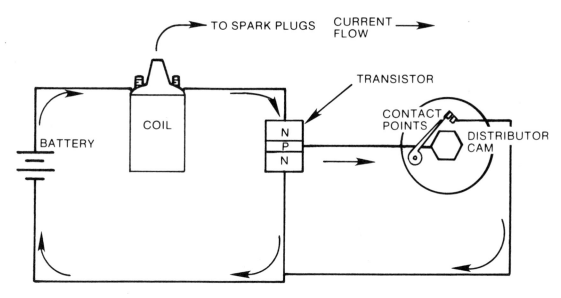

FIGURE I-23   Early Electronic Ignition System

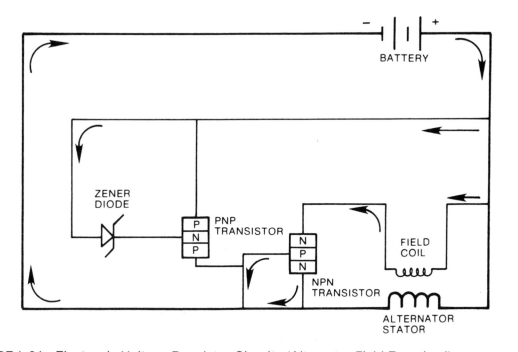

FIGURE I-24   Electronic Voltage Regulator Circuits (Alternator Field Energized)

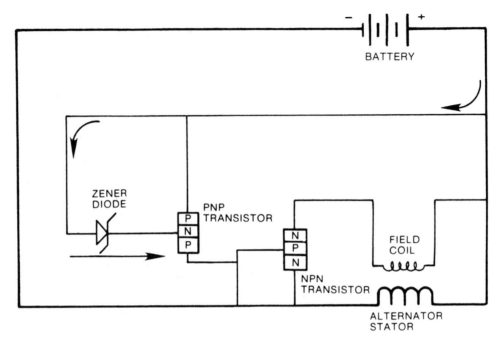

FIGURE I-25   Current Flow Through Zener Diode to PNP Transistor

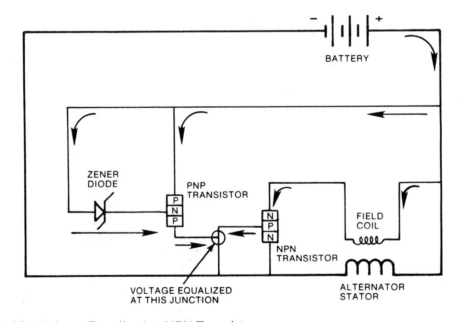

FIGURE I-26   Voltage Equalized at NPN Transistor

2. The zener diode breaks down and allows current to flow through it to the center section of a PNP transistor. See Figure I-25.

3. The PNP transistor turns on and sends current to the center section of the NPN transistor. This equalizes voltage at the junction and stops the current flow out of the NPN transistor section. See Figure I-26.

4. The NPN transistor shuts off.

5. The alternator field circuit is grounded through the NPN transistor; so, it shuts off also.

6. The alternator stops producing electricity.

7. Voltage drops below the zener diode set point.

8. The zener diode shuts off current.

9. The PNP transistor turns off.

10. The NPN transistor turns on.

11. Current flows in the field circuit.

12. The alternator begins producing electricity.

13. Go back to Step 1 to begin the process again.

This process repeats many times per second. An actual voltage regulator would have many additional parts, such as capacitors to protect the regulator from voltage surges and resistors to reduce current flow to and from the transistors. These systems may use conventional electronic components or integrated circuits. This is the first automotive use of what is called a *control loop.* Figure I-27 is an example of the control loop concept in which the alternator voltage, zener diode, transistors, and field current are constantly affecting each other.

INTEGRATED CIRCUITS

An integrated circuit is simply a large number of diodes, transistors, and other electronic components such as resistors and capacitors, all mounted on a *single* piece of semiconductor material. This type of system has a tremendous size advantage, for it is extremely small. Circuitry that used to take up entire rooms can now fit in your pocket. The principles of semiconductor operation remain the same in integrated circuits—only the size has changed.

An example of an integrated circuit is the electronic ignition system in many late model cars. The points and distributor cam are replaced by a stationary pick-up coil and a toothed wheel attached to the distributor shaft. Rotation of the distributor shaft sets up a magnetic field in the pick-up coil. This produces a

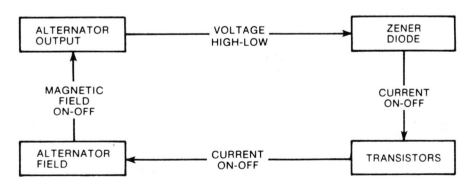

FIGURE I-27   Simple Control Loop

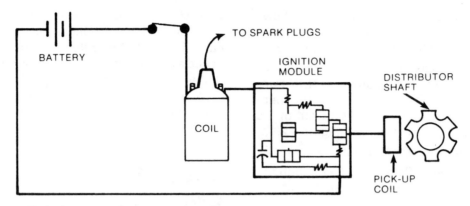

FIGURE I-28   Typical Modern Electronic Ignition System

low-voltage current, which is fed into an ignition module inside of the distributor. The current is amplified by a series of transistors and is used to shut off coil current, which completes its circuit through the ignition module. This causes coil secondary discharge in the same manner as a conventional ignition system. A typical modern electronic ignition system is shown in Figure I-28.

The key to **this** system is the ignition module. The entire **module** is about the size of a wristwatch. All operations in the module are performed by integrated circuits.

The increasingly small size of integrated circuits is very important to automobiles. This means that electronics is no longer confined to the simple tasks of rectifying alternator current or switching the coil on and off. Enough transistors, diodes, and other solid-state components can be installed in a car to make logic decisions and issue commands to other areas of the engine. This is the foundation of electronic engine control systems.

GLOSSARY

ALTERNATING CURRENT (AC)   Current that changes its direction of flow (cycles) a certain number of times per second. Most home and industrial electricity is 60 cycle AC.

AMPERES   The number of electrons flowing in a circuit. Amperes are used to measure current.

ATOM   The basic unit of an element, with a nucleus of protons and neutrons and rings of electrons arranged in shells. The number of protons and electrons varies with every element.

CHARGE   A property of matter which is caused by an excess or lack of electrons. Charges are either positive or negative.

CHEMICAL REACTION   A process by which excess electrons are created by the reaction of various chemicals. A battery operates by chemical reaction.

COMPLETE CIRCUIT   A circuit in which current can flow.

COMPOUND   A combination of two or more elements.

CONDUCTOR   A substance offering an easy (low resistance) path for electricity.

CURRENT   The flow of electrons in a circuit, measured in amperes or amps.

DIODE   A semiconductor device that will permit current to flow in one direction only. Used as a current rectifier in alternators.

DIRECT CURRENT (DC)   Current that flows in one direction only. Used in automobiles and other battery-operated devices.

ELECTROMAGNET A wire (usually in a coil) that produces a magnetic field when current flows through it.

ELECTRON A negatively charged particle in all atoms. Electrons are located in the electron shells around the nucleus.

ELEMENT A substance made up of identical atoms.

GROUND In automobiles, the side of a circuit connected to the vehicle body and frame. Most commonly referred to as a negative or positive ground.

INDUCTION Producing electricity by moving a conductor (such as a wire) through a magnetic field.

INERT An element or compound that is inactive both electrically and chemically.

INSULATOR A substance offering a difficulty (high resistance) path for electricity.

INTEGRATED CIRCUIT A device that contains many transistors, diodes, and other electronic parts, mounted on a single piece of semiconductor material.

JUNCTION The places where N-type and P-type materials are joined. Junctions are where the change from conductor to insulator, or vice versa, takes place.

MAGNET A device that attracts iron (or other ferrous metals) due to the field set up by the position of its electrons.

MAGNETIC FIELD The space around any type of magnet; this area is filled with magnetic force.

MAGNETIC INDUCTION The process of creating electricity by moving a wire (or many wires) in relation to a magnetic field. An alternator is an example.

MODULE A single unit containing many interrelated electronic parts, as in an electronic control module.

NEGATIVE Any part of a circuit that has a surplus of electrons when compared to another part of the same circuit.

NEUTRON A particle located in the atom's nucleus; it has no charge and no effect on electrical theory.

NPN A transistor arrangement in which a piece of P-type material is sandwiched between two pieces of N-type material.

N-TYPE MATERIAL Semiconductor material that contains an excess of electrons and has a negative charge.

NUCLEUS The center of the atom; it contains protons and neutrons.

OHMS A measurement of the amount of resistance in a circuit.

OHM'S LAW A mathematical formula expressing the relationship between amperage, voltage, and resistance in a circuit. Can be expressed as:

$$\text{Voltage} = \text{Amperage} \times \text{Resistance}$$

$$\text{Amperage} = \frac{\text{Voltage}}{\text{Resistance}}$$

$$\text{Resistance} = \frac{\text{Voltage}}{\text{Amperage}}$$

OPEN CIRCUIT A circuit in which current cannot flow.

PARALLEL CIRCUIT A circuit in which current may split up and flow in different amounts through each parallel part.

PATH OF LEAST RESISTANCE The easiest (shortest) way for electric current to complete its circuit. Electricity will always follow the path of least resistance.

PICK-UP COIL An electronic ignition device that reacts to a moving magnetic field by producing a small current flow, which is sent to the ignition module.

PNP A transistor arrangement in which a piece of N-type material is sandwiched between two pieces of P-type material.

P-TYPE MATERIAL Semiconductor material that lacks electrons and has a positive charge.

POLARITY The condition that determines the direction in which current flows in a circuit. One side of the circuit will always be positive, and the other side negative. Current will flow from negative to positive since the negative side always has an excess of electrons.

POSITIVE Any part of a circuit that has a lack of electrons when compared to another part of the same circuit.

PROTON A particle in all atoms; it is located in the nucleus of the atom and has a positive charge.

RECTIFY To change alternating current into direct current.

RESISTANCE Opposition to the flow of electrons, measured in ohms.

SEMICONDUCTOR A device that can function as either a conductor or an insulator, depending on its construction and current flow conditions.

SERIES CIRCUIT A circuit in which the same amount of current flows through all parts.

SERIES-PARALLEL CIRCUIT A circuit combining features of both series and parallel circuits. The entire current may run through some sections and be split in others.

SHELL Energy levels outside of the atom's nucleus. They contain electrons in various amounts, depending upon the kind of atom.

SHORT CIRCUIT A circuit that bypasses part of its intended path.

SIGNAL AMPLIFIER A transistor that controls a large amount of current through the action of the small amount of current at its center section.

TRANSISTOR A semiconductor switch.

VOLTAGE Electrical pressure caused by an imbalance of electrons. Sometimes called electromotive force (EMF).

ZENER DIODE A diode that will not allow current to flow in its reverse direction until a certain voltage is reached; at that point the diode breaks down and allows current to flow. Used in electronic voltage regulators.

COMPUTERS ON CARS

Objectives

After studying this chapter, you will be able to:

- name the major components that make up a microcomputer.
- list the cause or causes of CO, HC, and NO_x production.
- describe a stoichiometric air/fuel ratio.
- identify the air/fuel ratio needed to make a three-way catalytic converter work most effectively.
- explain why almost all cars currently being built for use in this country are equipped with a computerized engine control system.
- describe the difference between open-loop and closed-loop operation.
- explain the importance of performing a thorough prediagnostic inspection.

WHY COMPUTERS? _____

In 1963 positive crankcase ventilation systems were universally installed on domestic cars as original equipment. People in the service industry felt that dumping all of those crankcase gases in the induction system would plug up the carburetor and be harmful to the engine. Everyone in the service industry knew that all of the crankcase gases in the induction system would plug up the carburetor and ruin the engine. Engine life doubled. In 1968 exhaust emission devices were universally applied to domestic cars. Compression ratios began to go down, spark control devices denied vacuum advance during certain driving conditions, thermostat temperatures went up, air

pumps were installed, heated air intake systems were used during warm-up, and air/fuel ratios began to become leaner. Through the seventies evaporation control systems, exhaust gas recirculation systems, and catalytic converters were added. Although some of the emissions systems actually tended to improve driveability and even fuel mileage, for the most part driveability and fuel mileage suffered in order to achieve a dramatic reduction in emissions.

In 1973 and 1974, when domestic car fuel economy was at its worst, we also experienced an oil embargo and an energy shortage. The federal government responded by establishing fuel mileage standards in addition to the already established emissions standards. By the late 1970s, the car manufacturers were hard pressed to meet the evermore stringent emissions and mileage standards; and the standards set for the eighties looked impossible, Figures 1–1 and 1–2. To make matters worse, the consumer was getting into the picture, too. Not only did the consumer's car get poor fuel mileage, it had poor *driveability* (idled rough, often hesitated or stumbled during acceleration if the engine was not fully warmed up, and had little power). What made the situation so difficult was that the three demands—lower emissions, better mileage, and better driveability—were largely in opposition to each other using the technology available at that time, Figure 1–3.

What was needed was a much more precise way to control engine functions, or *engine calibration*. The automotive industry had already looked to microprocessors. In 1968 Volkswagen introduced the first large-scale production, computer-controlled electronic fuel injection system, an early version of the Bosch D-Jetronic system. In 1975 Cadillac introduced a computer-controlled electronic fuel injection (EFI) system. In 1976 Chrysler introduced a computer-controlled electronic spark control system, the Lean-Burn system. In 1977 Oldsmobile introduced a computer-controlled electronic spark control system that they called MISAR (microprocessed sensing automatic regulation). All of these systems had three things in common: they all controlled only one engine function, they all used an *analog* computer, and none of them started any landslide movement toward computer controls.

By the late seventies, the electronics industry had made great strides with *digital* microprocessors. These small computers with their comparatively low cost, compact size and weight, great speed, and application flexibility proved to be the answer for which the industry

EMISSION REQUIREMENTS (GRAMS/MILE) (PASSENGER CARS)

MODEL YEAR	HYDROCARBON (HC) CALIFORNIA	FEDERAL	CARBON MONOXIDE (CO) CALIFORNIA	FEDERAL	OXIDES OF NITROGEN (NOx) CALIFORNIA	FEDERAL
1978	0.41	1.5	9.0	15.0	1.5	2.0
1979	0.41	0.41	9.0	15.0	1.5	2.0
1980	0.39	0.41	9.0	7.0	1.0	2.0
1981	0.39	0.41	7.0	3.4	0.7	1.0
1982	0.39	0.41	7.0	3.4	0.4	1.0
1983	0.39	0.41	7.0	3.4	0.4	1.0
1984	0.39	0.41	7.0	3.4	0.4	1.0
1985	0.39	0.41	7.0	3.4	0.7	1.0
1986	0.39	0.41	7.0	3.4	0.7	1.0
1960 (No Control)	10.6		84		4.1	

FIGURE 1–1   Federal and California emissions standards compared to emissions from a typical car in 1960. *Courtesy of GM Product Service Training.*

FUEL ECONOMY STANDARDS

MODEL YEAR	MPG	TOTAL IMPROVEMENT OVER THE 1974 MODEL YEAR
1978	18.0	50%
1979	19.0	58%
1980	20.0	67%
1981	22.0	83%
1982	24.0	100%
1983	26.0	116%
1984	27.0	125%
1985	27.5	129%
1986	26.0*	116%
1987	26.0	116%
1988	26.0	116%

*Reduced from 27.5% by the federal government in 1986

FIGURE 1-2   Federally imposed fuel economy standards

was looking. Probably the most amazing feature of the digital computer is its speed. To put it into perspective, one of these computers controlling functions on an eight-cylinder engine running at 3,000 RPM can send the spark timing command to fire a cylinder; reevaluate input information about engine speed, coolant temperature, engine load, barometric pressure, throttle position, air/fuel mixture, spark knock, and vehicle speed; make a new decision about air/fuel mixture, spark timing, whether or not to turn on the EGR valve, canister purge valve, and the torque converter clutch; then take a short nap before sending the commands, all before the next cylinder fires.

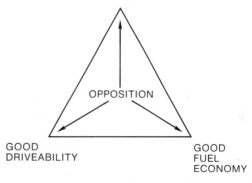

FIGURE 1-3   Opposition of exhaust emissions, driveability, and fuel economy to each other

At last the automotive manufacturers had the technology to precisely monitor and control, to instantly and automatically adjust, while driving, enough of the engine's calibrations to make the vehicle comply with the government's demands for emissions and fuel economy and still satisfy the consumer's demands for better driveability. This brought about what is probably the first real revolution in the automotive industry's recent history; other changes have been evolutionary by comparison. After some experimental applications in 1978 and 1979 and some limited production in 1980, the whole domestic industry was using comprehensive computerized engine controls in 1981. At the same time, General Motors produced more computers than anyone else in the world and used half of the world's supply of computer parts.

How Computers Work

Contrary to some "informed" opinions and numerous suggestions from movies, computers cannot think for themselves; when properly programmed, however, they can carry out explicit instructions with blinding speed and almost flawless consistency.

Communication Signals. A computer uses voltage values as communication signals, thus voltage is often referred to as a *signal* or a *voltage signal*. There are two types of voltage signals: analog and digital, Figure 1–4. An *analog signal's voltage* is continuously variable within a given range and time is used for the voltage to change. An analog signal is generally used to convey information about a condition that changes gradually and continuously within an established range. Temperature-sensing devices usually give off an analog signal. *Digital signals* also vary but not continuously, and time is not needed for the change to occur. Turning a switch on and off creates a digital signal; voltage is either there or it is not. Digital signals are often referred to as square wave signals.

Binary Code. A computer converts a series of digital signals to a binary number made up of 1s and 0s; voltage above a given value converts to 1, and zero voltage converts to 0. Each 1 or 0 represents a *bit* of information. Eight bits equal a *byte* (sometimes referred to as a *word*). All communication between the microprocessor, the memories, and the interfaces is in binary code, with each information exchange being in the form of a byte. If a series of digital signals converts to a binary number of 01111010, the numerical value (in $base_{10}$, which we are most familiar with) can easily be derived, as shown in Figure 1–5A. A power is assigned to each place in the binary number, starting from the right and working to the left. The rightmost place is given a power of 1, with the power doubling with each successive place to the left. Each digit in the binary number is multiplied by its respective power. The products are then added together to yield the $base_{10}$ numerical value. As you will see in Figure 1–5B, the largest value that an 8-bit computer can communicate is 255. Most automotive computers have 8-bit microprocessors, but as the computers take on more responsibility and have to communicate more information more rapidly, 16-bit units will become commonplace.

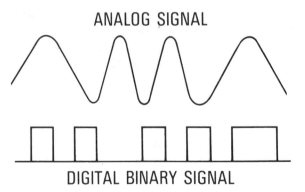

ANALOG SIGNAL

DIGITAL BINARY SIGNAL

FIGURE 1–4   Graphic illustration comparing analog and digital voltage signals. *Courtesy of GM Product Service Training.*

- Analog signal continously runable
- On and off signal within a given range

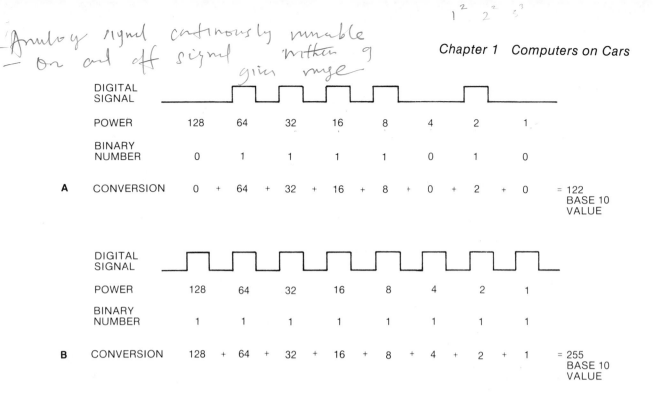

FIGURE 1-5   Binary numbers

Interface. The microprocessor is the heart of a computer, but it needs several support functions, one of which is *interface.* A computer has an input and an output interface circuit, Figure 1–6. The interface has two functions: it protects the delicate electronics of the microprocessor from the higher voltages of the circuits attached to the computer, and it translates input and output signals. The input interface translates all analog input data to binary code; most sensors produce an analog signal. It is sometimes referred to as *A/D,* analog to digital. The output interface, *D/A,* translates digital signals to analog for any controlled functions that need an analog voltage.

Memories. The microprocessor of a small computer does the calculating and makes all of the decisions, or data processing, but it cannot store information. The computer is therefore equipped with information storage capability called *memory.* The computer actually has three memories: read-only memory (ROM),

programmable read-only memory (PROM), and random access memory (RAM), Figure 1–7.

The ROM contains permanently stored information that instructs (programs) the microprocessor on what to do in response to input data. The microprocessor can read information from the ROM but cannot put information into it. The ROM unit is soldered into the computer and is not easily removed.

The PROM differs from the ROM in that it plugs into the computer and is more easily removed and reprogrammed or replaced with one containing a revised program. It contains program information specific to different vehicle model calibrations.

Some of the information stored in both the ROM and the PROM is stored in the form of look-up tables similar to the tables one might find in the back of a chemistry or math textbook. These tables enable the computer to interpret and make decisions in response to sensor data. For example, a piece of information (a

27

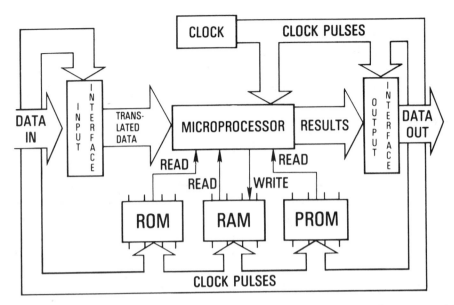

FIGURE 1-6   Interaction of the microprocessor and its support system. *Courtesy of GM Product Service Training.*

voltage value of 2.4 volts, about half throttle) is received from the throttle position sensor. This information plus information from the engine speed sensor is compared to a table for spark advance. This preprogrammed table tells the computer what the spark advance should be

![MICROPROCESSOR diagram with ROM, RAM, PROM]

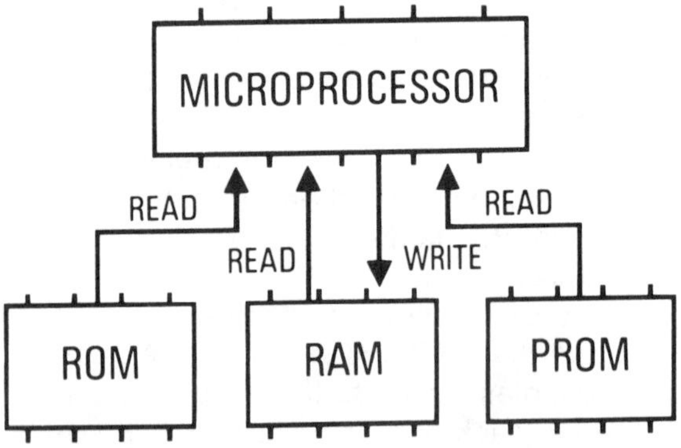

FIGURE 1-7   The three memories within a computer. *Courtesy of GM Product Service Training.*

for that throttle position and engine speed. The computer will then modify this spark advance value, by consulting other tables, with information concerning engine temperature and atmospheric pressure. These tables are sometimes referred to as *maps.*

The RAM is where temporary information is stored. It can be both read from and written into. If, for example, data comes in from a sensor and will be used for several different decisions, such as manifold vacuum information, the microprocessor will record it in RAM until it is updated. In the example presented in the above paragraph, the information concerning throttle position and engine speed came from the RAM.

In most automotive computers, the computer tests at least some of the input and output signals to see that the circuit that sent the signal or received the command is working properly. The computer knows what the voltage will be if the circuit were open and what the voltage will be if it were shorted. It watches to see that the voltage remains between those voltage values. On some circuits, the oxygen sensor circuit for instance, the computer expects to see the voltage go up and down, crossing a certain value. If during its continuous testing it sees something wrong (a fault), it will record information in RAM concerning the fault. The RAM is also soldered in place and not easily removed.

There are two kinds of RAMs: volatile and nonvolatile. A volatile RAM (sometimes called *keep-alive memory*) is erased when disconnected from its power source. In automotive applications a volatile RAM is usually connected directly to the battery by way of a fuse or fusible link so that when the ignition is turned off, the RAM is still powered. A nonvolatile RAM does not lose its stored information if its power source is disconnected. Vehicles with digital display odometers usually store mileage information in a nonvolatile RAM.

The terms ROM, PROM and RAM are fairly standard throughout the computer industry; however, all car manufacturers do not use the same terms in reference to their computer's memory. For instance, instead of using the term PROM, Ford calls their equivalent unit an engine calibration assembly. Another variation of a PROM is an E-PROM. An E-PROM has a housing with a transparent top that will let light through. If ultraviolet light strikes the E-PROM, it reverts back to its unprogrammed state. To prevent de-programming or memory loss, a cover must protect it from light. The cover is often a piece of tape. An E-PROM is sometimes used where ease of changing stored information is important. General Motors uses it as a PROM on some applications because it has more memory capacity.

Another variation of RAM is KAM (Keep Alive Memory). Often used as a subsection of the RAM, KAM is usually powered by a fuse straight from the battery rather than by the ignition or an ignition-controlled circuit. Therefore, when the ignition is turned off, the KAM

remains powered and its stored information is retained. Other types of memory that are unique to specific manufacturers will be introduced in appropriate chapters.

Clock Pulses. In order to maintain an orderly flow of information into, out of, and within the computer, a quartz crystal is used to produce a continuous, consistent time pulse, Figure 1–8. The pulse acts as a kind of rapid cadence by which information is transmitted. During the space or time interval between each pulse, one bit of binary code information is communicated from one part of the computer to another. This method provides an orderly flow of information.

The clock pulse is often referred by the term *baud rate.* The baud rate identifies the rate of the pulses. For example, a computer with a baud rate of 5000 could transmit 5000 bits of binary code information per second. Baud rates in automotive computers have gone up considerably since those used in the early eighties. The computers used by General Motors in the early eighties had a baud rate of 160 (Some of those computers were still used on some applications in the late eighties.), while their P-4 electronic control module introduced in the late eighties has a baud rate of 8192 with development underway to make them even faster. Ford introduced a computer in 1988 with a baud rate of 12500.

Data Links. When computers communicate with other electronic devices such as control panels, modules, some sensors, or other computers in the form of digital signals, they communicate through circuits called *data links.* Some data links transmit data in only one direction, although others transmit in both directions. What makes a data link different from an ordinary circuit is the manner in which it is controlled. Some control must be used so that the devices at each end of the data link know when to transmit and when to receive. Several control methods can be used, but to discuss them here would go beyond the goals of this

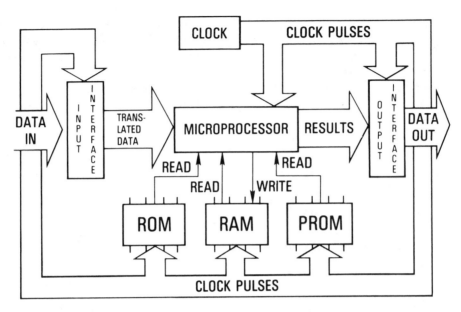

FIGURE 1-8   Clock pulses within a computer. *Courtesy of GM Product Service Training.*

text and the interest of most automotive students and technicians.

EMISSIONS, MILEAGE, AND DRIVEABILITY

The computer of a computerized engine control system has a mission: to reduce emissions, improve mileage, and improve driveability. The priority varies, however, under some driving conditions. For example, during warm-up, driveability has a higher priority than mileage; and at full throttle, some systems give the performance aspect of driveability a higher priority than emissions or mileage.

There are three major emissions that share the greatest concern for control on an automobile: hydrocarbon (HC), carbon monoxide (CO), and oxides of nitrogen (NO_x).

1. *HC.* Gasoline is a hydrocarbon compound. When it is burned properly, the hydrogen and carbon atoms separate; each mixes with oxygen to form water (H_2O) and carbon dioxide (CO_2), Figure 1-9. If for any reason the gasoline in the cylinder fails to burn, it is pumped into the exhaust system as a raw HC molecule. There can be several reasons for all or part of the HC in the cylinder not to burn:

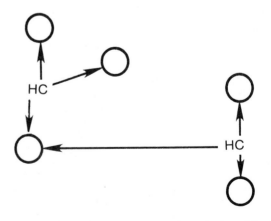

FIGURE 1-9   Chemical reaction during ideal combustion

[HG
GO
NOx] — Incomplete combustion
Too much fuel
— high temp.

Chapter 1 Computers on Cars

- Any electrical malfunction, which prevents the spark plug from igniting the fuel mixture.
- A lack of mechanical integrity, which allows fuel to escape from the cylinder.
- An overly lean mixture, which hampers flame propagation.
- An exceedingly rich mixture, which uses up the oxygen before all of the HC has been even partially burned.
- A low cylinder surface temperature, which rapidly draws heat from those HC molecules in contact with it and thus prevents them from reaching or maintaining ignition temperature. A low cylinder surface temperature can be caused by a low coolant (low temperature or stuck open thermostat) or an overadvanced ignition timing.

2. *CO.* During combustion each carbon atom tries to take on two oxygen atoms (CO_2). If, however, there is a deficiency of oxygen in the combustion chamber, some carbon atoms are only able to combine with one oxygen atom and thus produce CO. An overly rich air/fuel ratio is the only cause of CO production.

3. *NO_x.* Air is made up of about 80% nitrogen and 20% oxygen. Under normal conditions nitrogen and oxygen do not chemically unite. When raised to a high enough temperature, however, they unite to form a nitrogen oxide compound, NO. In the presence of atmospheric oxygen, NO rapidly becomes nitrogen dioxide, NO_2. These compounds are grouped into a family of compounds referred to as oxides of nitrogen, NO_x. Oxide of nitrogen compounds start to form at 1,120° C (2,040° F), with production increasing as temperatures go higher.

Catalytic Converter

Controlling the air/fuel ratio and the spark timing, keeping the coolant temperature near

---

Other Exhaust Emissions From Gasoline Engines

Aldehydes
Ammonia
Carboxylic acids
Inorganic solids
    lead (from leaded fuel)
    soot
Sulfur oxides (from fuel impurities)

93° C (200° F) or more, and controlling intake air temperature and combustion temperature dramatically reduce exhaust emissions. Even these measures, however, do not eliminate them. The catalytic converter is the single most effective device for controlling exhaust emissions. Placed in the exhaust system, the *catalyst* agents cause low-temperature oxidation of HC and CO and reduction of NO_x, yielding H_2O, CO_2, O_2, and nitrogen. Placing the converter downstream in the exhaust system allows the exhaust gas temperature to have dropped significantly before the gas enters the converter and thus prevents the production of more NO_x in the oxidation process. The catalysts that are most often used, *platinum* and *palladium* for oxidation and *rhodium* for reduction, are coated over a porous material, usually aluminum oxide in pellet form or aluminum oxide in a honeycomb structure. Aluminum oxide's porosity provides a tremendous amount of surface area on which the catalyst material can be applied to allow maximum exposure of the catalyst to the gases.

There have been three different types of catalytic converters used on automobiles: a single-bed, two-way, introduced on cars in the mid-seventies; a single-bed, three-way; and a dual-bed, three-way. The three-way converters were introduced on cars in the late seventies.

A single-bed converter has all of the catalytic materials in one chamber, Figure 1–10. A dual-bed converter has a reducing chamber for

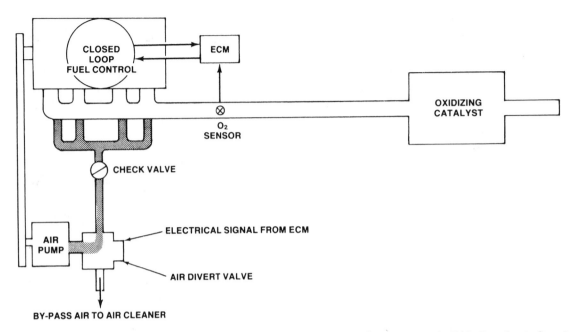

FIGURE 1–10   Single-bed, three-way catalytic converter. *Courtesy of GM Product Service Training.*

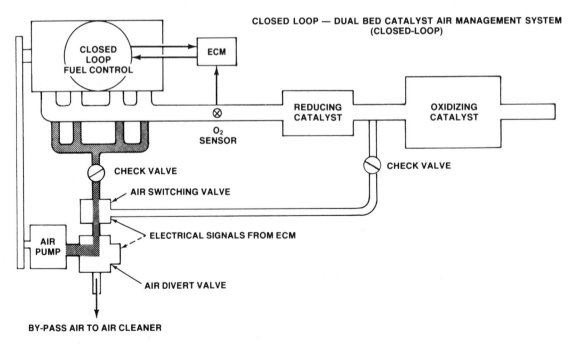

FIGURE 1–11   Dual-bed, three-way catalytic converter. *Courtesy of GM Product Service Training.*

ing any driving condition. For example, following a cold start, it can be advanced for driveability. During light load operation with a partially warmed-up engine, it can be retarded slightly to hasten engine warm-up and reduce exhaust emissions. During acceleration or wide-open throttle operation, it can be adjusted for maximum torque.

Idle Speed Control. Electronic idle speed control can more nearly maintain the best compromise between a speed at which the engine is prone to stall and one that wastes fuel and produces an annoying lurch as the automatic transmission is pulled into gear. It can keep idle speed from yo-yoing up and down as the air-conditioning clutch cycles on and off and can be designed to prevent dieseling. It can also adjust idle speed in an effort to compensate for low-charging system voltage, low engine temperature, and engine overheating.

Torque Converter Clutch Control. Many engine control computer systems control a clutch in the torque converter. This function eliminates converter slippage and heat production and thus contributes to fuel economy. On some manual transmission applications, the computer operates a *shift light,* which alerts the driver when to upshift in order to obtain maximum economy.

EGR Valve Control. The EGR valve has not been the best thing to happen to fuel economy or driveability. Many earlier EGR systems had the EGR valve open anytime the engine was off idle and short of full throttle; any control of it is an improvement. Most engine control computer systems control when the EGR valve is turned on and leave it off for more of the time when combustion temperatures are not high enough to produce significant amounts of NO_x. Some systems even control how much it opens *(modulate it).*

Air-conditioning Control. Some computerized engine control systems feature an air-conditioning clutch cutout function that disengages the air-conditioning clutch during heavy throttle operation. This function contributes somewhat to driveability.

Turbocharger Boost Control. To increase performance, each manufacturer is offering at least one turbocharged engine option. In most cases, the computer controls the amount of boost that the turbocharger can develop. It is worth noting that each of the manufacturers are using an air-to-air intercooler on selected turbo applications. This lowers manifold air temperature by 120° to 150° F and allows for even more aggressive turbo boost and spark advance.

Closed Loop, Open Loop

Closed Loop. Made possible by the development and use of the oxygen sensor, a feature of all comprehensive computerized engine control systems is the ability to operate the engine in a closed-loop mode. The term *closed loop* describes an intimate, triangular relationship between the oxygen sensor, the computer, and the fuel-metering control device. The oxygen sensor tells the computer what the air/fuel mixture is. The computer sends a command to the fuel-metering control device to adjust the air/fuel ratio toward a stoichiometric air/fuel ratio (14.7 to 1 at sea level). The adjustment usually causes the air/fuel ratio to cross stoichiometric, and it starts to move away from 14.7 in the opposite direction. The oxygen sensor sees this and reports it to the computer. The computer again issues a command to adjust back toward 14.7 to 1. With the speed at which this cycle occurs, the air/fuel ratio never gets very far from 14.7 to 1 in either direction. This cycle repeats continuously. No other sensor in the system has any input into the closed-loop relationship.

Closed loop is the most efficient operating mode, and the computer is programmed to keep the system in closed loop as much as possible. There are, however, criteria that must be met before any of the systems can go into closed loop.

35

- The oxygen sensor must reach operating temperature (around 315° C [600° F]).
- The engine coolant temperature must reach a criterion temperature (varies somewhat but tends to be around 65° C (150° F).
- A predetermined amount of time must elapse from the time the engine was started (this varies from a few seconds to one or two minutes).

There are other operating conditions, such as hard acceleration or, in some cases, prolonged idle, that force the system out of closed loop.

Open Loop. *Open loop* is used during periods of time when a stoichiometric air/fuel ratio is not appropriate such as during engine warm-up or at wide-open throttle (WOT). During this operational mode, the computer uses input information concerning coolant temperature, engine load, barometric pressure, and engine speed to determine what the air/fuel ratio should be. Once the necessary information is processed, the computer sends the appropriate command to the mixture control device. The command does not change until one of the inputs changes. In this mode the computer does not use oxygen sensor input and therefore does not know if the command it sent actually achieved the most appropriate air/fuel ratio for the prevailing operating conditions. As a result of this weakness of the open-loop mode, the computer puts the system into closed loop as soon as possible and keeps it there for as much of the time as possible. Certain failures within the system prevent it from going into closed loop or cause it to drop back out of closed loop.

ATTITUDE OF THE TECHNICIAN _____

Historically many automotive technicians had a negative attitude about the changes that the manufacturers continuously put into cars. Changes mean always having to learn something new, having to cope with a new procedure, and so on. There is another way to look at it, though: every time new knowledge or a new skill is required, it represents a new opportunity to get ahead of the pack. Most of us are in this business to make money; the more you know that other people have not bothered to learn yet, the more money you can make. Successful, highly paid automotive technicians earn the money they get because of what they know. Their knowledge enables them to produce more with the same effort.

Computerized automotive control systems represent the newest and probably the most significant and most complicated development ever to occur in the history of automotive service. (It is complicated, not because it is more difficult, but because it involves more processes.) We have seen only the beginning. It is likely that before long, every major component and function on the car will be electronically controlled. Those who appreciate the capability of these systems and learn everything they can about them will be able to service them and do very well while other people scratch their heads and complain.

Approaching Diagnosis

Diagnosing problems on complicated electronic systems takes a little different approach than many automotive technicians are used to. The flat-rate pay system has encouraged many of us to take shortcuts whenever possible; however, shortcuts on these systems get you into trouble. Use the service manual and follow procedures carefully. It is very useful to spend some time familiarizing yourself with the diagnostic guides and charts that manufacturers present in the service manual or manuals you will be using.

Prediagnostic Inspection. Although most computerized engine control systems feature some degree of self-diagnostic capability, none of them monitors spark plugs, spark plug wires, valves, vacuum hoses, PCV valves, and other engine support and emission-control components that are not a part of the computer system. A seemingly unrelated part such as a spark plug wire can have a direct impact on the performance of the system. For example, a shorted spark plug wire prevents its cylinder from firing. The unburned oxygen coming out of that cylinder, if it is on the same side as the oxygen sensor on a V-type engine, causes the oxygen sensor to mistakenly see a lean condition. The computer responds by enriching the air/fuel mixture. Replacing the oxygen sensor, half or all of the other sensors, the computer, the headlights, and the front bumper does not solve the problem. Replacing the shorted spark plug wire and possibly cleaning the spark plug does solve the problem.

The computer knows the engine operating parameters (throttle position, atmospheric pressure, engine speed, temperature, load, etc.), and with this information it can calculate exactly what the spark timing should be. It continuously makes calculations and sends its spark timing commands. It does not, however, know what base timing is; and the command it sends out is added to base timing. If base timing is incorrect, a spark plug will fire at the wrong time in spite of the computer.

Having a closed thermostat stick or a restricted radiator causes the engine to overheat, possibly causing the computer to think that the coolant temperature-sensing device or its connecting circuit is shorted and setting a fault code in its diagnostic memory. The coolant-sensing device, however, is only reporting what it sees; the fault is in the cooling system.

Do not overlook the possibility that someone has replaced a closed 195° thermostat with one that opens at 160° or that the charcoal canister is saturated with gasoline or that it is being purged when it should not be. Do not overlook the possibility of loose or shorted electrical wires or of cracked or misrouted vacuum hoses. Any such problem, although it may not be directly related to the computer system, can cause a driveability problem and can in some cases cause the computer to mistakenly report a problem in one of its circuits. The service manual for each vehicle with an engine control computer system should provide instructions for making a prediagnostic inspection, because the system's diagnostic procedures assume that all such unrelated, or perhaps it would be better to say semirelated, systems are functioning properly.

All of the engine components, especially engine support components, and some non-engine components such as the transmission are more related in their influence on each other than they have ever been before computer control systems. It is extremely important to remember that in order for the computer control system to work properly, all other engine-related systems must be operating to manufacturer's specifications. If routine maintenance, such as changing oil, PCV valve, spark plugs, spark plug wires, is due, it is advisable to do it before beginning diagnostic procedures.

Excessive HC Production

Overadvancing ignition timing by 6° can cause HC production to go up by as much as 25% and NO_x production to increase by as much as 20% to 30%. A thermostat that opens at 160° F can cause HC production to go up as much as 100 to 200 parts per million.

REVIEW QUESTIONS _____

1. Name two types of voltage signal.
2. The term *square wave* refers to what?
3. Most sensors create what type of signal?
4. How many different characters are in the binary code?

5. Data coming into the automotive-type microcomputer is first processed by _____ .

6. What does the abbreviation ROM stand for and what is its function?

7. Name at least two ways that a PROM is different from a ROM.

8. Name at least two ways that a RAM is different from a ROM.

9. Name two types of RAMs.

10. What is the function of the clock pulse in a computer?

11. Name at least three things that cause excessive HC from the exhaust.

12. What causes high CO production?

13. What causes high NO_x production?

14. What air/fuel ratio is necessary for the three-way catalytic converter to work effectively?

15. Name two different types of three-way catalytic converters.

16. Describe the condition(s) that must exist before the computer system can function properly and before a diagnostic procedure should be begun.

ASE-type Questions. (Actual ASE test questions will probably not be this product specific.)

17. Technician A says that computerized engine control systems are intended to reduce exhaust emissions. Technician B says that computerized engine control systems are designed to improve fuel mileage and driveability. Who is correct?
    a. A only
    b. B only
    c. both A and B
    d. neither A nor B

18. A car with a driveability complaint is being checked. During the inspection it is found that base timing is off by 4° and one spark plug is faulty. Also the engine control computer's self-diagnostic test shows an oxygen sensor code. Which of the three faults identified are most likely related?
    a. the timing and the oxygen sensor code
    b. the oxygen sensor code and the spark plug
    c. all three
    d. none of them

19. Technician A says that a fault code indicating a coolant temperature sensor problem could be the result of a faulty sensor or one of its circuit connections. Technician B says that it could be the result of a faulty thermostat. Who is correct?
    a. A only
    b. B only
    c. both A and B
    d. neither A nor B

20. Technician A says that for a car to use a three-way catalytic converter, it must also have some way to test and control its air/fuel mixture. Technician B says that if a car has a three-way catalytic converter, the converter can be either a single-bed or a dual-bed design. Who is correct?
    a. A only
    b. B only
    c. either A or B
    d. neither A nor B

GLOSSARY _____

A/D A converter circuit that translates an analog signal into digital quantities.

ANALOG A signal that continuously varies within a given range and with time being used for the change to occur.

BAUD RATE In a stream of binary signals, one baud is one bit per second. The term "baud" is in honor of the Frenchman Baudot, who is responsible for much of the early development in unit data transmission.

BIT   A single piece of information in a binary code system; somewhat comparable to a single letter within a word or a single digit within a number.

BYTE   Eight pieces of binary code information strung together to make a complete unit of information.

CATALYST   An agent that causes a chemical reaction between other agents without being changed itself.

CLOSED LOOP   Describes an intimate, signal-response, triangular relationship between the O_2 sensor, the ECM, and the M/C solenoid.

D/A   A converter circuit that translates digital signals to analog voltage values.

DIGITAL   Using values (numbers) expressed as digits with specific meanings to represent the variables in a set of operations.

DRIVEABILITY   Those factors, including ease of starting, idle quality, acceleration without hesitation, and so on, that affect the ease of driving and reliability.

ENGINE CALIBRATION   Adjustments or settings that produce desired results such as spark timing, air/fuel mixture, EGR control, and so on.

INTERFACE   Circuitry that converts both input and output information to analog or digital signals as needed and filters external circuit voltage to protect computer circuits.

MICROPROCESSOR   A processor contained on an integrated circuit, processor (central processing unit) that makes the arithmetic and logic decisions in a microcomputer.

MODULATE   To find a position between two extremes; between on and off.

OPEN LOOP   An operational mode in which the air/fuel mixture is calculated based on coolant temperature, engine speed, and engine load without benefit of the oxygen sensor.

PALLADIUM (Pd)   A silver-white, ductile, metallic element; number 46 on the Periodic Table.

PLATINUM (Pt)   A heavy, grayish white, ductile, metallic element; number 78 on the Periodic Table.

PROM   Programmable read-only memory.

RAM   Random access memory.

RHODIUM (Rh)   A white, hard, ductile, metallic element; a member of the noble metals family; number 45 on the Periodic Table.

ROM   Read-only memory.

STOICHIOMETRIC   In automotive terminology it refers to an air/fuel ratio in which all combustible materials are used with no deficiencies or excesses; 14.7 parts air to 1 part fuel, by weight, at sea level.

COMMON COMPONENTS

Objectives

After studying this chapter, you will be able to:

- state the advantages and disadvantages of a multipoint fuel injection system.
- define the terms *pulse width* and *duty cycle.*
- describe how an oxygen sensor works.
- describe how a thermistor is used as a temperature-measuring sensor.
- describe how a potentiometer is used as a position-measuring sensor.
- describe the operation of a Hall-effect switch.
- describe a sensor that is commonly used to measure air pressure.
- describe how a simple switch is used as a sensor.
- define the term *actuator.*

Among the different car manufacturers, vehicle models, and years, there are many different computerized engine control systems. Although there are significant differences between the various systems, when compared they are actually more alike than different. Some of the more common components and circuits will be discussed in this chapter to avoid needless duplication when discussing the specific systems. As you read the following chapters, you may find it useful to refer to this chapter to clarify how a particular component or circuit works.

COMMON FEATURES

Computers

The comprehensive automotive computer is a small, highly reliable, solid-state digital computer protected inside a metal box. It receives information in the form of voltage signals from several sensors and other input sources. With this information, which the computer reads several thousand times per second, it can make decisions about engine calibration functions such as air/fuel mixture, spark timing, EGR application, and so forth. These decisions appear as commands sent to the *actuators*—the solenoids, relays, and motors that carry out the output commands of the computer. Commands usually amount to turning on or off an actuator. In most cases the ignition switch provides voltage to the actuators either directly or through a relay. The computer controls the actuator by using one of its internal solid-state switches (transistors) to ground the actuator's circuit. In this way the transistor only has to deal with the low-voltage side of the circuit (the voltage is dropped and most of the heat is produced in the actuator, outside the computer).

WARNING: No attempt should be made to open the computer's metal housing. The housing protects the computer from static electricity. Opening it or removing any circuit boards from it in any other than carefully controlled laboratory conditions will likely result in damage to some of its components.

Location. On most systems the computer is located inside the passenger compartment. This protects it from the harsh, high-temperature environment in the engine compartment.

Engine Calibration. Because of the wide range of vehicle sizes and weights, engine and transmission options, axle ratios, and so forth, most manufacturers use a PROM, sometimes called an *engine calibration unit,* to calibrate the computer to a specific vehicle, Figure 2–1. For example, the vehicle's weight affects the load on the engine; therefore, to optimize ignition timing, ignition timing calibration must be programmed for that vehicle's weight. In a given model year, a manufacturer can have well over a hundred different vehicle models, if counting different engine and transmission options, but can use less than a dozen different computers. This is made possible by using an engine calibration unit to program the computer to the specific calibration needs of each vehicle. The calibration unit is usually removable.

Five-Volt Reference

With few exceptions all of the systems use 5 volts to operate their sensors. Within the elec-

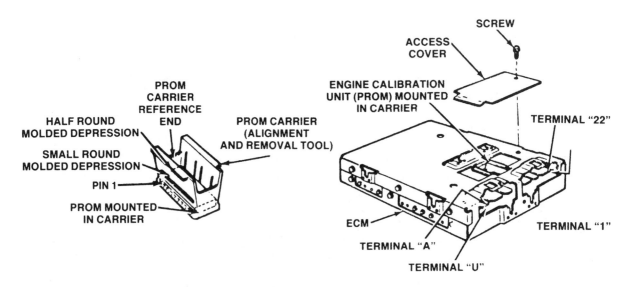

FIGURE 2–1   ECM and PROM. *Courtesy of GM Product Service Training.*

tronics industry, 5 volts has been almost universally adopted as a standard for information transmitting circuits. This voltage value is high enough to provide reliable transmission and low enough not to damage the tiny circuits on the chips in the computer.

Fuel Injection

There are two types of fuel injection systems used with comprehensive computerized engine control systems: single point and multipoint. They each use an intermittent or timed spray to control fuel quantity.

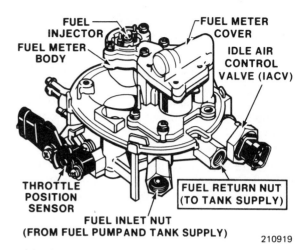

FIGURE 2-2   TBI unit. *Courtesy of GM Product Service Training.*

Constant Spray and Timed Spray Injection

There are two basic methods of controlling the quantity of fuel introduced by a fuel injection system: the constant spray injection and the timed spray injection. Historically these have been port injection systems (placing an injector in the center of the intake manifold, above the throttle blades, is a comparatively new development). As the term *constant spray* implies, in this system each injector sprays a continuous stream of fuel into each intake port; and the quantity of fuel introduced is controlled by a variable fuel pressure. Fuel pressure is controlled by an airflow meter. The Bosch K-Jetronic is the most common example of this type of system.

In timed spray systems, fuel pressure is constant and the injectors are turned on and off. Fuel quantity is controlled by how long the injectors are turned on, and the injectors are controlled by an electronic module (a single-function computer system). The Bosch D-Jetronic and L-Jetronic systems are the best known examples of this type of system.

Single-point Injection. Single-point injection is often referred to as *throttle body injection.* Single point means that fuel is introduced into the engine from one location. This system uses an intake manifold similar to what would be used with a carbureted engine, but the carburetor is replaced with a throttle body unit, Figure 2-2. The throttle body unit contains one or two solenoid-operated injectors that spray fuel directly over the throttle blade (or blades). Fuel under pressure is supplied to the injector. The throttle blade is controlled by the throttle linkage just as in a carburetor. The computer controls voltage pulses to the solenoid-operated injector, which opens and sprays fuel into the throttle bore. The amount of fuel introduced is controlled by the length of time the solenoid is energized. This is referred to as *pulse width.* The amount of air introduced is controlled by the opening of the throttle blade.

The EFI system is characterized, especially on smaller engines, by excellent throttle response and good driveability. Experience has shown, however, that the system is best suited for engines with small cross-sectional area manifold runners that at low speeds will keep the fuel mixture moving at a higher velocity.

This reduces the tendency for the heavier fuel particles to fall out of the airstream.

Multipoint Injection. Multipoint injection is often referred to as *port injection* and means that fuel is introduced into the engine from more than one location. This system uses an injector at each intake port. Fuel is sprayed directly into the port, just on the manifold side of the intake valve.

The multipoint injection system provides the most advanced form of fuel control yet developed. It offers the following advantages:

- Spraying precisely the same amount of fuel directly into the intake port of each cylinder eliminates the unequal fuel distribution so inherent when already mixed air and fuel are passed through an intake manifold.
- Because there is no concern about fuel condensing as it passes through the intake manifold, there is less need to heat the air or the manifold.
- Because there is no concern about fuel molecules falling out of the airstream while moving through the manifold at low speeds, the cross-sectional area of the manifold runners can be larger and thus offer better cylinder-filling ability (volumetric efficiency, VE) at higher speeds.
- Most of the manifold-wetting process is avoided; some wetting still occurs in the port areas. If fuel is introduced into the intake manifold, some will remain on the manifold floor and walls especially during cold engine operation and acceleration. Fuel metering has to allow for this fuel in order to avoid an overlean in the cylinders. It has to be accounted for again during high-vacuum conditions because it will then begin to evaporate and go into the cylinders.

In general port fuel injection provides better engine performance and excellent driveability while maintaining or lowering exhaust emission levels and increasing fuel economy.

Fuel Injection History

Fuel injection for gasoline engines dates at least as far back as the middle 1930s. It has captured the interest of automotive engineers and performance enthusiasts for years by promising to be a superior method of fuel induction. Only recently, however, has its technology evolved far enough (see **Measuring Air Mass** in this chapter) to enable fuel injection to surpass the carburetor's ability to deliver performance, driveability, economy, and emissions reduction under all driving conditions.

Fuel Pressure Regulators. All fuel injection systems that are part of a comprehensive computerized engine control system use pulse width as the primary means of controlling fuel metering. A pressure regulator is used to provide a constant pressure to the injector. An electric fuel pump supplies under pressure to the regulator.

The fuel pressure regulator is located on the fuel rail or manifold and allows excess fuel to return to the tank. It has a diaphragm with fuel pressure on one side and atmospheric pressure on the other side, Figure 2–3. A diaphragm spring pushes the diaphragm against fuel pressure. As the diaphragm moves in the direction the spring is pushing it, a flat disc in the center of the diaphragm closes off the fuel return passage. This allows less fuel to return to the tank and causes fuel pressure to go up, closer to fuel pump pressure. The increased fuel pressure moves the diaphragm against the spring. This allows more fuel to escape through the return line and thus causes fuel pressure to drop. The diaphragm will always find a bal-

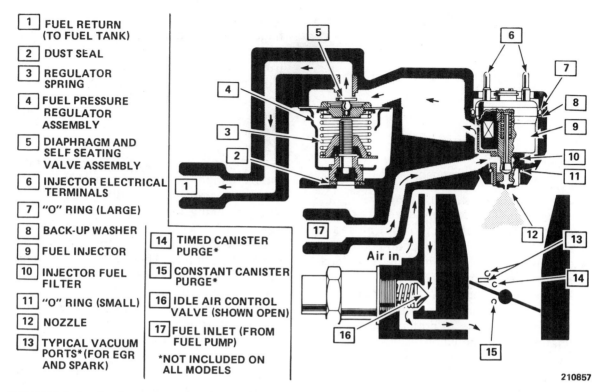

1	FUEL RETURN (TO FUEL TANK)
2	DUST SEAL
3	REGULATOR SPRING
4	FUEL PRESSURE REGULATOR ASSEMBLY
5	DIAPHRAGM AND SELF SEATING VALVE ASSEMBLY
6	INJECTOR ELECTRICAL TERMINALS
7	"O" RING (LARGE)
8	BACK-UP WASHER
9	FUEL INJECTOR
10	INJECTOR FUEL FILTER
11	"O" RING (SMALL)
12	NOZZLE
13	TYPICAL VACUUM PORTS*(FOR EGR AND SPARK)
14	TIMED CANISTER PURGE*
15	CONSTANT CANISTER PURGE*
16	IDLE AIR CONTROL VALVE (SHOWN OPEN)
17	FUEL INLET (FROM FUEL PUMP)

*NOT INCLUDED ON ALL MODELS

210857

FIGURE 2–3   Fuel pressure regulator. *Courtesy of Cadillac Motor Car Division.*

anced position that will keep fuel pressure at the desired value.

On some single-point injection systems, the spring side of the diaphragm is exposed to atmospheric pressure. Changes in atmospheric pressure produce a slight change in fuel pressure. Some multipoint injection systems connect manifold vacuum to the spring side of the diaphragm. Changes in atmospheric pressure or in manifold pressure change the total pressure on the spring side of the diaphragm and therefore the fuel pressure. The intent, however, is not to increase or decrease fuel delivery but to maintain a constant pressure drop across the injector. For example, when manifold pressure goes up, the injector has to spray fuel into a higher pressure environment. In order for the correct amount of fuel to be delivered during the pulse width, fuel pressure

must be increased in proportion to the manifold pressure increase.

SENSING DEVICES

Exhaust Oxygen Sensors

One of the major components of all comprehensive computerized engine control systems is the oxygen sensor, Figure 2–4. The amount of free oxygen in the exhaust stream is a direct result of the air/fuel ratio. A rich mixture yields little free oxygen; most of it is consumed during combustion. A lean mixture, on the other hand, yields considerable free oxygen because all of the oxygen is not consumed during combustion. With an oxygen sensor in or near the exhaust manifold, measuring the amount of oxygen in the exhaust reveals the air/fuel ratio burned in the cylinder.

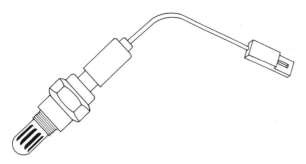

FIGURE 2-4   Oxygen sensor

The heart of the oxygen sensor is a hollow ceramic body *(zirconium dioxide)* closed at one end, Figure 2-5. The inner and outer surfaces of the ceramic body are coated with separate, thin, gas-permeable films of platinum. The platinum coatings serve as electrodes. The outer electrode surface is covered with a thin, porous, ceramic layer to protect against contamination from combustion residue. This body is placed in a metal shell similar to a spark plug shell except that the shell has a louvered nose that encloses the closed end of the ceramic

body. When the shell is screwed into the exhaust pipe or manifold, the louvered end extends into the exhaust passage. The outer electrode surface contacts the shell; the inner electrode connects to a wire that goes to the computer. Ambient oxygen is allowed to flow into the hollow ceramic body and to contact the inner electrode.

When the zirconium dioxide reaches 300° C (570° F), it becomes oxygen-ion conductive. If the oxygen in the exhaust stream is less than that in the ambient air, a voltage is generated. The greater the difference, the greater the voltage, Figure 2-6. Voltage generated by the oxygen sensor will normally range from a minimum of 0.1 volt (100 millivolts) to a maximum of 0.9 volt (900 millivolts). The computer has a preprogrammed value, called a *set-point,* that it wants to see from the oxygen sensor during closed-loop operation. The set-point is usually between 0.45 volt and 0.5 volt and equates to the desired air/fuel ratio. Voltage values below the set-point are interpreted as lean, and values above it are interpreted as rich.

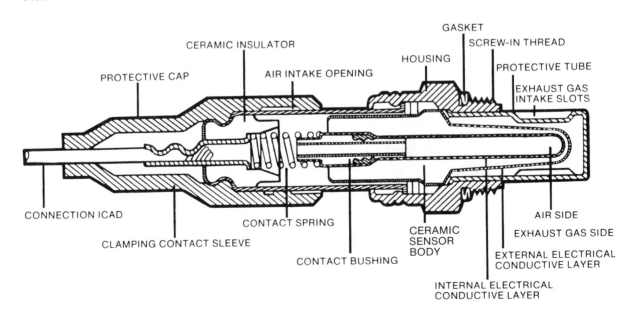

FIGURE 2-5   Exhaust gas oxygen sensor. *Courtesy of Ford Motor Co.*

By the mid to late 1980s, most manufacturers had begun using heated oxygen sensors on some engine applications. The operation of the sensor itself is no different; however, a heat element is placed inside it to keep it hot enough. This is largely due to a tendency to move the sensor a little lower in the exhaust system so that the exhaust sampled by the sensor is more of an average of what all of the cylinders are producing. Placing the oxygen sensor lower in the exhaust pipe means that it is exposed to less heat due to the rapid loss of heat through the pipe wall. Two additional wires are attached to the oxygen sensor to power the heat element. In most cases, one powers the heat element when the ignition is turned on and the other goes to a fixed ground. The circuit is not part of the computer control system.

Oxygen sensors are subject to contamination under certain conditions. Contaminating either electrode surface (the outside electrode is the most vulnerable) will shield it from oxygen and adversely affect its performance. Prolonged exposure to rich fuel mixture exhaust can cause it to be carbon fouled. This can sometimes be burned off by operating in a lean condition for two or three minutes. The oxygen sensor will also be fouled by exposure to:

- exhaust from leaded fuel.
- the vapors from some silicone-based gasket-sealing compounds.
- the residues of coolant leaking into the combustion chamber.

Fouling by these materials will probably require that the oxygen sensor be replaced.

Thermistors

A *thermistor* is a resistor made from a semiconductor material. Its electrical resistance changes greatly and predictably as its temperature changes. At $-40°$ C ($-40°$ F) a typical thermistor can have a resistance of 100,000 ohms. At $100°$ C ($210°$ F) the same thermistor

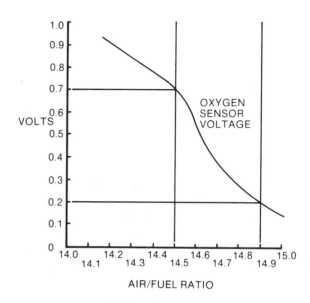

FIGURE 2-6   Relationship of oxygen sensor voltage to air/fuel ratio (approximate)

can likely have a resistance between 100 and 200 ohms. Even small changes in temperature can be observed by monitoring the thermistor's resistance. This characteristic makes it an excellent means of measuring the temperature of such things as water and air.

There are negative temperature coefficient (NTC) and positive temperature coefficient (PTC) thermistors. The resistance of the NTC thermistor goes down as its temperature goes up, although the resistance of the PTC type goes up as its temperature goes up. Most thermistors, including those used as temperature sensors on automotive computer systems, are the NTC type. A good example is an engine coolant temperature sensor, common to just about all computerized engine control systems. The coolant temperature sensor consists of a thermistor in the nose of a metal housing, Figure 2-7. The housing screws into the engine, usually in the head, with its nose extending into the water jacket so that the thermistor element will be the same temperature as that of

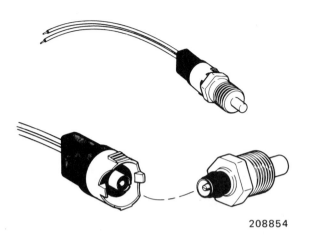

FIGURE 2-7   Coolant temperature sensor.
Courtesy of GM Product Service Training.

the coolant. The computer sends a regulated voltage signal (reference voltage, usually 5 volts) through a fixed resistance and then on to the sensor, Figure 2-8. A small amount of current flows through the thermistor (usually shown as a resistor symbol with an arrow diagonally across it) and returns to ground. This is a voltage divider circuit (current flows through the first resistance unit then has an alternate

path in parallel with the second resistance unit) and is commonly used as a temperature-sensing circuit. Because the resistance of the thermistor changes with temperature, the voltage drop across it changes also. The computer monitors the voltage drop across the thermistor and, using preprogrammed values, converts the voltage drop to a corresponding temperature value.

Potentiometers

A potentiometer is another application of a voltage divider circuit. It has a movable center contact or *wiper* that taps or senses the voltage between two resistance units, Figure 2-9. The potentiometer is usually used to measure either linear or rotary motion. If we were going to use a potentiometer to measure throttle position, we would probably attach it to one end of the throttle shaft. As the throttle is opened, the rotation of the throttle shaft moves the wiper, which slides along the wire-wound resistor. The computer sends a reference voltage to one terminal of the potentiometer, point A. If the wiper is positioned near point A (wide-open throttle on most applications), there will be a low voltage drop between points A and B (low

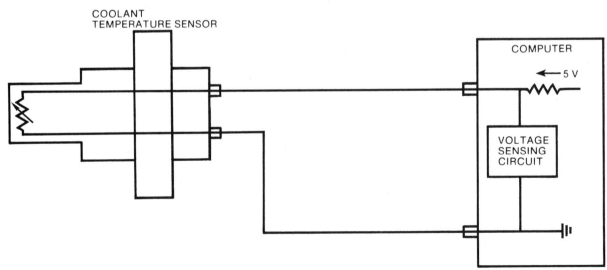

FIGURE 2-8   Temperature-sensing circuit

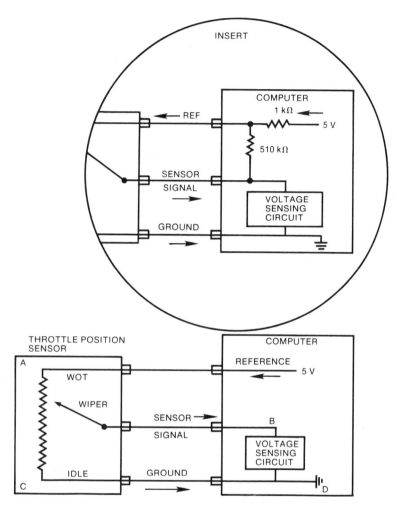

FIGURE 2-9   Potentiometer as throttle position sensor

resistance) and a high voltage drop will exist between points B and C. When the wiper is positioned near point C (idle position on most applications), there is a high voltage drop between points A and B and a low drop between points B and C. The computer monitors the voltage drop between points B and C (C should be the same as D) and interprets a low voltage, 0.5 for example, as idle position. A high voltage, around 4.5, will be interpreted as wide-open throttle (WOT). Voltages between these values will be interpreted as a proportionate throttle position.

The insert in Figure 2-9 shows a slight variation in the throttle position sensor circuit. Some applications use this variation to enhance diagnostic capability. The high resistance connection between the reference voltage circuit and point B causes near reference voltage to appear at the sensor signal terminal if an open circuit occurs in the harness that connects the sensor to the computer.

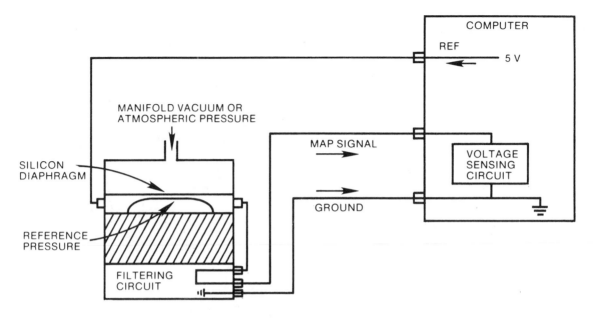

FIGURE 2-10   Silicon diaphragm pressure sensor

Pressure Sensors

Intake manifold pressure is a direct response to throttle position and engine speed, with throttle position being the most significant factor. The greater the throttle opening, the greater the manifold pressure becomes (lower vacuum). At WOT manifold pressure is nearly 100% of atmospheric pressure. Intake manifold pressure can therefore be translated as engine load and is a critical factor in determining many engine calibrations. Barometric pressure also impacts on manifold pressure and in most systems is considered when calculating engine calibrations such as air/fuel mixture and ignition timing.

Intake manifold pressure (usually a negative pressure or *vacuum*) and/or barometric pressure are, on most systems, measured with a silicon diaphragm that acts as a resistor, Figure 2-10. The resistor/diaphragm (about 3 millimeters wide) separates two chambers. As the pressure on the two sides of the resistor/diaphragm varies, it flexes. The flexing causes the

resistance of this semiconductor material to change its electrical resistance. The computer applies a reference voltage to one side of the diaphragm. As the current crosses the resistor/diaphragm, the amount of voltage drop that occurs depends on how much the diaphragm is flexed. The signal is passed through a filtering circuit before it is sent to the computer as a DC analog signal. The computer monitors the voltage drop and, by using a look-up chart of stored pressure values, determines exactly what the pressure is to which the diaphragm is responding.

There are two slightly different pressure sensor designs that use the *piezoresistive* silicon diaphragm: the absolute pressure and the pressure differential sensors, Figure 2-11. In one design the chamber under the diaphragm is sealed and contains a fixed reference pressure. The upper chamber is exposed to either intake manifold pressure or to atmospheric pressure. If the upper chamber is connected to the intake manifold, the sensor functions as a *manifold absolute pressure sensor.* Absolute, as used in MAP, refers to a sensor that compares a varying pressure to a fixed pressure. The output signal (return voltage) from this sensor increases as pressure on the variable side of the diaphragm increases (wider throttle opening). If the upper chamber is exposed to atmospheric pressure, the sensor functions as a *barometric pressure sensor.*

The differential pressure sensor combines the functions of both the MAP and the BARO sensors. Instead of using a fixed pressure, one side of the resistor/diaphragm is exposed to barometric pressure and the other side is connected to intake manifold pressure. The output signal is the result of subtracting manifold vacuum from barometric pressure. Opposite to the MAP sensor, however, as manifold pressure decreases (higher vacuum), output voltage increases.

Other types of pressure sensors are discussed in Chapters 7 and 8.

Measuring Air Mass

Among most manufacturers' computerized engine control system offerings is at least one port fuel injection or multipoint injection sys-

PRESSURE SENSOR DESIGN

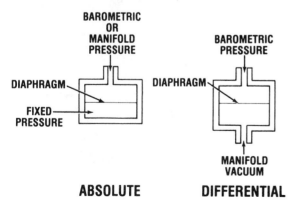

ABSOLUTE   DIFFERENTIAL

FIGURE 2-11 Two types of silicon diaphragm pressure sensors. *Courtesy of GM Product Service Training.*

tem, and it appears almost certain that multi-point injection will quickly become the predominate fuel-metering system. Although a multipoint injection system offers several distinct advantages over other types of fuel-metering systems (discussed earlier in this chapter), it has one shortcoming. A multipoint injection system provides less opportunity for the fuel to evaporate than a system where the fuel is introduced into the center of the intake manifold does, and only the fuel that is evaporated before combustion occurs is useful. In the past this problem was probably most apparent as a lean stumble when accelerating from idle and was the result of the high manifold pressure that comes from the sudden increase in throttle opening. With the throttle open, atmospheric pressure forces air into the manifold faster than the engine can use it. The turbulence produced during the intake and compression strokes (and the increased oxygen density on turbocharged applications) helps to atomize the fuel and hasten its evaporation. The condition is only completely overcome, however, by spraying in additional fuel to compensate for the failure of the heavier hydrocarbon molecules to evaporate. The capability of the digital microprocessor provides the ability to calculate just the right amount of fuel to avoid a lean stumble without sacrificing economy or emissions.

Given an engine at normal operating temperature and warm intake air, there is one precise amount of fuel that should be injected into the intake port, for each manifold pressure value within the operating range, that will provide an evaporated 14.7 to 1 air/fuel ratio in the combustion chamber with the lowest possible leftover unevaporated HC. The industry has long been able to identify and meter that amount of fuel for each manifold pressure value. The problem has been that the correct amount of fuel is so critical that if the system were adjusted to perfection one day, it probably would not be right the next day because the air's ability to evaporate fuel varies in response to changes in atmospheric pressure, humidity, and temperature (the air's mass).

In order to make the multipoint injection system work as well as it does now, the manufacturers had to find a way to determine the air's mass and meter fuel accordingly. Two methods have been used successfully to determine a value equivalent to air's mass: the mass airflow sensor and the speed-density formula. These are both discussed in the **Inputs** section of Chapter 5.

Hall-Effect Switches

A *Hall-effect switch* is frequently used to sense engine speed and crankshaft position. It provides a signal each time a piston reaches top dead center, and its signal serves as the primary input on which ignition timing is calculated. Because of the frequency with which this signal occurs and the critical need for accuracy, the Hall-effect switch is a popular choice because it and its related circuitry provide a digital signal. On some applications a separate Hall-effect switch is used to monitor camshaft position as well. Systems that do not use a Hall-effect switch, use either the magnetic pick-up unit in the distributor as the crankshaft speed and position information source or its equivalent outside the distributor.

A Hall-effect switch consists mainly of a permanent magnet, a *gallium arsenate crystal* with its related circuitry, and a shutter wheel, Figure 2–12. The permanent magnet is mounted so that a small space is between it and the gallium arsenate crystal. The shutter wheel, rotated by a shaft, alternately passes its vanes through the narrow space between the magnet and the crystal, Figure 2–13. When a vane is between the magnet and the crystal, the vane intercepts the magnetic field and thus shields the crystal from it. When the vane moves out, the gallium arsenide crystal is invaded by the magnetic field. A steady current is being passed through the crystal from end to end. When the magnetic lines of force from the permanent magnet invade the crystal, the current flow across the crystal is distorted. This results in a weak voltage potential being produced at the crystal's top and bottom surfaces, negative at one surface and positive at the other. As the shutter wheel turns, the crystal provides a weak high/low voltage signal.

This signal is usually modified before it is sent to the computer. For example, as shown in Figure 2–13, the signal is sent to an amplifier, which strengthens and inverts it. The inverted signal (high when the signal coming from the Hall-effect switch is low) goes to a *Schmitt trigger* device, which converts the analog signal to a digital signal. The digital signal is fed to the base of a switching transistor. The transistor is switched on and off in response to the signals generated by the Hall-effect switch assembly.

When the transistor is switched on, current flows through it from another circuit to ground. The other circuit can come from the ignition switch or from the computer, as shown in Figure 2–13. In either case it has a resistor between its voltage source and the transistor. A voltage-sensing circuit in the computer connects to the

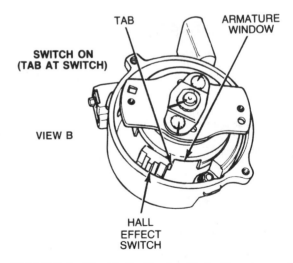

FIGURE 2-12   Hall-effect switch. *Courtesy of Ford Motor Co.*

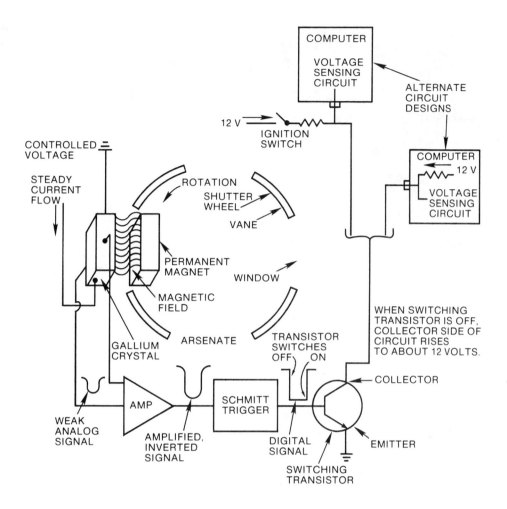

FIGURE 2-13   Hall-effect switch circuit

circuit between the resistor and the transistor. When the transistor is on and the circuit is complete to ground, a volt drop occurs across the resistor. The voltage signal to the voltage-sensing circuit is less than 1 volt. When the transistor is switched off, there is no drop across the resistor; and the voltage signal to the voltage-sensing circuit is near 12 volts. The computer monitors the voltage level and can determine by the frequency at which it rises and falls what engine speed is. Each time it

rises, the computer knows that a piston is approaching top dead center.

Detonation Sensors

To optimize performance and fuel economy, ignition timing must be adjusted so that combustion will occur during a specific number of crankshaft degrees beginning at top dead center of the power stroke. If it occurs any later, less pressure will be produced in the cylinder. If it occurs any sooner, detonation will

FIGURE 2–14   Knock sensor

occur. Due to variations in fuel quality, cylinder-cooling efficiency, machining tolerances and their effect on compression ratios, and so forth, a preprogrammed spark advance schedule can result in spark knock under certain driving conditions. In order to be able to provide an aggressive spark advance and avoid spark knock, many systems use a detonation sensor, often referred to as a *knock sensor.* It alerts the computer when spark knock occurs so that the timing can be retarded.

The knock sensor, Figure 2–14, screws into some section of the engine where it will be subjected to the high-frequency vibration caused by the spark knock. Most knock sensors contain a piezoelectric crystal. *Piezoelectric* refers to a characteristic of certain materials whereby they produce an electrical signal in response to physical stress or experience stress in response to an electrical signal. The most commonly used piezoelectric material produces an oscillating voltage signal in response to pressure or vibration. The oscilla-

tions occur at the same frequency as the spark knock (5,000 to 6,000 hertz [Hz], or cycles per second). Although most vibrations within the engine and many elsewhere in the vehicle cause the knock sensor to produce a signal, the computer only responds to signals of the correct frequency.

Switches

Most systems have several simple switches that are used to sense and communicate information to the computer. The information includes whether or not the transmission is in gear, the air-conditioning is on, the brakes are applied, and so on. The same type of circuit is usually used for each switch regardless of what condition the switch is measuring.

As an example, let's take a switch used to tell the computer whether or not the automatic transmission is in gear, Figure 2–15. The ignition switch feeds 12 volts through 1 k ohms of resistance to the P/N switch. Let's assume that when the transmission is in park or neutral, the switch is closed and thus allows current to go to ground. During this condition the only significant resistance between the voltage source and ground is the 1 k-ohm resistor, and all but a tiny fraction of the 12 volts is dropped across it. The voltage signal to the computer is near zero. When the transmission is shifted into gear, the switch opens. Now the only path to ground is through the 1 k-ohm resistor and then the 10 k-ohm resistor in the computer. As the current goes through the 1 k-resistor, about 1 volt is dropped and the remaining 11 volts is dropped across the 10 k-ohm resistor. The computer recognizes that when voltage is present at the sensing circuit, the transmission is in gear.

E-cells

Some selected systems use an E-cell to indicate to the computer when a specified amount of engine operation has occurred. The E-cell actually measures a specific amount of ignition on time, which equates to an estimated number of vehicle miles. Its purpose is to alert the computer when the predetermined mileage has occurred so that the computer can adjust specific calibrations to compensate for engine wear.

The E-cell contains a silver cathode and a gold anode. Ignition voltage is applied to the E-cell through a resistor. The passage of current through the cell controls a signal to the computer. As current passes through the cell, a chemical reaction takes place; the silver on the cathode is attracted to the gold anode. After a certain amount of on time, the silver cathode is completely deplated and the circuit opens.

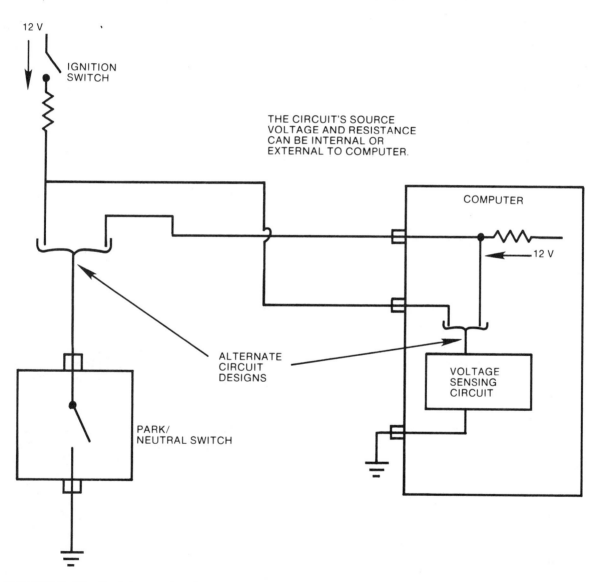

FIGURE 2–15   Park/neutral switch

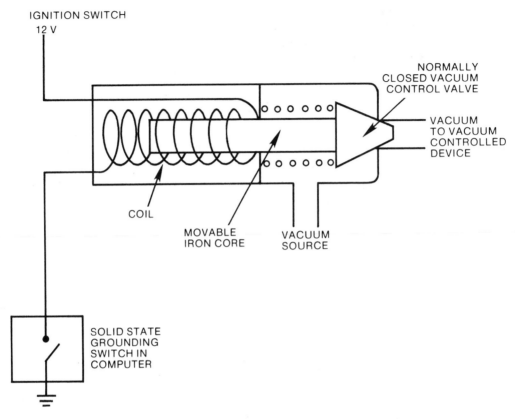

IGNITION SWITCH
12 V

NORMALLY
CLOSED VACUUM
CONTROL VALVE

VACUUM
TO VACUUM
CONTROLLED
DEVICE

COIL

MOVABLE
IRON CORE

VACUUM
SOURCE

SOLID STATE
GROUNDING
SWITCH IN
COMPUTER

FIGURE 2-16   Typical solenoid-controlled vacuum valve

How long this takes is controlled by the amount of current allowed to flow through the E-cell (how much resistance is in the circuit).

ACTUATORS _____

Solenoids

The most common actuator is a valve powered by a solenoid. The valve usually controls vacuum, but it can also be a valve designed to control fuel vapor, airflow, oil or water flow, and so forth. A solenoid consists of a coil of wire with a movable iron core, Figure 2-16. When the solid-state switching circuit in the computer grounds the solenoid circuit, the coil becomes energized. The magnetic field developed by the current passing through the coil windings pulls the iron core so that it occupies the entire length of the coil. This movement of the core opens the valve attached to its other end.

Duty-cycled and Pulse Width Solenoids. In some cases the switching circuit that drives the solenoid turns on and off rapidly. If the solenoid circuit is turned on and off (cycled) ten times per second, then the solenoid is *duty cycled*. To complete a cycle, it must go from off to on to off again. If it is turned on for 20% of each tenth of a second and off for 80%, it is said to be on a 20% duty cycle. In most cases where a solenoid is duty cycled, the computer has the ability to change the duty cycle to achieve a desired result. The best example of this is the mixture control device on many computer-controlled carburetors, where the duty cycle will frequently change to control the air/fuel mixture.

If a solenoid is turned on and off rapidly but there is no set number of cycles per second at which it is cycled, the on time is referred to as *pulse width.* For example, let's assume that the computer has decided that the EGR valve should be open but that it should be open to only 60% of its capacity. The computer can energize a solenoid valve such as the one in Figure 2–16 to allow vacuum to the EGR valve. To control or *modulate* the amount of vacuum to the EGR valve, the computer issues pulse width commands to a solenoid that when energized opens a port that allows atmospheric pressure to bleed into the EGR vacuum line,

Figure 2–17. Because the computer wants the EGR valve open to 60% of its capacity, it selects pulse widths from a look-up chart to appropriately weaken the vacuum signal to the EGR valve. Some literature refers to pulse width as *dithering.*

Relays

A *relay* is a remote control switch that allows a light-duty switch such as an ignition switch, a blower motor switch, a starter switch, or a driver switch in a computer to control a device that draws a relatively heavy current load. The most common example in automotive computer application is a fuel pump relay, Figure 2–18. When the computer wants the fuel pump on, it turns on its driver switch (transistor). The transistor applies 12 volts to the relay

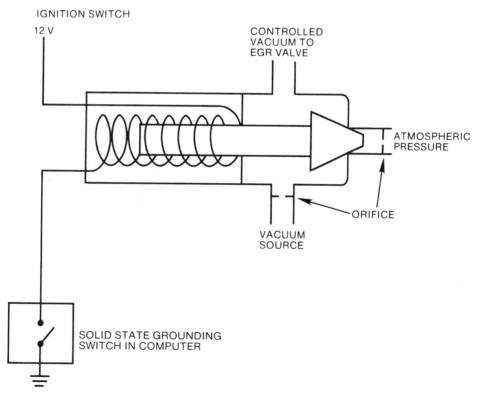

FIGURE 2–17   Typical duty-cycled vent valve

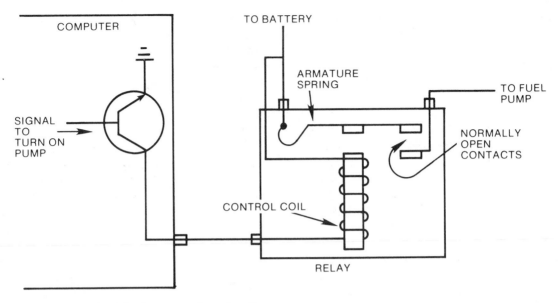

FIGURE 2–18   Typical fuel pump relay circuit

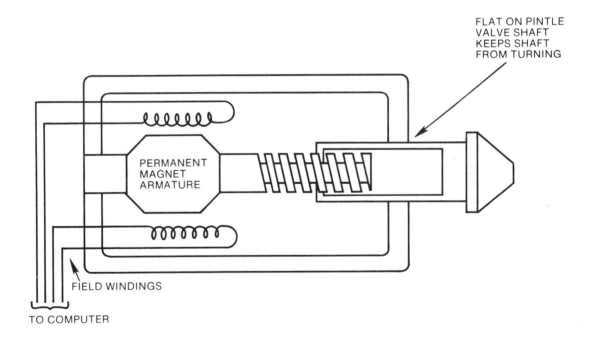

FIGURE 2–19   Typical stepper motor

coil, which develops a magnetic field in the core. The magnetic field pulls the armature down and closes the normally open contacts. The contacts complete the circuit from the battery to the fuel pump. On many systems the fuel pump relay is one of the few examples where the computer supplies the voltage to operate a device instead of the ground connection.

Electric Motors

When electric motors are used as actuators, there are two unique types used.

Stepper Motors. A stepper motor, as used in these applications, is a small electric motor powered by voltage pulses. It contains a permanent magnet armature and usually either two or four field coils, Figure 2–19. A voltage pulse applied to one coil or pair of coils causes the motor to turn a specific number of degrees. The same voltage pulse applied to the other coil or pair of coils makes the motor turn an equal number of degrees in the opposite direction.

The armature shaft usually has a spiral on one end and this spiral connects to whatever the motor is supposed to control. As the motor turns one way, the controlled device, a *pintle valve* for instance, is extended. As it turns the other way, the valve is retracted. The computer can apply a series of pulses to the motor's coil windings in order to move the controlled device to whatever location is desired. The computer can also know exactly what position the valve is in by keeping count of the pulses applied. The stepper motor has been used to control air/fuel mixture and idle speed.

Permanent Magnet Field Motors. Some systems use this type of motor to control idle speed, Figure 2–20. It is a simple, reversible, DC motor with a wire-wound armature and a permanent magnet field. The polarity of the voltage applied to the armature winding determines the direction in which the motor spins. The armature shaft drives a tiny gear

drive assembly that either extends or retracts a plunger. The plunger contacts the throttle linkage; and when it is extended, the throttle blade opening is increased. When it is retracted, the throttle blade opening is decreased. The computer has the ability to apply a continuous voltage to the armature until the desired idle speed is reached.

DIAGNOSIS AND SERVICE

Special Tools

A good service manual that applies specifically to the vehicle being serviced is essential. It should contain all of the necessary charts and diagnostic procedures and should list the tools necessary for diagnostic operations.

Most systems have a data link used in the factory to test the system's operation as it leaves the assembly line. Several companies

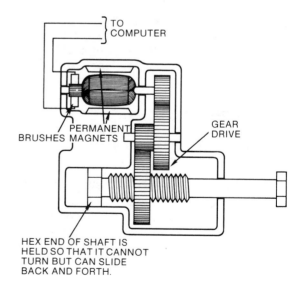

FIGURE 2–20   Permanent magnet field reversible DC motor

have designed and are selling test equipment that can connect to the data link and get service code information from the computer. In most cases other information such as sensor voltage values and the state of position of some actuators can also be obtained. This information is either shown as a digital readout on the test instrument or is printed on a piece of paper. This type of tool is very helpful and in some cases is essential.

High-impedance Digital Volt-Ohmmeter (DVOM). When diagnosing most systems, a DVOM is required for two reasons.

1. Many of the readings to be taken require a higher degree of accuracy than is obtained by reading a needle position on a scale.

2. Many of the circuits that will be tested have high resistance and very low voltage and amperage values. Applying a standard meter with its relatively low internal resistance to such a circuit can easily constitute a short to the circuit. In other words the meter's circuit offers an easier electrical path than the circuit or portion of the circuit being tested. A *high-impedance* meter has a minimum of 10 million ohms of internal resistance; a standard meter has only about 10 thousand.

Service Tips

Oxygen Sensor. When handling or servicing the oxygen sensor, several precautions should be observed.

- A boot is often used to protect the end of the sensor where the wire connects. It is designed to allow air to flow between it and the sensor body, Figure 2–21. The vent in the boot and the passage leading into the chamber in the oxygen-sensing element must be open and free of grease or other contaminants.
- Those with soft boots must not have the boot pushed onto the sensor body so far as to cause it to overheat and melt.

- No attempt should be made to clean the sensor with any type of solvent.
- Do not short the sensor lead to ground, and do not put an ohmmeter across it. Any such attempt can permanently damage the sensor. The single exception is that a digital high-impedance (10 megohm minimum) voltmeter can be placed across the sensor lead to ground (the sensor body).
- If an oxygen sensor is reinstalled, its threads should be inspected and if necessary cleaned with an 18 mm spark plug thread chaser. Before installation its threads should be coated with an anti-seize compound, preferably liquid graphite and glass beads. Replacement sensors may already have the compound applied.

PERSONAL SAFETY: It is often recommended that the oxygen sensor be removed while the engine is hot to reduce the possibility of thread damage. If doing so wear leather gloves to prevent burns on the hand.

Vacuum Leaks. Vacuum leaks can have far-reaching effects. Examine Figure 2–22. Assume that the diaphragm ruptured in the primary vacuum break (or it could be a vacuum hose cracked or left off anywhere in that circuit). The results would be as follows:

- The primary vacuum break would not work.
- The secondary vacuum break would not work (rich condition, possibly stalling during warm-up).
- The heated air system would not work (poor cold driveability).
- The vacuum sensor would report "no vacuum" to the computer, which would in

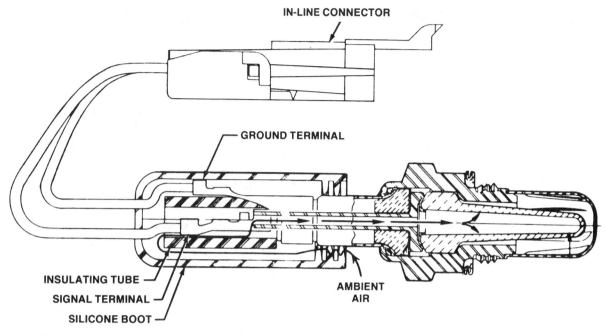

IN-LINE CONNECTOR

GROUND TERMINAL

INSULATING TUBE

SIGNAL TERMINAL

SILICONE BOOT

AMBIENT AIR

FIGURE 2–21   Oxygen sensor. *Courtesy of GM Product Service Training.*

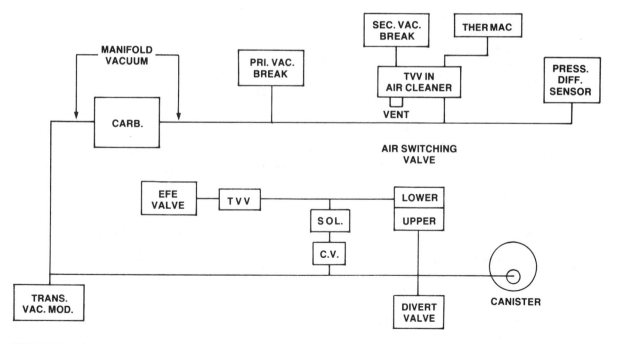

MANIFOLD VACUUM

PRI. VAC. BREAK

SEC. VAC. BREAK

THERMAC

TVV IN AIR CLEANER

VENT

PRESS. DIFF. SENSOR

CARB.

AIR SWITCHING VALVE

EFE VALVE

TVV

SOL.

LOWER

UPPER

C.V.

CANISTER

TRANS. VAC. MOD.

DIVERT VALVE

FIGURE 2–22   Typical vacuum circuit. *Courtesy of GM Product Service Training.*

turn issue erroneous spark and fuel mixture commands.

- A vacuum sensor code would probably be set.

Vacuum leaks downstream from the airflow sensor on a multipoint injection system cause poor idle if the engine idles at all.

Charcoal Canister. Charcoal canister saturation can often be the result of fuel tank overfilling. Slowly filling the tank right to the top of the filler neck causes the antioverfill chamber in the tank to fill and thus leaves no room for fuel expansion. When the fuel does expand, it is pushed up and into the charcoal canister. Canister purging then dumps so much hydrocarbon into the intake manifold that the system may not be able to compensate, even in closed loop. Many driveability problems can be linked to charcoal canister problems; don't overlook it in diagnostic procedures.

Avoid Damaging the Computer. When servicing a computerized engine control system vehicle, observe the following in order to avoid producing a voltage spike that could damage the computer:

- Avoid starting the vehicle with the aid of a battery charger.
- Turn the ignition off before connecting jumper cables.
- Turn the ignition off before disconnecting the battery or computer connectors.
- Replace any system actuator with less than specified resistance or that has a shorted clamping diode. A *clamping diode* is a diode placed across an electrical winding to suppress voltage spikes that are produced when the winding is turned on and off.
- Be sure that the winding of any electrical accessory added to the vehicle has sufficient voltage spike containment capability. A resistor of 200- to 300-watt capacity can be used instead of a diode.
- Use caution when testing the charging system, and never disconnect the battery cables with the engine running.

REVIEW QUESTIONS _____

1. What does the term *duty cycle* mean?
2. How is duty cycle different from pulse width?
3. How does the oxygen sensor know if the air/fuel mixture is rich?
4. What type of thermistor has higher resistance at lower temperatures?
5. Diagram a voltage divider circuit.
6. Diagram a potentiometer circuit.
7. Describe how a Hall-effect switch turns on and off a switching transistor.
8. Describe how the Hall-effect switch's switching transistor signals the computer.
9. What information does the Hall-effect switch provide for the computer?
10. Name a resistance device commonly used to measure air pressure.
11. Why is 5 volts most commonly used as a reference voltage for sensor operation?
12. Describe how a simple switch is used as a sensing device.
13. What is an actuator?
14. Describe the operation of a solenoid.
15. Describe the operation of a relay.
16. Describe how to reverse the direction of rotation of a permanent magnet field motor.
17. For what is a permanent magnet field motor usually used?

ASE-type Questions. (Actual ASE test questions will probably not be as product specific.)

18. Technician A says that a rich exhaust stream contains little free oxygen because most of the available oxygen was combined with other elements during combustion. Technician B says that little free oxygen is present in a rich exhaust stream because there was not enough oxygen in

the air/fuel mixture from the beginning. Who is correct?
a. A only
b. B only
c. both A and B
d. neither A nor B

19. The voltage reading from the oxygen sensor of a car with a computerized engine control system is consistently around 0.7 volt. This indicates that the system is operating in a _____ condition.
a. rich
b. lean
c. normal
d. abnormal

20. Technician A says that a potentiometer is usually used to monitor throttle position. Technician B says that a thermistor is usually used to monitor throttle position. Who is correct?
a. A only
b. B only
c. both A and B
d. neither A nor B

21. Technician A says that an oxygen sensor can be fouled by fuel additives such as lead and antioxidation agents. Technician B says that an oxygen sensor can be fouled by use of a gasket compound containing silicon and by fuels containing alcohol. Who is correct?
a. A only
b. B only
c. both A and B
d. neither A nor B

GLOSSARY _____

ACTUATOR   Any mechanism, such as a solenoid, relay, or motor, that when activated causes a change in the performance of a given system or circuit.

CLAMPING DIODE   A diode placed across a coil or winding to trap the voltage spikes produced when the device is turned on or off.

DUTY CYCLE   The portion of time during each cycle when an electrical device is turned on, when the device is cycled at a fixed number of cycles per unit of time.

GALLIUM ARSENATE CRYSTAL   A semiconductor material that changes its conductivity when exposed to a magnetic field.

HALL-EFFECT SWITCH   A magnetic switching device often used to signal crankshaft position and speed to the computer.

Hz (HERTZ)   Cycles per second.

IMPEDANCE   The total circuit resistance including resistance and reactance.

PIEZOELECTRIC   A crystal that produces a voltage signal when subjected to physical stress such as pressure or vibration.

PIEZORESISTIVE   A semiconductor material whose electrical resistance changes in response to changes in physical stress.

PINTLE VALVE   A valve with a stem, some part of which is cone shaped. The cone-shaped portion of the stem fits against a seat to close the passage through the valve.

POTENTIOMETER   A three-terminal variable resistor that acts as a voltage divider.

PULSE WIDTH   The portion of time during a cycle when an electrical device is turned on, when the rate at which the cycles occur varies.

SCHMITT TRIGGER   A device that trims an analog signal and converts it to a digital signal.

THERMISTOR   A temperature-sensing device containing a coil of wire whose resistance changes dramatically as the temperature changes. Most thermistors have a negative temperature coefficient (the resistance goes up as the temperature goes down).

ZIRCONIUM DIOXIDE (ZrO_2)   A white crystalline compound that becomes oxygen-ion conductive at about 315° C (600° F).

3 GENERAL MOTORS' COMPUTER COMMAND CONTROL

Objectives

After studying this chapter, you will be able to:

- describe torque converter clutch operation.
- explain how a dwellmeter is used to evaluate air/fuel mixture control.
- describe the operation of the mixture solenoid in a computer command control system.
- describe how idle speed is controlled in a computer command control system.
- state when the air management system directs air to the exhaust manifold, catalytic converter, atmosphere.
- describe what a trouble code is and how it is obtained from the computer.

The computer command control (CCC) system was General Motors' first widely used comprehensive computerized engine control system. It was first introduced in mid-1980. Beginning with the 1981 model year, all of General Motors' carbureted passenger cars have made use of the CCC system. It should be noted that the CCC system varies slightly from one engine application to another and from one model year to another. In 1982 four different CCC systems were used. Most 1982 engines used the full function system, which was a slightly updated version of the 1981 system. Beginning in 1982 Oldsmobile 5-liter engines were equipped with a limited control system; the T car (Chevrolet Chevette and Pontiac

From EFC to CCC

General Motors introduced its first electronic closed-loop fuel control system in 1978. This single function system, called electronic fuel control (EFC), was limited to selected four-cylinder engines. In 1979 GM introduced its first comprehensive engine control system on selected applications: computer-controlled catalytic converter (C-4). The system was improved for 1981 production and was renamed the computer command control. Someone at GM decided they did not want their engine control system named after a Ford transmission.

T-1000) used a minimum function system. The Chevrolet and GMC S truck (S-10 and S-15, introduced in 1982) with the California emissions package and a four-cylinder gas engine used an imported electronic control module (ECM) that was slightly different from the other CCC systems. There are also slight variations within each of these systems.

Although all of this can seem a bit bewildering, it need not be. All versions of the CCC system are much more alike than different. Once the basic system and function are understood and a good service manual is at hand for specific directions and specifications, you will find the CCC system very manageable.

ELECTRONIC CONTROL MODULE (ECM)

The engine control computer used in the CCC system and in all other General Motors systems is called an electronic control module, Figure 3-1. The inputs it receives and the functions it controls are shown in Figure 3-2. The ECM is located in the passenger compartment, usually near the glove compartment or behind the passenger kick panel. The PROM (programmable read-only memory) is located under an access cover, Figure 3-3. It is responsible for fine-tuning engine calibrations for each specific vehicle and when plugged in becomes a functioning part of the ECM.

Note: The PROM can easily be installed backwards, which will result in electrical damage to it. Consult a service manual or the **System Diagnosis and Service** section of this chapter.

Keep Alive Memory (KAM)

The ECM monitors its most important sensor and selected actuator circuits. If a fault such as an open circuit occurs during normal operation, a fault code will be stored in the KAM. The code can be retrieved at a later time to assist the technician in locating the problem. This feature is especially useful because it enables the ECM to report a fault that occurred recently but is no longer present, as well as those that are still present. The KAM receives battery power at all times so that the code can be retained in memory.

OPERATING MODES

The two basic modes of operation are closed loop and open loop.

Closed Loop

Before the system can go into closed loop, the coolant must reach a temperature of approximately 65° C (150° F), the oxygen sensor must reach at least 300° C (570° F), and a predetermined amount of time must elapse following engine start-up. This time is programmed into the PROM and varies with engine application. It can be as little as a few seconds on some engines or as much as a couple of minutes on others.

FIGURE 3-1   ECM. *Courtesy of GM Product Service Training.*

Inputs	Codes	ECM	Codes	Outputs
Coolant Temperature Sensor	14, 15		23	Fuel Mixture
Vacuum Sensor	34		42	Electronic Spark Timing
Barometric Pressure Sensor (if used)	32	ROM and PROM process information (inputs) and issue commands (outputs).		Electronic Spark Retard (if used)
Throttle Position Sensor	21			Idle Speed
Distributor Reference (crank position—engine speed)	12, 41			AIR Management
				EGR
Oxygen Sensor	13, 44, 45			Canister Purge
Vehicle Speed Sensor	24			Torque Converter Clutch (or shift light on manual transmission)
Ignition On				
Air Conditioner On/Off				
Park/Neutral Switch		RAM monitors indicated circuits, sets codes, and reads codes out when put in diagnostics.		Air Conditioner
System Voltage				Early Fuel Evaporation
Transmission Gear Position Switch(es)				Diagnosis (check engine light)
Spark Knock Sensor (if used)	43			(test terminal)
EGR Vacuum Indicator Switch (only on 1984 and later models)	53			(serial data)
Idle Speed Control Switch	35			

FIGURE 3-2   Overview of computer command control. Inputs and outputs with code numbers are monitored for faults. Those without are not.

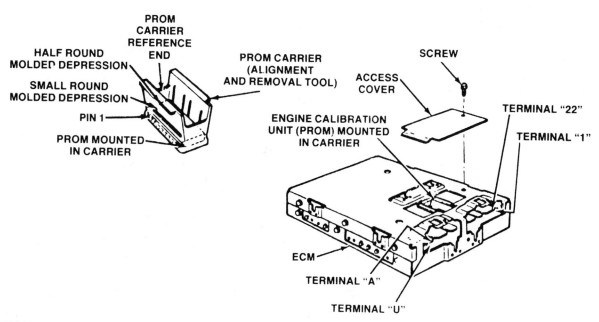

FIGURE 3-3   ECM and PROM. *Courtesy of GM Product Service Training.*

The system drops out of closed loop if the vehicle is operated at or near wide-open throttle (WOT), if the coolant temperature drops below the criterion value, or if certain system component failures occur.

Note: Nineteen eighty-two and later T cars (Cheverolet Chevette and Pontiac T-1000) used a minimum function CCC system. These systems use a coolant temperature switch instead of a coolant temperature sensor. As the engine reaches operating temperature, the switch closes and thus puts the system into closed loop (assuming the O_2 sensor is hot and the proper time has expired). If the engine then cools down and the switch opens, the system stays in closed loop.

Open Loop

This is an operating mode that includes several suboperational modes.

Start-up Enrichment. This provides a rich command to the mixture control solenoid for a short time after each engine start. The duration depends on engine temperature, as measured by the coolant temperature sensor, when the engine is started. If the engine is started hot, this mode will be shorter than if it is started cold.

Blended Enrichment. This mode occurs during engine warm-up (engine coolant and/or the oxygen sensor are not up to proper operating temperature). In this mode the ECM controls the air/fuel ratio by using information from sensors concerning coolant temperature, throttle position, manifold pressure, barometric pressure, and engine speed. As engine temperature comes up, the air/fuel ratio is adjusted leaner by the ECM. In this open-loop mode, the ECM attempts to adjust the air/fuel ratio using fixed, programmed instructions in its read-only memory. If the air/fuel ratio is not correct because of a carburetor malfunction, vacuum

leak, and so forth, the ECM will not recognize the problem and will not respond to it. In other words in open loop the ECM does what it is programmed to do, but the engine still does not run correctly unless other engine components function properly.

Power Enrichment. This mode occurs when the vehicle is operated at or near wide-open throttle. The ECM sends a steady power enrichment signal to the mixture control solenoid. This provides the rich mixture required by an engine when manifold vacuum is near zero.

Shutdown. This mode turns off the M/C solenoid, and thus creates a full, rich operating condition, when the reference pulse to the ECM indicates 200 RPM or less and when battery voltage to the ECM is less than 9 volts.

Limp In. This mode allows the engine to continue operating in spite of most major failures, such as an ECM failure, that can occur within the system. This is a *fail soft* mode that provides no fuel control (full rich or full metered) and no spark advance.

INPUTS

Coolant Temperature Sensor (CTS)

With the exception of some T car applications (Chevrolet Chevette and Pontiac T-1000), all CCC systems use a CTS, Figure 3–4, as described in Chapter 2. It is most often located near the thermostat housing and is the single most important sensor in the system. Its input impacts on just about every decision the computer makes. (This statement can generally be applied to all systems.)

Some 1982 and later T cars are equipped with a minimum function system and use a coolant temperature switch instead of a CTS. The switch is open while the coolant is below a specified temperature and closes as the coolant approaches normal operating temperature. In this system the ECM only knows whether the coolant is below or above a specific temperature.

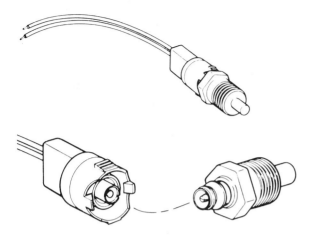

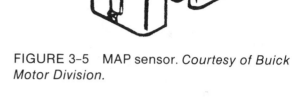

FIGURE 3-5   MAP sensor. *Courtesy of Buick Motor Division.*

FIGURE 3-4   Coolant temperature sensor. *Courtesy of GM Product Service Training.*

Note: The MAP and BARO sensors used on 1980 and early 1981 vehicles were round metal units, both mounted on the same bracket, under the hood.

Pressure Sensors

There are three different types of pressure sensors used on various CCC applications. They are all of the piezoresistive silicon diaphragm type and are nearly identical in appearance, Figure 3-5. Each can, however, be identified by a colored plastic insert at its electrical harness connector.

Manifold Absolute Pressure (MAP) Sensor. The MAP sensor has either a red or orange (orange for turbo engines) connector. The MAP sensor is often used in combination with a BARO sensor.

Barometric Pressure (BARO) Sensor. The BARO sensor has a blue connector. On some applications it is located under the hood and near the MAP sensor. On others it is inside the passenger compartment and under the instrument panel.

Pressure Differential (VAC—for vacuum) Sensor. The VAC sensor has a gray or black connector and is used by itself.

Oxygen (O₂) Sensor

The O_2 sensor used on 1981 and earlier models had two wires and used a vented silicone boat to cover and protect its open end, Figure 3-6. The second wire served as a back-up ground; the sensor shell is already grounded to the manifold. Beginning in 1982 the ground

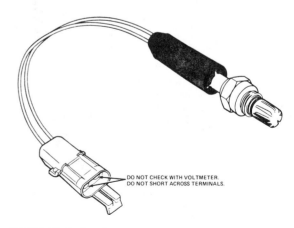

DO NOT CHECK WITH VOLTMETER.
DO NOT SHORT ACROSS TERMINALS.

FIGURE 3-6   Oxygen sensor, early style (two wire). *Courtesy of GM Product Service Training.*

wire was not used, and a metal boot replaced the silicone boot. The silicone boot melted if it was pushed too far onto the O₂ sensor toward the exhaust manifold.

Note: There is an O₂ sensor ground wire that runs from the ECM to the engine block (not connected to the O₂ sensor).

Throttle Position Sensor (TPS)

The TPS is a potentiometer mounted in the bowl section of the carburetor, Figure 3-7. When the throttle is depressed, the accelerator pump lever forces the TPS plunger down. This action moves the potentiometer's wiper. The TPS is the only sensor in the CCC system that can be adjusted. Because engine vacuum and

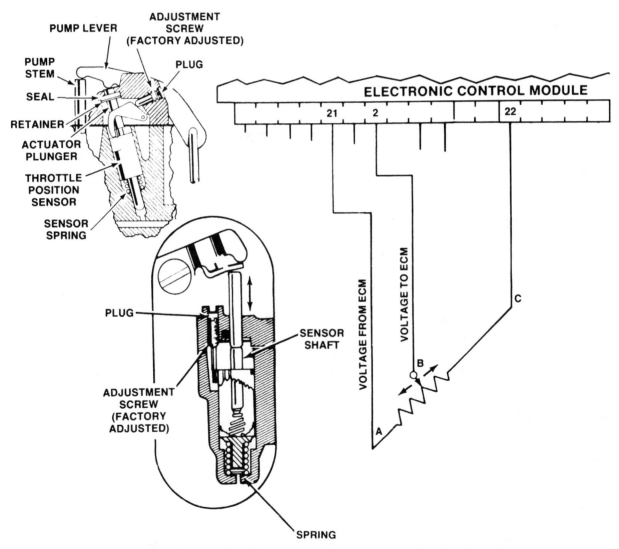

FIGURE 3-7   Throttle position sensor. *Courtesy of GM Product Service Training.*

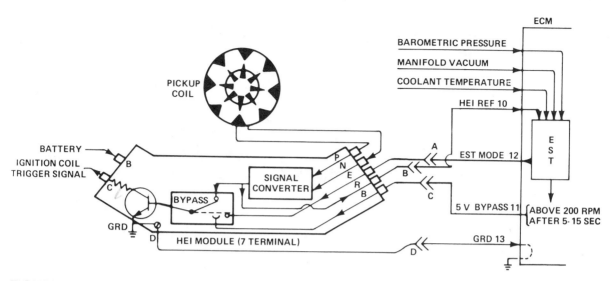

FIGURE 3-8   Electronic spark control circuit. *Courtesy of GM Product Service Training.*

---

throttle position are so closely related, the ECM expects to see the readings from these two sensors closely synchronized.

Distributor Reference Pulse (REF)

The REF can also be abbreviated as DIST or REF pulse. This signal tells the ECM what engine speed is and what position the crankshaft is in. The REF is obtained by tapping into the pickup coil circuit inside the HEI module, Figure 3-8.

The pickup coil signal is generated as the distributor shaft and timer core turn, moving the points of the timer core past the points of the magnetic pickup unit. Each time the magnetic field strengthens or weakens as a result of

the points aligning and misaligning, a voltage signal is produced in the pickup coil. A signal converter in the HEI module changes the analog signal produced by the pickup coil to a digital signal. On HEI applications prior to CCC, the pickup coil signal was used to signal the HEI module when to turn off the main switching transistor, which turned off the primary ignition circuit. On CCC applications the only time that the pickup coil signal is used by the module is during starting and some system failure conditions. During normal operation the pickup coil signal is used only as a REF signal to the ECM. This is explained further in the **Outputs** section of this chapter.

From 1982 to 1985, a Hall-effect switch, Figure 3-9, was mounted between the pickup coil and the rotor on the Chevrolet, 3.8 liter (229 cid.), odd-firing V6. It was used to produce greater timing accuracy. The odd-firing engine is especially susceptible to signal converter error. Notice in Figure 3-10 that the pickup is still there. It, however, is only used to start the engine. Notice also that terminal R of the HEI module is not used (on applications without a

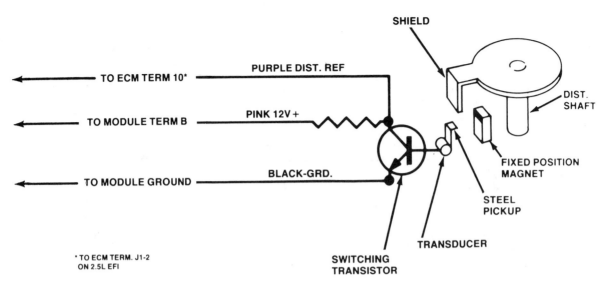

FIGURE 3-9   Hall-effect switch. *Courtesy of GM Product Service Training.*

Hall-effect switch, terminal R feeds the REF signal to the ECM) and that the REF signal comes to the ECM directly from the Hall-effect switch. Once the engine has started, the solid-state switching device, illustrated in Figure 3-10 as a relay, has disconnected the pickup coil from the main switching transistor and connected in its place the EST lead from the HEI terminal E. The ECM is now in charge of spark timing.

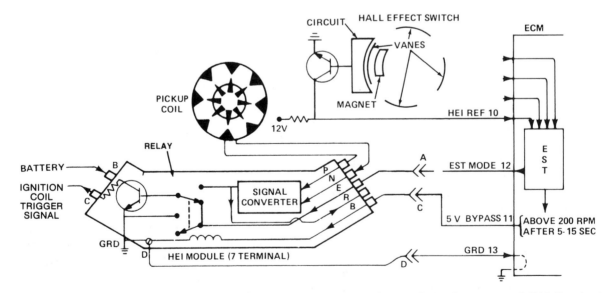

FIGURE 3-10   Electronic spark control system with Hall-effect switch. *Courtesy of GM Product Service Training.*

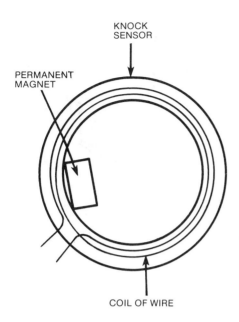

FIGURE 3–11   Early-style knock sensor

FIGURE 3–12   Late-style knock sensor

Detonation Sensor

The detonation sensor (knock sensor) is part of the electronic spark control (ESC) system that has been made a subsystem of CCC on some applications. A similar spark control system, also called ESC, has been a stand-alone system for several years on non-CCC vehicles.

On early editions of CCC and prior to it, a knock sensor other than the piezoelectric sensor was used. The early style, Figure 3–11, looked similar to the later one, Figure 3–12, but used a small permanent magnet confined in a chamber. A coil was formed by wrapping many turns of fine wire around the outside of the chamber wall. The coil and chamber was then enclosed in the center housing. When detonation occurs the vibration it produces causes the magnet to vibrate. This vibration then causes the magnet to rapidly and repeatedly detach and reattach from the chamber wall and thus results in an induced voltage in the coil around the chamber. This voltage signals the ECM that detonation is occurring.

On many General Motors engines, the knock sensor is screwed into the block in the crankcase area to reduce its sensitivity. If the knock sensor were mounted in the upper part of the engine, a clicking valve lifter could often cause the sensor to produce its "knock occurring" signal.

Vehicle Speed Sensor (VSS)

Two types of VSS are used on CCC applications. The earlier type consists of a *light-emitting diode* (LED) and a phototransistor. Both are housed in a plastic connector that plugs into the back of the speedometer housing, near the speedometer cable, Figure 3–13. The LED is powered by the ignition switch on 1982 and later models and by the ECM on earlier models. When the ignition is turned on, the LED directs its infrared light beam, invisible to the human eye, toward the back of the speedometer cup, which is painted black. The drive magnet, which spins with the speedometer cable, has a reflective surface. As the drive magnet moves into the light from the LED, the

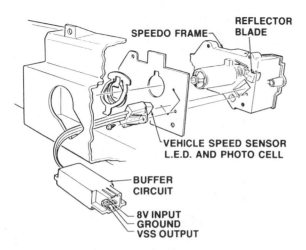

FIGURE 3-13   Vehicle speed sensor. *Courtesy of GM Product Service Training.*

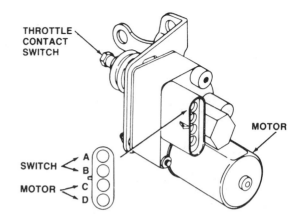

FIGURE 3-14   Idle speed control motor. *Courtesy of GM Product Service Training.*

light is reflected back to the phototransistor. Every time the light strikes the phototransistor, a voltage is produced. These voltage pulses are fed to a buffer switch.

The signal from the VSS is an analog signal that varies in amplitude as well as frequency as vehicle speed changes. The buffer switch modifies the raw VSS signal to 2,002 digital pulses per mile. This information is used by the ECM to determine when to apply the torque converter clutch.

Some later models that feature an electronically operated speedometer and no speedometer cable use a pulse generator device as a VSS. This device mounts on the transmission where the speedometer cable drive gear would be if it were used. This transmission-driven device rotates a magnet near a coil. As each pole of the magnet swings by the coil, a weak voltage signal, similar to a magnetic pickup unit in a distributor, is produced in the coil. This signal produces an analog voltage signal that is proportional to the speed of the car. These signals are fed to a buffer, which converts them to 4,004 digital pulses per mile. These signals are sent to the ECM, which can calculate vehicle speed from the signal frequency.

Ignition Switch

The ignition switch is one of the two power supplies to the ECM. When the ignition is turned on, the ECM initializes (starts its programmed routine) and is ready to function. The ignition switch also powers most of the actuators that the ECM controls. The other power supply comes from the battery by way of a fuse. It powers the diagnostic memory. These power supplies are also the inputs by which the ECM monitors system voltage.

Park/Neutral (P/N) Switch

The P/N switch consists of another pair of contacts being added to the neutral safety switch that has been on cars with automatic transmissions for years. It provides a signal to the ECM indicating whether the transmission is in gear or not. This information is used to help control engine idle speed. On some applications the ECM will not issue any spark advance or EGR commands unless the transmission is in gear.

Air-conditioning (A/C) Switch

On CCC cars equipped with air-conditioning, the A/C switch is connected by a wire to the ECM. When the A/C is turned on or off, the ECM is informed and uses this information in its control of idle speed.

Idle Speed Control (ISC) Switch

At closed throttle the throttle lever, being pulled by the return spring, presses against the ISC plunger. The pressure exerted on the ISC plunger causes a set of contact points at the base of the plunger to close, Figure 3–14. This signals the ECM to be in charge of idle speed. Opening the throttle allows the contacts to open. This signals the ECM not to control idle speed, Figure 3–15.

Transmission Switches

Most transmissions used in CCC-equipped vehicles have one or more hydraulically operated electric switches screwed into the valve body. These switches provide the ECM with signals that tell it what gear the transmission is in. This information enables it to more effectively control torque converter clutch operation.

EGR Vacuum Diagnostic Control Switch

Beginning in 1984 most CCC systems included a vacuum-operated switch that tied into the vacuum hose between the EGR valve and the EGR control solenoid, Figure 3–16. When vacuum is applied to the EGR valve, the switch closes. If the ECM detects a closed EGR diagnostic switch during starting, idle, or any other time it has not commanded the EGR to be applied, it will turn on the check engine light and set a fault code in its diagnostic memory. If it sees an open switch during any time it has commanded the EGR to be applied, it will turn on the check engine light and set the same code.

OUTPUTS

Mixture Control (M/C) Solenoid

In a CCC system, the M/C solenoid effectively replaces the full power system that would otherwise be used in the carburetor of a non-CCC vehicle. It, however, has much more articulate control of air/fuel ratio over a much wider range of engine operating conditions than the conventional full power system does, Figures 3–17, 3–18, and 3–19. The M/C solenoid is powered by the ignition switch. It is duty cycled by a solid-state grounding switch in the ECM, Figure 3–20. When the ECM grounds the M/C solenoid circuit, the solenoid is turned on and drives the solenoid plunger down. When the plunger is driven down, it drives a metering rod(s) down into the main metering jet(s) to restrict fuel flow. In some cases the end of the solenoid plunger itself provides the restriction. This produces a lean air/fuel mixture. When the ECM opens the circuit, the solenoid is turned off and the restriction is removed. This allows

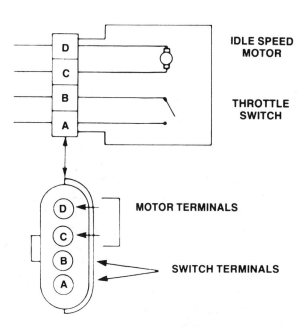

FIGURE 3–15   Idle speed control motor connector. *Courtesy of GM Product Service Training.*

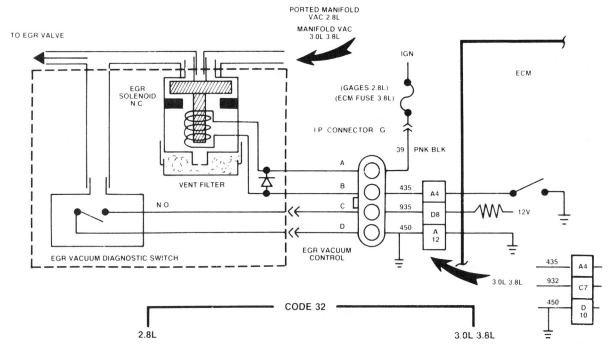

FIGURE 3-16   EGR diagnostic switch circuit. *Courtesy of GM Product Service Training.*

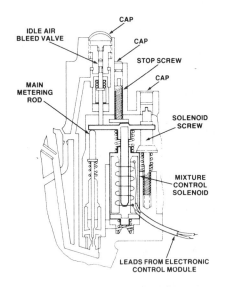

FIGURE 3-17   M/C solenoid, Rochester Dual-jet or Quadrajet. *Courtesy of GM Product Service Training.*

more fuel to flow into the main metering system and thus produces a rich mixture.

The ECM is programmed to cycle the M/C solenoid ten times per second. It cannot vary from ten cycles per second except in shutdown mode, when it is turned off. The ECM can, however, vary the duty cycle anywhere from 90% to 10% (hold the solenoid down for 90% of the cycle or 10% of the cycle, Figure 3–21).

Measuring the Duty Cycle. The relative time that the M/C solenoid is turned on or off can be measured with a dwellmeter, Figure 3–22. The wire that runs from the M/C solenoid to the ECM has a special connector (dwell test connector) for this purpose. A dwellmeter, when connected to the dwell test connector and to ground and set on the six-cylinder (60°) scale, indicates in degrees the relative amount of time that the M/C solenoid is turned on, Figure 3–23.

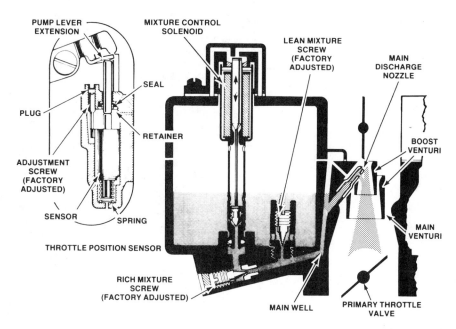

FIGURE 3–18   M/C solenoid, Rochester Varajet. *Courtesy of GM Product Service Training.*

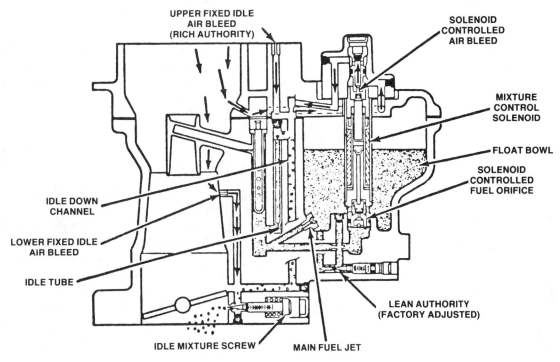

FIGURE 3–19   M/C solenoid, Holley 6510 C. *Courtesy of GM Product Service Training.*

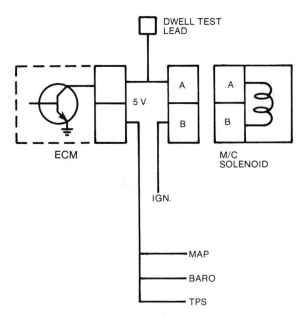

FIGURE 3-20   M/C solenoid circuit

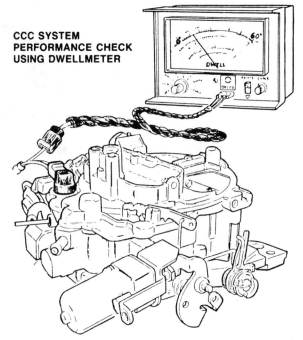

FIGURE 3-22   Dwell test lead. *Courtesy of, GM Product Service Training.*

Dwell Readings. During start-up enrichment or WOT operation, a dwell reading of 6° is expected. During engine warm-up (open loop) an initial reading around 10° is considered normal, depending on how cold the engine is.

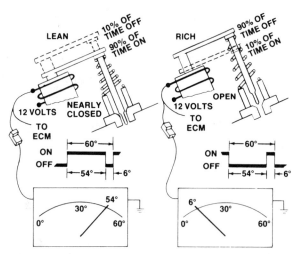

FIGURE 3-21   M/C solenoid duty cycle. *Courtesy of GM Product Service Training.*

As the engine warms up, the dwell reading should steadily increase. In open loop the ECM selects an air/fuel ratio based on engine temperature, engine load, barometric pressure, throttle position, and engine speed. This produces the best results in terms of exhaust emissions, fuel mileage, and driveability. The fuel mixture, as indicated by the dwell reading, remains constant until some of the input information changes or until the system goes into closed loop. When the coolant reaches a temperature of about 150° F, the O_2 sensor reaches about 600° F, and the required time has elapsed, the ECM puts the system into closed loop. At that moment the dwellmeter needle begins wagging. It usually wags, or varies, over a range of 5° to 10°. The dwell reading is now taken by selecting the center or average of the range over which it is sweeping.

The dwell is varying because of the way the ECM responds to the O_2 sensor to control the

RELATIONSHIP OF DWELLMETER READINGS TO MIXTURE CONTROL SOLENOID CYCLING

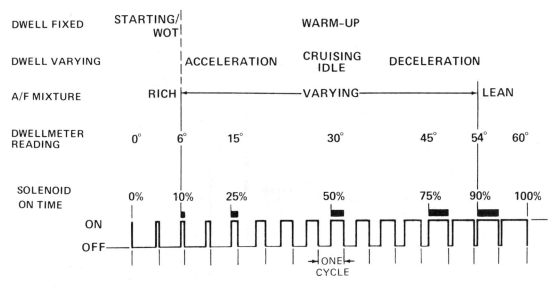

FIGURE 3-23   Dwell readings. *Courtesy of GM Product Service Training.*

air/fuel ratio. When the O_2 sensor sees a lean condition, lean being anything leaner than 14.7 to 1, it reports "lean" to the ECM. The ECM responds with a rich command to the M/C solenoid. This produces a rich condition, which the O_2 sensor immediately reports. This in turn evokes a lean command. In this way the air/fuel ratio constantly bounces back and forth across 14.7 to 1. It most often remains within a range of about 14.5 to 14.9 to 1.

At the same time that the M/C solenoid is cycling or stroking the main metering system, it is also stroking an idle air bleed valve in the idle speed–low speed system. The idle air bleed valve, Figure 3-24, is designed as a variable restriction to the amount of air that enters the idle speed–low speed circuit. When the M/C solenoid is down, the idle air bleed valve plunger drops and allows maximum air into the idle speed–low speed circuit (lean mixture). When the M/C solenoid plunger is up, the idle air bleed valve plunger is driven up and allows

minimum air into the idle speed–low speed circuit (rich mixture). With this arrangement the M/C solenoid is controlling the air/fuel ratio in whatever circuit is being used.

Electronic Spark Timing (EST)

The EST function of the ECM eliminates the vacuum and centrifugal advance mechanisms of earlier distributors, Figure 3-25. In order to achieve this, the HEI module has been modified to work with the ECM. The ECM now controls timing by signalling the HEI module when to turn the primary ignition circuit off. The HEI module only responds to these spark timing commands from the ECM once the engine has started.

Note: In the following discussion and in the accompanying illustrations, the HEI module will be treated as though it has a mechanical bypass relay with a double set of contact points. This is done to make its

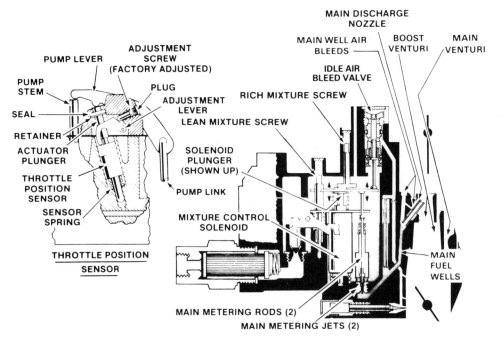

FIGURE 3-24 Rochester Dualjet or Quadrajet. *Courtesy of GM Product Service Training.*

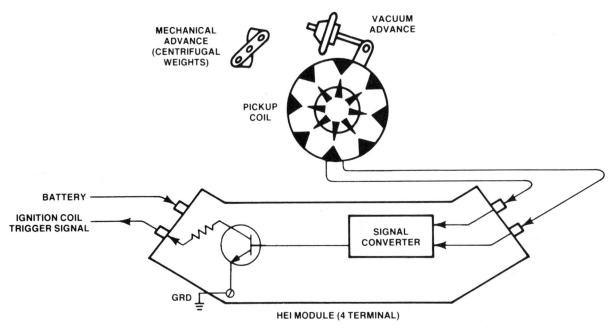

FIGURE 3-25 HEI circuit with mechanical and vacuum advance. *Courtesy of GM Product Service Training.*

operation easier to explain. Solid-state components are actually used to achieve the functions discussed.

During cranking, the bypass relay is in the de-energized or module mode. In this mode the main switching transistor of the HEI module is connected to the pickup coil, Figure 3-26. The pickup coil signal, after it is converted from an analog to a digital signal, turns the switching transistor on and off. This same signal is sent to the ECM as the HEI reference pulse.

When the ECM sees a REF pulse signal of about 200 RPM or more for about five seconds, it decides that the engine is running. It then sends a 5-volt signal through the bypass wire to the bypass relay. This causes the double contacts to move. The upper contact disconnects the base of the main switching transistor from the pickup coil and connects it to the EST wire. The lower contact simultaneously disconnects the EST wire from ground. The system is now in EST mode, and timing is being controlled by the ECM. The ECM considers barometric pressure, manifold pressure, coolant temperature,

engine speed, and crankshaft position and then sends to the HEI module the optimum spark timing command.

Note: Some Chevrolet Chevette and Pontiac T-1000 CCC systems do not have EST.

Electronic Spark Control (ESC)

Engines that have strong potential for spark knock are equipped with ESC. The ESC actually becomes a subsystem of the CCC system when it is used on a CCC vehicle. It has its own electronic module that works in conjunction with the ECM. Its function is to retard ignition timing when detonation (uncontrolled, rapid burning of the fuel charge) occurs. Two slightly different versions of ESC systems have been used.

The 1981 system, Figure 3-27, passes the EST and the bypass commands from the ECM through the ESC module on their way to the HEI module. In the event of spark knock, the detonation sensor notifies the ESC module

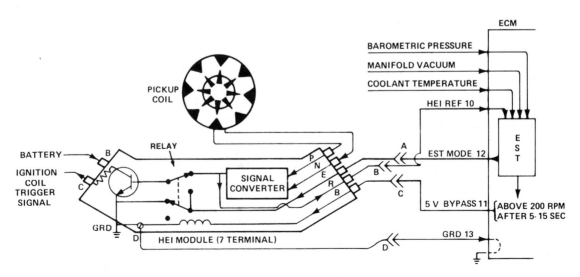

FIGURE 3-26   HEI circuit with EST. *Courtesy of GM Product Service Training.*

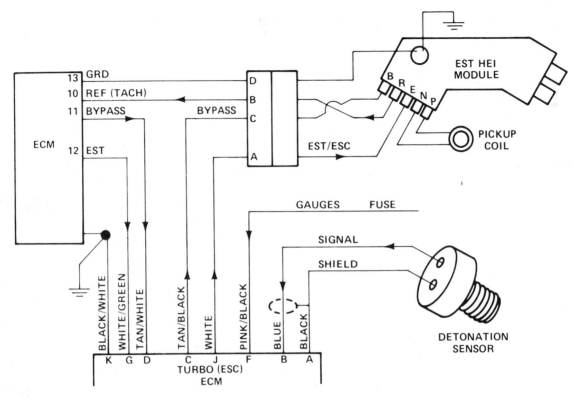

FIGURE 3-27   ESC circuit, pre-1982. *Courtesy of GM Product Service Training.*

that detonation is occurring. The ESC module then modifies the EST command. Spark timing is retarded about 4° per second until the detonation clears up. When the detonation sensor no longer "hears" detonation occurring, the ESC module begins restoring the spark advance it took away. The spark advance is restored at a slower rate than it was removed.

In the 1982 and later version of ESC, Figure 3-28, the EST and bypass commands do not pass through the ESC module. Instead, when the knock sensor produces a voltage signal that indicates that detonation is occurring, the ESC module sends a request to the CCC ECM for it to retard timing. It works as follows. Ignition voltage is applied to terminal F of the ESC

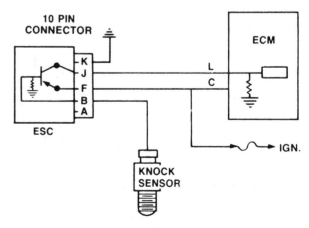

FIGURE 3-28   ESC circuit, 1982 and later. *Courtesy of GM Product Service Training.*

module. The module feeds this voltage, reduced to about 10 volts, to terminal J. It is then carried to terminal L of the ECM. When a detonation signal is applied to terminal B of the ESC module, the module reduces the voltage at terminal J to less than 1 volt. The ECM retards timing until voltage reappears at terminal L. Either of the ESC systems can easily be tested by tapping on the intake manifold with a metal object while watching the timing with a timing light or timing meter. Be sure to consult the service manual for specific directions.

Air Management Valve

As mentioned earlier making the three-way catalytic converter work at maximum efficiency is one of the major goals of the CCC system. It achieves this primarily by controlling the air/fuel ratio. On applications using a dual-bed, three-way catalytic converter, the system achieves making the catalytic converter work at maximum efficiency by the timely introduction of air from the air pump into the oxidizing side of the converter. The ECM-controlled air management system is responsible for controlling air injection into the exhaust system.

CCC vehicles equipped with dual-bed, three-way converters also have a dual air management valve. Those with single-bed, three-way converters use a single air management valve, Figures 3–29 and 3–30.

Single Valve. The single valve system can direct air to one of two places. During engine warm-up air is directed to the exhaust manifold. This will:

- reduce HC and CO by burning it in the exhaust system.
- raise O_2 sensor temperature more quickly.

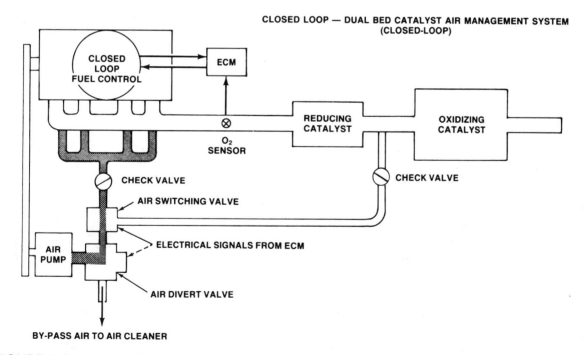

FIGURE 3–29   Air management with dual-valve and dual-bed catalyst. *Courtesy of GM Product Service Training.*

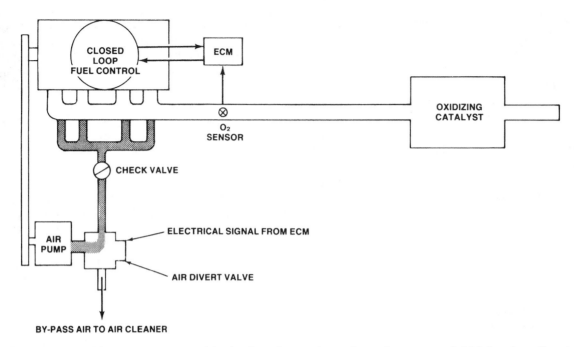

FIGURE 3-30   Air management with single valve and catalyst. *Courtesy of GM Product Service Training.*

- raise the catalytic converter temperature more quickly; the converter needs to be between 205° C (410° F) and 260° C (500° F) to operate effectively.

As the coolant temperature approaches 65° (150° F), the temperature at which the CCC system goes into closed loop, the ECM commands the divert valve to go into *divert mode.* Depending on the particular application, the air is diverted to either the air cleaner or to the atmosphere. To continue pumping air into the exhaust manifold after the system goes into closed loop is misleading to the O_2 sensor. CCC systems with single-bed converters are always in divert mode except during warm-up.

Dual Valve. On vehicles equipped with a dual-bed converter, air can be directed to one of three places. During engine warm-up the divert valve directs air to the air switching valve, Figure 3-29. The switching valve directs the air to the exhaust manifold. As the engine ap-

proaches closed-loop temperature, the ECM commands the switching valve to direct air to the oxidizing bed of the catalytic converter. Air being pumped into the oxidizing bed of the converter achieves the following:

- It prevents the additional oxygen from flowing across the O_2 sensor. This would cause it to give an incorrect signal to the ECM.
- It lowers the temperature in the exhaust manifold. Continuing to pump oxygen into the exhaust manifold after the engine has reached normal operating temperature can produce additional NO_x.
- It makes the oxidizing bed of the catalytic converter operate at maximum efficiency without interfering with the efficiency of the reducing bed.

During WOT operation and deceleration, the ECM commands the divert valve to dump

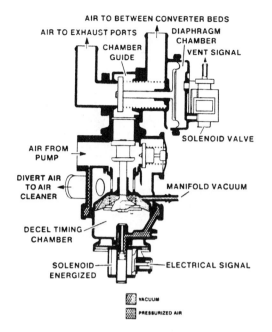

FIGURE 3-31 Divert mode (deceleration). *Courtesy of GM Product Service Training.*

the air, or use the divert mode, because there is no other safe place to put it. Either of these driving conditions results in an increased amount of HC and CO in the exhaust system. If during these conditions air were to continue being pumped into any part of the exhaust system, the converter would quickly overheat. Certain failures within the CCC system, such as in the M/C solenoid circuit or EST, also cause the ECM to put air management into divert mode.

Air Management Valve Designs. Many different valve designs have been used on various CCC system applications; thus, a discussion of each different valve is impractical. Most of them, however, are more alike than different. They all direct airflow by moving an internal valve to open or block a passage. The valves can be moved by a vacuum diaphragm, an electric solenoid, or a combination of the two. Figures 3-31, 3-32, and 3-33 show a typical dual valve in different modes of operation.

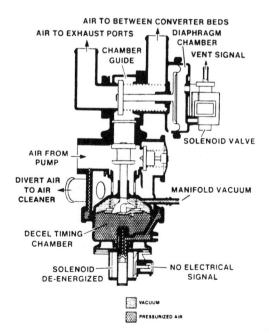

FIGURE 3-32 Divert mode (ECM commanded). *Courtesy of GM Product Service Training.*

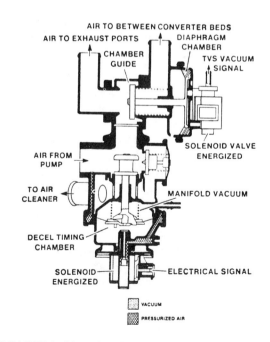

FIGURE 3-33 Air directed to exhaust ports. *Courtesy of GM Product Service Training.*

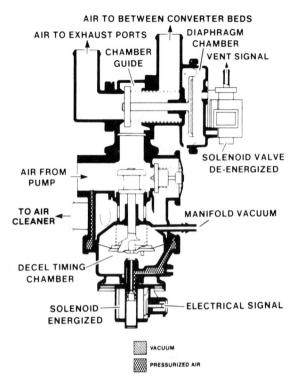

FIGURE 3-34   Warm engine mode (closed loop). *Courtesy of GM Product Service Training.*

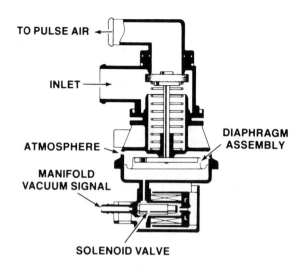

FIGURE 3-35   Pulsair shutoff valve. *Courtesy of GM Product Service Training.*

In Figure 3-31 the vehicle is decelerating. The high vacuum signal has pulled the diaphragm of the divert valve (lower valve) up. The passage to the switching valve (upper valve) is blocked, and air is diverted to the air cleaner. The air cleaner serves as a silencer. Diverting the injection pump air to it has no significant effect on air/fuel ratio.

In Figure 3-32 the ECM has put the air management system into divert mode by de-energizing the solenoid. This allows pump air pressure to be routed to the decel timing chamber. With the help of the manifold vacuum on the other side of the diaphragm, the pressure forces the diaphragm and divert valve up.

Figure 3-33 shows the air management valve directing air to the exhaust manifold during engine warm-up. Manifold vacuum is nor-

mally not strong enough to pull the divert valve diaphragm up; and the ECM is keeping the divert solenoid energized, which blocks pump air pressure from entering the decel timing chamber. The spring holds the divert valve down and allows air to flow up to the switching valve. The switching valve solenoid, energized by the ECM, allows vacuum to be applied to the upper diaphragm. This pulls it and the attached switching valve to the right to block the converter air passage and thus opens the exhaust port passage.

In Figure 3-34 the engine has warmed up and the CCC system is ready to go into closed loop. The ECM has de-energized the switching valve solenoid, which blocks vacuum to the diaphragm. The spring moves the valve to the left, opening the converter passage and blocking the exhaust port passage.

All air management valves used with CCC systems have a pressure relief valve. At high engine speeds, the pressure relief valve exhausts excess air to reduce system pressure. This can easily be mistaken for a divert mode function.

Pulsair System. Some vehicles use a *Pulsair* system instead of an air injection system. Between each exhaust pulse in the exhaust manifold, a low pressure develops. This low pressure can be used to syphon air into the exhaust ports. The air is drawn through a tube having a check valve. The check valve blocks reverse airflow during the exhaust pulse.

Pulsair Shutoff Valve. When a Pulsair system is combined with the CCC system, the ECM uses a Pulsair shutoff valve to control air flowing into the exhaust system, Figure 3–35.

During engine warm-up the solenoid is energized and vacuum is applied to the diaphragm. The valve is pulled down and air is allowed to flow into the exhaust system. With the exception of how air is made to flow, this system is very similar to the single-valve air management system.

Idle Speed Control (ISC)

Idle Speed Control Motor. In order to achieve the established goals for exhaust emissions, fuel economy and driveability, an articulate idle speed control is necessary. This is achieved by using a permanent magnet field reversible electric motor that attaches to the side of the carburetor, Figure 3–36.

The ECM normally has a fixed idle speed that it wants to maintain during idle. It moves the throttle as necessary with the ISC motor; however, it commands a higher idle speed in response to any one of the following three conditions:

- During closed choke idle, the fast idle cam holds the throttle blade open enough to lift the throttle linkage off of the ISC plunger. This allows the ISC switch to open so that the ECM does not monitor idle speed. As the choke spring allows the fast idle cam to fall away and the throttle returns to warm idle position, the ECM notes the still low coolant temperature and commands a slightly higher idle speed.

- If the engine starts to overheat, the ECM commands a higher idle speed to increase coolant flow.
- If system voltage falls below a predetermined value, the ECM commands a higher idle speed in order to increase generator speed and output.

During warm idle if the automatic transmission is pulled into gear (forward or reverse), the park/neutral switch signals the ECM of the impending load on the engine. It commands the ISC motor to extend the plunger. The throttle blade opens at about the same time that the transmission engages, and no appreciable change in engine speed occurs. If the air-conditioning is turned on, the ECM receives a signal from the air-conditioning switch. The ECM commands a wider throttle blade opening to compensate for the additional load of the air-conditioning compressor.

Starting in 1982 models with smaller engines were equipped with an ISC relay. The ISC relay uses a solid-state, fixed time delay device to keep the ISC motor activated in a retract mode after the key is turned off. This allows the throttle to completely close each

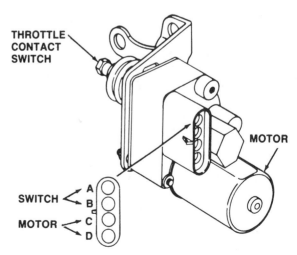

FIGURE 3–36   ISC motor assembly. *Courtesy of GM Product Service Training.*

time the engine is turned off and thus prevents dieseling.

There is no idle speed adjustment capability on CCC systems using an ISC motor. Attempting to adjust idle speed by adjusting the plunger screw results in the ECM moving the plunger to compensate for the adjustment. Idle speed does not change until the ISC plunger motor uses up all of its plunger travel trying to compensate for the adjustment, at which point the system is completely out of calibration. Proper calibration can be restored, however, by following the *minimum* and *maximum authority* adjustment procedures as outlined in the service manual. When idle speed driveability problems occur, the ISC system is usually responding to or being affected by the problem not causing it.

Note: Be sure the ignition is turned off before connecting or disconnecting the ISC motor; otherwise, damage to the ECM can occur.

Idle Stop Solenoid. Some engine applications do not use an ISC motor. Instead they use an idle stop solenoid. The idle stop solenoid is essentially the same as an antidiesel solenoid, which was used for years and dates back to the first emission control systems. When the ignition is turned on, the solenoid is energized and extends its plunger to hold the throttle lever off of the idle stop screw. The extended plunger provides the curb idle position for the throttle blade. When the ignition is turned off, the plunger retracts and allows the throttle blade to more fully close to prevent dieseling. The same device can be connected to the A/C switch. In this case the idle stop screw provides curb idle, and the solenoid increases idle speed slightly when the A/C is turned on.

Idle Load Compensator (ILC). Beginning in 1982 the Oldsmobile limited control CCC system used an ILC instead of an ISC motor, Figure 3-37. The ILC is not ECM controlled; it is controlled by engine vacuum. The ILC is very similar in construction to a vacuum advance unit on an older style distributor. It contains a diaphragm with a spring pushing against one side. The other side attaches to a plunger that sticks out of the unit in the form of an adjustable screw. The head of the plunger screw acts as the throttle stop to control idle speed. Mani-

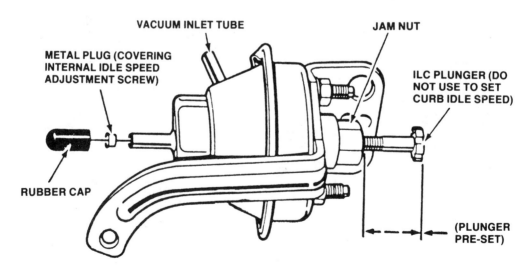

FIGURE 3-37   Idle load compensator. *Courtesy of GM Product Service Training.*

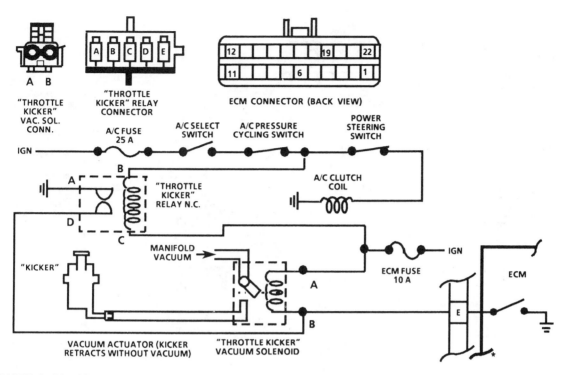

FIGURE 3-38   Vacuum-operated throttle kicker circuit (2.8 L). *Courtesy of GM Product Service Training.*

fold vacuum is applied to the spring side of the diaphragm. When the automatic transmission is pulled into gear, the load on the engine causes engine vacuum to go down slightly. This allows the spring to push the diaphragm and plunger a little farther and thus opens the throttle blades slightly wider. The same thing occurs for any load or condition that causes engine vacuum to diminish.

Idle speed adjustments can be made to the ILC; however, the adjustment is made at the vacuum side of the diaphragm not at the plunger screw. The idle adjusting screw in the stem on the back of the unit changes the preload on the diaphragm spring. Refer to an appropriate service manual for adjustment procedures

Throttle Kicker. Some later model CCC applications use either a vacuum or electronic *throttle kicker*. The functions of a throttle kicker are explained in Figures 3–38 and 3–39.

Torque Converter Clutch (TCC)

The TCC improves fuel mileage by eliminating hydraulic slippage and heat production in the converter once a cruise speed has been achieved.

Clutch. The major component of the TCC system is the clutch itself. It becomes a fourth element added to a conventional torque converter, Figure 3–40. The clutch plate, which is splined to the turbine, has friction material bonded to its engine side and near its outer circumference. The converter cover has a ma-

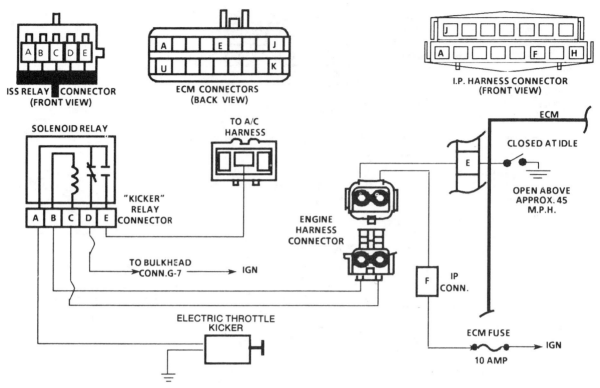

FIGURE 3-39 Electrically operated throttle kicker circuit. *Courtesy of GM Product Service Training.*

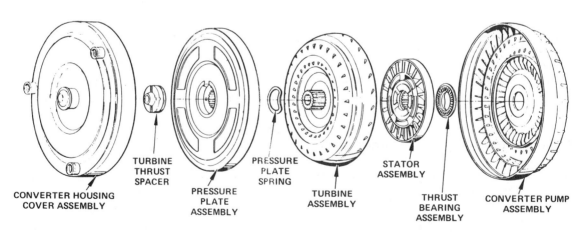

FIGURE 3-40 Torque converter with TCC. *Courtesy of GM Product Service Training.*

chined surface just inside where the converter drive lugs are attached. This machined surface mates up to the friction material on the disc when the clutch is applied. When hydraulic pressure is applied to the turbine side of the disc, the disc is forced against the converter cover. The friction between the clutch's friction material and the cover causes the complete assembly to turn as a solid unit. The pressure source is converter feed oil coming out of the pressure regulator valve in the transmission valve body. In non-TCC applications converter feed oil is used to charge and cool the converter by circulating through the converter and then the transmission cooler in the radiator. On TCC applications, however, converter feed oil must pass through a converter clutch apply valve on its way to the converter and as it returns from the converter. The apply valve, a small valve in the transmission, controls the direction of oil flow through the torque converter, Figure 3–41.

With the apply valve in its at-rest position and transmission in neutral, or the car moving at low speed, converter feed oil is directed into the converter through the release passage by way of the hollow turbine shaft. This oil is fed into the converter between the converter cover

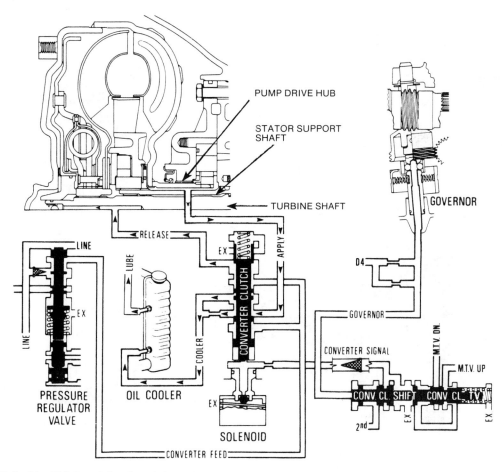

FIGURE 3–41   TCC apply circuit in release position. *Courtesy of GM Product Service Training.*

and the clutch disc. The oil forces the disc to move away from the cover and thus releases the clutch. Converter feed oil flows over the circumference of the disc and circulates through the converter. It exits through the apply passage, a passage between the pump drive hub and the stator support shaft.

Two criteria must be met before the clutch is applied. The transmission must be ready hydraulically; and the ECM must be satisfied with engine temperature, throttle position, engine load, and vehicle speed. When the transmission is in the right gear (this varies from transmission to transmission), hydraulic pressure is supplied at one end of the clutch apply valve,

Figure 3–42. This hydraulic pressure (converter apply signal) has the potential to move the apply valve into the apply position. The apply valve is moved, however, only if the ECM is ready for the clutch to be applied.

The ECM controls the position of the clutch apply valve with a solenoid that can open or close an exhaust port at the converter apply signal end of the apply valve. If the solenoid is not energized, the exhaust port is open; and converter apply signal oil exhausts from the signal end of the apply valve as fast as it arrives. Sufficient pressure does not develop to move the apply valve. Converter feed oil continues to flow into the converter through the

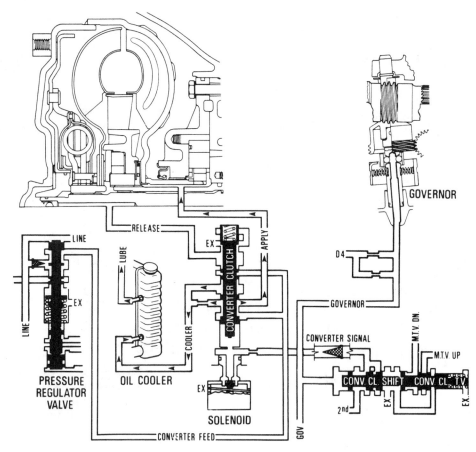

FIGURE 3–42   TCC apply circuit in apply position. *Courtesy of GM Product Service Training.*

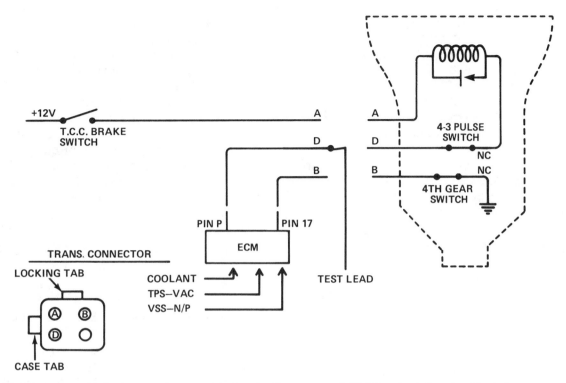

FIGURE 3-43   TCC electronic control circuit. *Courtesy of GM Product Service Training.*

release passage, Figure 3-41. When the solenoid is energized, the exhaust port is blocked. The converter apply signal oil develops pressure, and the apply valve is moved and held in the apply position. Converter feed oil now flows into the converter through the apply passage, Figure 3-42. The clutch is applied and prevents oil from exhausting through the release passage, and static pressure exists.

Figure 3-43 shows a typical TCC solenoid control circuit. The power source is the ignition switch, by way of the gauge fuse. The brake switch is in series on the high-voltage side of the circuit. Anytime the brakes are applied, the solenoid is de-energized. After the current passes through the apply solenoid, it arrives at the 4-3 pulse switch, used only on automatic transmissions with fourth-speed overdrive. The 4-3 pulse switch momentarily disengages the

---

TCCs and Four-speed Automatic Transmissions

On vehicles using a four-speed automatic transmission, the TCC must be applied while the transmission is in fourth gear. Any failure that results in the clutch not applying during prolonged fourth-gear operation can result in transmission damage as a result of oil overheating. This can occur in a relatively short period of steady-state fourth-gear operation.

TCC to allow a smooth 4-3 downshift. It is closed at all other times. From there the current goes to the ECM, where the circuit is grounded when the ECM is ready to apply the TCC. When the transmission goes into fourth gear, the fourth-gear switch opens to alert the ECM that the transmission is in overdrive. In overdrive the ECM holds the TCC on through a much wider range of throttle position than it does in third or second gear. To put it in engineering language, the ECM holds the TCC on through a wider throttle position *window.* The term *window* refers to a range within a range. For instance, if the TPS signal to the ECM were 0.5 volt at idle and 4.5 volts at WOT, the ECM might select 2.0 to 3.5 as the window in which it would keep the TCC applied.

The TCC test lead, which comes from the low-voltage side of the circuit, can be used to tell when the ECM has electrically applied the TCC circuit or to ground the circuit and thus override the ECM. Its use is covered in more detail under **System Diagnosis and Service** in this chapter.

Exhaust Gas Recirculation (EGR) Valve

The primary purpose of the EGR valve is to control oxides of nitrogen (NO_x) by lowering combustion temperature. It also is used to control detonation. The pressing need for better fuel economy and driveability combined with the development of computer-controlled ignition timing has pushed ignition timing to the ragged edge of detonation. Without the cool-

ing effect of EGR on combustion temperature, detonation occurs in most cases. Even so overuse of EGR causes serious driveability problems especially on cold engines. It makes sense then to have the ECM in charge of EGR operation. On CCC applications the EGR is turned on or off or on some later models is modulated (the amount of opening is controlled) by one or more ECM-controlled solenoids. The exact method and amount of EGR control varies considerably with engine and model year application. Generally, however, EGR systems controlled by CCC can be summarized by one of the following statements:

- The ECM operates a solenoid that when energized blocks vacuum. When not energized it passes ported vacuum to the EGR valve, Figure 3–44.
- The ECM operates a solenoid that when activated bleeds atmospheric pressure into the ported vacuum signal to the EGR valve. A bleed solenoid can be used to turn the EGR valve off or to partially close (modulate) it, Figure 3–45.
- The ECM operates two solenoids, one to block or pass ported vacuum to the EGR valve and the other as bleed solenoid to modulate it, Figure 3–46.
- The ECM operates two blocking solenoids in series. Each one is operated ac-

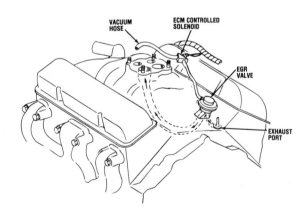

FIGURE 3–44   Typical EGR valve. *Courtesy of GM Product Service Training.*

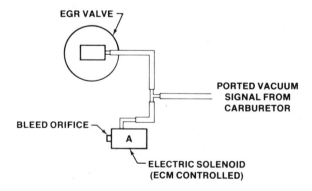

FIGURE 3–45   EGR bleed solenoid. *Courtesy of GM Product Service Training.*

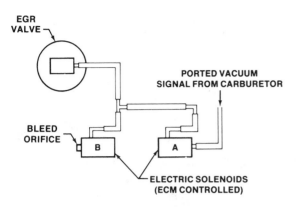

FIGURE 3-46   EGR bleed and control solenoids. *Courtesy of GM Product Service Training.*

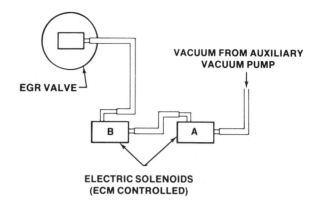

FIGURE 3-47   Two EGR control solenoids.
Courtesy of GM Product Service Training.

cording to different sets of engine calibration criteria, Figure 3-47.

- A pulse width modulating solenoid (rapidly turned on and off with the on-time
  variable) precisely controls the amount
  of vacuum that is allowed to the EGR
  valve, Figure 3-48.

One other EGR control system that can be
encountered on a CCC vehicle should be discussed at least briefly, the *aspirator-controlled
EGR,* Figure 3-49. The aspirator, mounted in
the air cleaner, contains a small venturi. Air
from the air pump is directed to the aspirator.
During periods of low engine vacuum, the aspi-

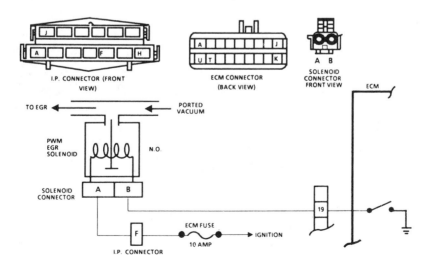

FIGURE 3-48   Pulse width modulated EGR control solenoid. *Courtesy of GM Product Service
Training.*

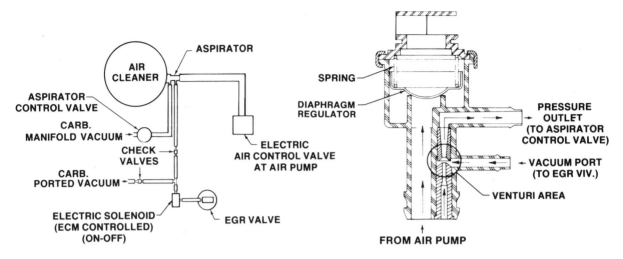

FIGURE 3-49   Aspirator-assisted EGR valve. *Courtesy of GM Product Service Training.*

rator control valve allows air to pass through the venturi portion of the aspirator to produce a vacuum that keeps the EGR valve open. When engine vacuum is high, the aspirator control valve closes to prevent air from passing through the aspirator venturi. The EGR valve is now operated by ported vacuum.

Note: Generally electrical failures in any of the EGR control systems result in the EGR valve being open.

Early Fuel Evaporation (EFE)

The EFE system applies heat to the intake manifold area beneath the carburetor to help evaporate the fuel and to keep it in a vapor state. CCC systems use one of two different types of EFE systems.

Exhaust Heat Type. An EFE valve (known years ago as a heat riser valve and located where the exhaust pipe connects to the exhaust manifold on one side of the engine) is closed by a vacuum motor, Figure 3-50. This forces exhaust gases to flow through a passage in the intake manifold beneath the space over

which the carburetor is mounted (plenum). Vacuum to the vacuum motor is controlled by a solenoid that is controlled by the ECM, Figure 3-51. This system varies from earlier, non-CCC, EFE systems in that they used a *vacuum control valve* (VCV) to control vacuum to the EFE actuator.

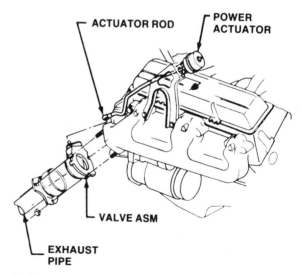

FIGURE 3-50   Vacuum-actuated EFE valve. *Courtesy of GM Product Service Training.*

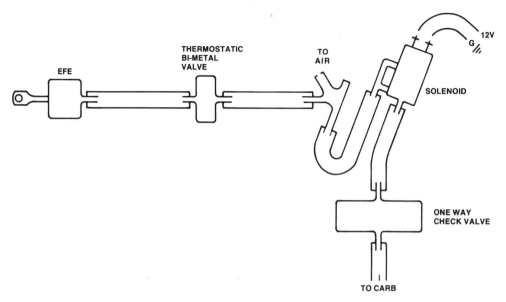

FIGURE 3–51   EFE vacuum control circuit. *Courtesy of GM Product Service Training.*

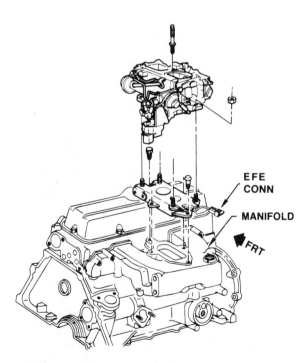

FIGURE 3–52   Electrically heated EFE valve. *Courtesy of GM Product Service Training.*

Electric Grid Type. The electric EFE uses a ceramic-encased heating element under the carburetor, Figure 3–52. The heater element is powered by a relay, Figure 3–53. When the ignition is on, voltage is available at one of the normally open relay contacts. Ignition voltage is also applied to the relay coil waiting for ground by the ECM. When coolant temperature is low, the ECM grounds the coil, the relay contacts close, and the heater element is powered. Electric EFE circuits vary slightly with engine application and model year.

Controlled Canister Purge (CCP)

On many CCC vehicles, the ECM operates a solenoid that controls purge vacuum to the *purge valve* on the charcoal canister. When the CCC system is in open loop, or before a predetermined time period has elapsed since startup, or below a specific RPM, the purge solenoid is energized and blocks purge vacuum. In closed loop, after a specified elapsed time, and above a specific RPM, the purge solenoid is de-energized and canister purging occurs.

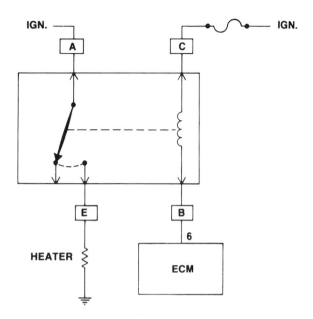

FIGURE 3-53   Electric EFE control circuit. *Courtesy of GM Product Service Training.*

Check Engine Light and Lamp Driver

All CCC systems use a check engine light. If the ECM sees a fault in one of the circuits that it monitors for malfunctions, it turns on the check engine light located on the instrument panel. This warns the driver that a malfunction exists. By grounding the test terminal (one of the terminals in the assembly line communication link, located under the dash), the ECM causes the check engine light to flash. These flashes indicate one or more codes that identify the circuit or circuits in which the fault exists. This is explained more fully under **System Diagnosis and Service** in this chapter.

In 1982 a remote lamp driver was added to the system. It is a separate module, located under the dash in most cases, that powers the check engine light and gives it greater reliability.

E-cell

Some Oldsmobile and Chevrolet eight-cylinder engine CCC applications incorporate an E-cell, often referred to as *green engine calibration unit.* The E-cell slightly modifies some engine calibrations during engine break-in. It is not necessary to replace the E-cell even if it fails prematurely.

SYSTEM DIAGNOSIS AND SERVICE

Self-diagnosis

The ECM monitors the major input sensors, the M/C solenoid, EST, and their respective circuits for proper operation. If the ECM sees a fault, such as an open, a short, or a voltage value that stays too high or too low for too long, in any of the circuits it monitors, it will turn on a check engine light on the instrument panel and record a code number in its diagnostic memory. Some later models have a service engine soon light and a service engine now light, either of which can be selected depending on the severity of the problem. The code number identifies the circuit in which the fault exists. The check engine light warns the driver that a fault exists. The check engine light should come on, however, anytime the ignition is on without the engine running.

Diagnostic Memory. A portion of the ECM's random access memory is devoted to diagnostic memory, sometimes referred to as long-term memory. The diagnostic memory enables the ECM to store code numbers, referred to as *fault codes,* that identify the type of fault and the circuit in which the fault exists. A technician can obtain the stored codes by either of two methods. Grounding the test terminal in the assembly line communication link (ALCL), often referred to as the assembly line diagnostic link (ALDL), puts the system into diagnostics, Figure 3-54. The ECM reads out its stored codes by flashing the check engine light. All codes are stored as two-digit numbers. Two quick flashes followed by a pause of

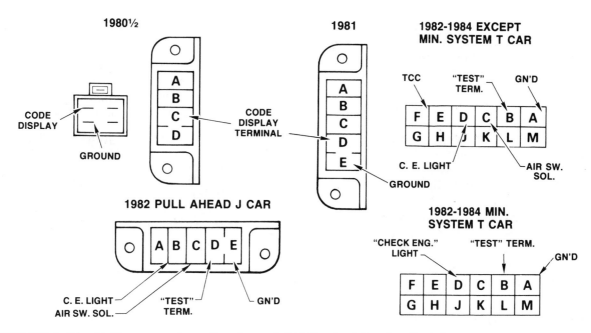

FIGURE 3-54   ALCL connectors. *Courtesy of GM Product Service Training.*

about two seconds represents 2; three more quick flashes represents the second digit—3. Figure 3-55 demonstrates a code 12 being given. Once the two-digit number is completed, a slightly longer pause separates it from the next two-digit number, or code. When a code is given, it is repeated twice before moving on to the next code. The second method of obtaining stored codes is to plug a special piece of test equipment, manufactured by several different companies and ranging in size from large engine analyzers to small handheld units, into the ALCL. With the ignition on, the test unit receives any stored codes from the ECM and displays them on its digital screen.

Note: The ALCL is most often located just under the instrument panel, somewhere between the steering column and the radio.

Note: Diagnostic procedures for 1987 and later are written to include the use of an

ALCL tool, often referred to as a "scanner," where it is applicable. Previously, diagnostic procedures were written around the use of a voltmeter to obtain such measurements as sensor readings.

Intermittent Faults. If the ECM sees that a fault has turned on the check engine light and

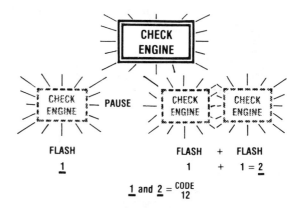

FIGURE 3-55   Check engine light in diagnostics. *Courtesy of GM Product Service Training.*

set a code in its memory, the check engine light will stay on until the ignition is turned off or until the ECM no longer sees the fault. If the perceived fault clears up, the check engine light will go off, but the code will still be stored in memory. If the fault condition does not appear again within the next fifty ignition cycles (starting the engine and turning it off), the code will automatically be erased.

Clearing the Memory. The code can also be erased by disconnecting the power supply from terminal R of the ECM for ten seconds. This can most easily be done by either pulling the fuse marked ECM from the fuse panel, disconnecting the negative battery lead, or on some later models disconnecting a fusible link near the positive battery terminal.

Behavior in Diagnostics

When the test terminal in the ALCL is grounded and the engine is not running, the check engine light should begin flashing code 12. Code 12 means no reference pulse is coming from the distributor, which is expected since the engine is not running. Any additional stored codes are displayed in numerical order after code 12.

Note: The ignition should always be on before grounding the test terminal; otherwise, the check engine light does not function properly.

There are some other noteworthy points about code 12. It is not a storable code (it is not stored in memory). It is most often used to indicate that the self-diagnosis function of the ECM is working. If you put the system into diagnostics with the engine off and do not get a code 12, the self-diagnosis function is not working properly and must be repaired before you can continue. The only time that the presence of code 12 indicates a fault is when the engine is running. In this case it means that the REF pulse is not coming in from the distributor.

While the codes are being displayed, the ECM also:

- sends a steady 30° dwell command to the M/C solenoid.
- energizes all ECM-controlled solenoids.
- pulses the ISC motor in and out.

If the test terminal is grounded while the engine is running, the ECM will:

- take out the open-loop timer, which maintains the time laps before the system can go into closed loop.
- take out some of the adaptive enrichment modes.
- send a 30° dwell command to the M/C solenoid if the system is in open loop and is not in an enrichment mode.
- cause EST to be set at a fixed spark advance.

While in diagnostics (test terminal grounded), the ECM does not store new codes.

The 1982 and later T car (Chevrolet Chevette and Pontiac T-1000) has no diagnostic memory. It can only display codes concerning faults that it can currently see. In order to receive fault codes from that minimum function system, the engine must be running when the test terminal is grounded.

Diagnostic Procedures

The diagnostic procedures are contained in several different charts or groups of charts. They are as follows:

- diagnostic circuit check
- customer complaint (or in newer manuals, driveability symptoms)
- system performance check
- diagnostic charts without trouble codes
- diagnostic charts with trouble codes
- diagnostic charts on related components

It is important to carefully follow the directions when any of the diagnostic procedures or flowcharts are being used. Failing to do so or

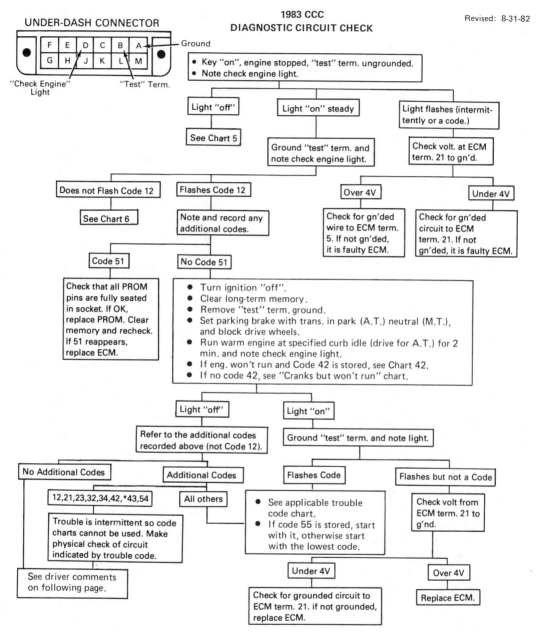

UNDER-DASH CONNECTOR

"Check Engine" Light

"Test" Term.

1983 CCC
DIAGNOSTIC CIRCUIT CHECK

Revised: 8-31-82

- Key "on", engine stopped, "test" term. ungrounded.
- Note check engine light.

Light "off"
See Chart 5

Light "on" steady
Ground "test" term. and note check engine light.

Light flashes (intermittently or a code.)
Check volt. at ECM term. 21 to gn'd.

Does not Flash Code 12
See Chart 6

Flashes Code 12
Note and record any additional codes.

Over 4V
Check for gn'ded wire to ECM term. 5. If not gn'ded, it is faulty ECM.

Under 4V
Check for gn'ded circuit to ECM term. 21. If not gn'ded, it is faulty ECM.

Code 51
Check that all PROM pins are fully seated in socket. If OK, replace PROM. Clear memory and recheck. If 51 reappears, replace ECM.

No Code 51
- Turn ignition "off".
- Clear long-term memory.
- Remove "test" term. ground.
- Set parking brake with trans. in park (A.T.) neutral (M.T.), and block drive wheels.
- Run warm engine at specified curb idle (drive for A.T.) for 2 min. and note check engine light.
- If eng. won't run and Code 42 is stored, see Chart 42.
- If no code 42, see "Cranks but won't run" chart.

Light "off"
Refer to the additional codes recorded above (not Code 12).

Light "on"
Ground "test" term. and note light.

No Additional Codes
12,21,23,32,34,42,*43,54
Trouble is intermittent so code charts cannot be used. Make physical check of circuit indicated by trouble code.

Additional Codes
All others

Flashes Code
- See applicable trouble code chart.
- If code 55 is stored, start with it, otherwise start with the lowest code.

Flashes but not a Code
Check volt from ECM term. 21 to g'nd.

See driver comments on following page.

Under 4V
Check for grounded circuit to ECM term. 21. if not grounded, replace ECM.

Over 4V
Replace ECM.

Before replacing an ECM, always check PROM for correct application and installation. Also, remove terminal(s) from ECM connector for circuit involved, clean terminal contact and expand it slightly to increase contact pressure and recheck to see if problem is corrected. In case of repeat ECM failure, check for a shorted solenoid or relay controlled by the ECM.

The system performance check should be performed after any repairs to the system have been made.

*It is possible to set a false Code 42 on starting, but the "Check Engine" light will not be "on". No corrective action is necessary.

FIGURE 3-56 Diagnostic circuit check. Charts 5 and 6, which are referred to in the diagram, are found in the manufacturer's service manual. *Courtesy of GM Product Service Training.*

taking shortcuts usually results in inaccurate conclusions.

Diagnostic Circuit Check, Figure 3-56. This chart, or procedure, should always be used before attempting to diagnose any ECM-controlled system. It is primarily responsible for sorting out why a code is stored. It also aids in discovering why the check engine light is not working properly if it is not. There can be several reasons for a code being stored:

- *Existing problem.* A fault can exist, in which case its corresponding code is called a *hard code.*
- *Intermittent problem.* A fault can develop and clear up by itself, in which case the resulting code is *intermittent.*
- *No real problem.* Someone previously working on the car can open a CCC system circuit that the ECM monitors, while the ignition is on, and cause a code to be set. Or a strong radio signal, such as can be encountered near an airport, can cause the code to set. This type of code is often referred to as a *phantom code.*

It is important that this type of code identification be made. The trouble code charts are written assuming that a fault exists. If one of the trouble code charts is used to pursue a phantom or intermittent code, the chart most often leads to an invalid conclusion because it is beginning with an invalid assumption, that a fault currently exists when in fact none does.

Driver Comment Chart/Driveability Symptoms Section, Figure 3-57. This chart or section, depending on the model year, acts as a guide to aid the technician in determining the cause of problems that either do not have trouble codes or have intermittent trouble codes.

System Performance Check, Figure 3-58. This procedure should always be used after repairing any part of the CCC system or any component that the ECM controls. It verifies that the heart of the system, fuel mixture control, is working properly. If the fuel mixture is not being controlled properly, it refers to another chart for further diagnosis.

A trouble code found in memory and shown by the Diagnostic Circuit Check to be intermittent should be dismissed as a phantom code when no driver complaint or performance problem is identified and the System Performance Check shows satisfactory results.

Diagnostic Charts without Trouble Codes. This is a series of charts contained in the service manual. Any of the charts within the series can assist in finding the cause of any one of several specific fault conditions that do not have a corresponding trouble code. Figure 3-59 is shown as an example. These charts should only be used when referred to by one of the preceding charts (diagnostic circuit check, driver comment, or system performance check) or when a specific condition addressed by one of them has been identified. As mentioned previously, improper use of the charts most often leads to mistakes, sometimes costly ones.

Diagnostic Charts with Trouble Codes. This is a series of charts contained in the service manual. These charts can be used to find the cause of a fault once the fault has been identified by a code stored in the ECM's memory and after the diagnostic circuit check has been performed to verify that the fault currently exists. Figure 3-60 is shown as an example.

Diagnostic Charts on Related Components. This series of charts, contained in the service manual, helps with the diagnosis of components and related circuits controlled by the ECM but not monitored by it for proper operation. These components and circuits include the air management valve and the torque converter clutch. Figure 3-61 is shown as an example.

Dwell Diagnosis

If the system is in closed loop and the dwell is varying between 10° and 50°, then we know that the system is able to maintain the

DRIVER COMMENT   Revised: 8-31-82

Engine performance troubles (stalling, detonation, surge, fuel economy, etc.)

IF THE "CHECK ENGINE" LIGHT IS NOT ON, NORMAL CHECKS THAT WOULD BE PERFORMED ON CARS WITHOUT THE SYSTEM SHOULD BE DONE FIRST.
IF GENERATOR OR COOLANT LIGHT IS ON WITH THE CHECK ENGINE LIGHT, THEY SHOULD BE DIAGNOSED FIRST.
INSPECT FOR POOR CONNECTIONS AT COOLANT SENSOR, M/C SOLENOID, ETC., AND POOR OR LOOSE VACUUM HOSES AND CONNECTIONS. REPAIR AS NECESSARY.

- Intermittent check engine light but no trouble code stored.
    - Check for a loose connection in the circuit from:
        - Ignition coil to ground and arcing at spark plug wires or plugs.
        - Bat. to ECM term's. C and R
        - ECM terms. A and U to engine ground
        - EST wires should be kept away from spark plug wires, distributor housing, coil and generator. Wires from ECM term. 13 to dist. and the shield around EST wires should be a good ground.
        - Open Diode across A/C Compressor Clutch.

- Loss of long term. memory
    - Grounding dwell lead for 10 seconds with "test" term. ungrounded with engine running should give code 23. This code should be retained in long term. memory after the engine is stopped and restarted. If it is not, ECM is at fault.

- Stalling, Rough Idle, or Improper Idle Speed
    - See idle speed control . . . Page 4A-60

- Denotation (Spark Knock)
    - Check — ESC performance, if applicable . . . Page 4A-55
                MAP or Vacuum Sensor output . . . Pages 2A-20 and 2A-21
                EGR Check . . . Pages 4A-56 and 4A-57
                TPS enrichment operation . . . Chart No. 4
                HEI operation . . . Page 4A-54

- Poor Performance and/or Fuel Economy and Surging
    - See — Carb. on car service — Divisional Service Manual
                EFE check . . . Pages 4A-58 thru 4A-60
                TCC operational check . . . Page 4A-63
                EST diagnosis . . . Page 4A-54
                ESC diagnosis if applicable . . . Page 4A-55

- Poor Full-Throttle Performance
    - See Chart 4 if equipped with TPS

- Intermittent No-Start
    - Incorrect pick-up coil or ignition coil. See "Cranks, But Won't Run" chart.
    - Intermittent ground connections on ECM.

- All Other Comments.
    - Make system performance check on warm engine (upper radiator hose hot)

The system performance check should be performed after any repairs to the CCC system have been made

FIGURE 3–57   Driver complaint chart. Pages and charts referred to in this illustration are found in the manufacturer's service manual. *Courtesy of GM Product Service Training.*

SYSTEM PERFORMANCE CHECK

Revised: 8-31-82

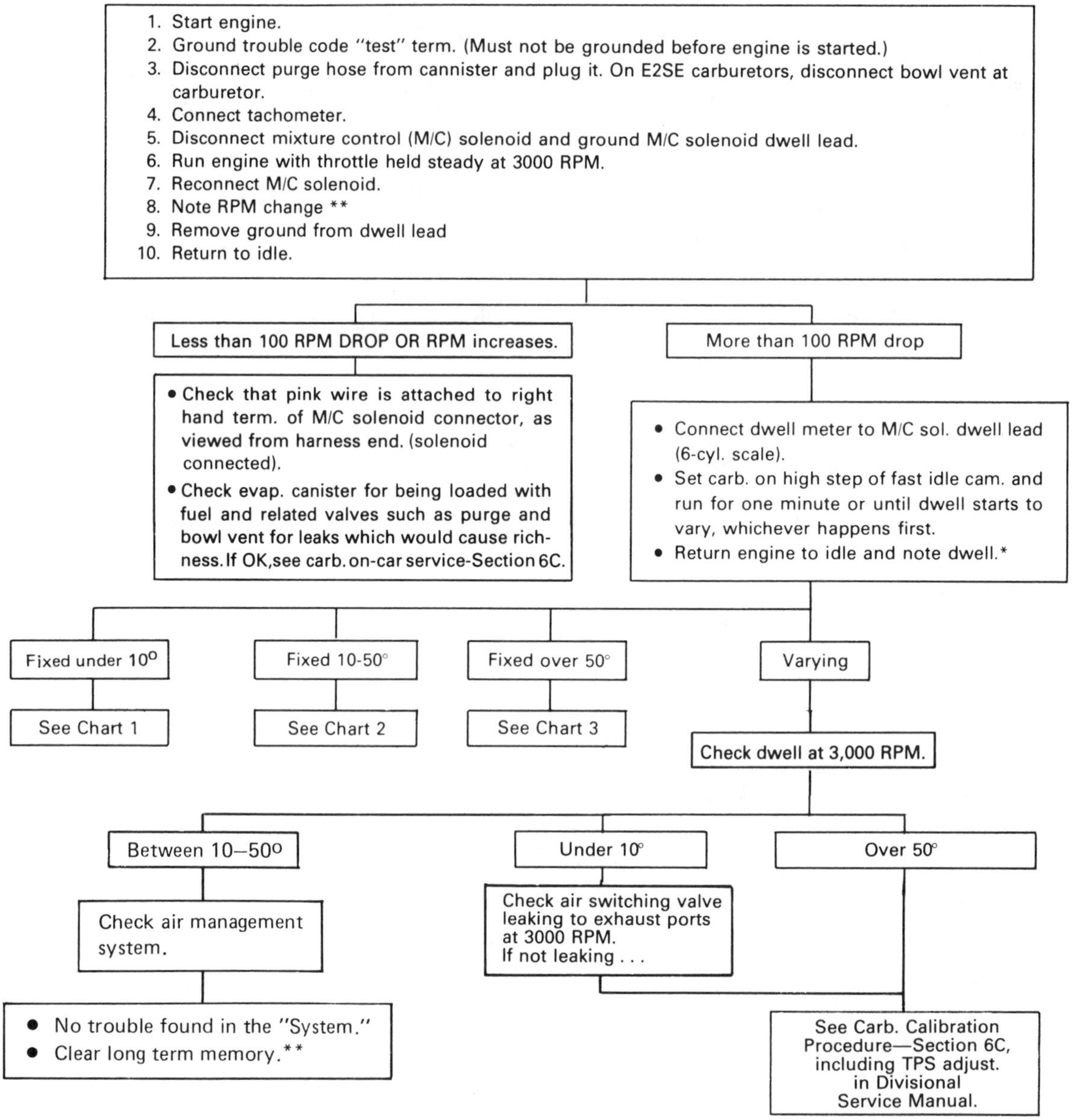

1. Start engine.
2. Ground trouble code "test" term. (Must not be grounded before engine is started.)
3. Disconnect purge hose from cannister and plug it. On E2SE carburetors, disconnect bowl vent at carburetor.
4. Connect tachometer.
5. Disconnect mixture control (M/C) solenoid and ground M/C solenoid dwell lead.
6. Run engine with throttle held steady at 3000 RPM.
7. Reconnect M/C solenoid.
8. Note RPM change **
9. Remove ground from dwell lead
10. Return to idle.

Less than 100 RPM DROP OR RPM increases.

- Check that pink wire is attached to right hand term. of M/C solenoid connector, as viewed from harness end. (solenoid connected).
- Check evap. canister for being loaded with fuel and related valves such as purge and bowl vent for leaks which would cause richness. If OK, see carb. on-car service-Section 6C.

More than 100 RPM drop

- Connect dwell meter to M/C sol. dwell lead (6-cyl. scale).
- Set carb. on high step of fast idle cam. and run for one minute or until dwell starts to vary, whichever happens first.
- Return engine to idle and note dwell.*

Fixed under 10°	Fixed 10-50°	Fixed over 50°	Varying
See Chart 1	See Chart 2	See Chart 3	Check dwell at 3,000 RPM.

Between 10–50°

Check air management system.

- No trouble found in the "System."
- Clear long term memory.**

Under 10°

Check air switching valve leaking to exhaust ports at 3000 RPM. If not leaking . . .

Over 50°

See Carb. Calibration Procedure—Section 6C, including TPS adjust. in Divisional Service Manual.

*Oxygen sensors may cool off at idle and the dwell change from varying to fixed. If this happens running the engine at fast idle will warm it up again.

**If car is equipped with an electric cooling fan, it may lower the RPM when it engages.

FIGURE 3-58   System performance check. Charts referred to in this illustration are found in the manufacturer's service manual. *Courtesy of GM Product Service Training.*

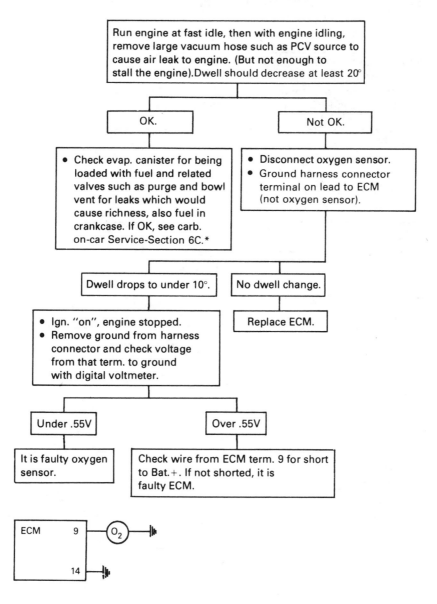

CHART #3

Revised: 4-21-82

(Rich Exhaust Indication)

DWELL FIXED OVER 50°

Run engine at fast idle, then with engine idling, remove large vacuum hose such as PCV source to cause air leak to engine. (But not enough to stall the engine).Dwell should decrease at least 20°

OK.

Not OK.

- Check evap. canister for being loaded with fuel and related valves such as purge and bowl vent for leaks which would cause richness, also fuel in crankcase. If OK, see carb. on-car Service-Section 6C.*

- Disconnect oxygen sensor.
- Ground harness connector terminal on lead to ECM (not oxygen sensor).

Dwell drops to under 10°.

No dwell change.

Replace ECM.

- Ign. "on", engine stopped.
- Remove ground from harness connector and check voltage from that term. to ground with digital voltmeter.

Under .55V

Over .55V

It is faulty oxygen sensor.

Check wire from ECM term. 9 for short to Bat.+. If not shorted, it is faulty ECM.

ECM   9   O₂

14

*Use divisional service manual.

FIGURE 3–59   Typical diagnostic chart. *Courtesy of GM Product Service Training.*

TROUBLE CODE 21   Revised: 2-15-81

THROTTLE POSITION SENSOR CIRCUIT

Check for stuck* or misadjusted** TPS Plunger Repair as necessary.
If OK, proceed:

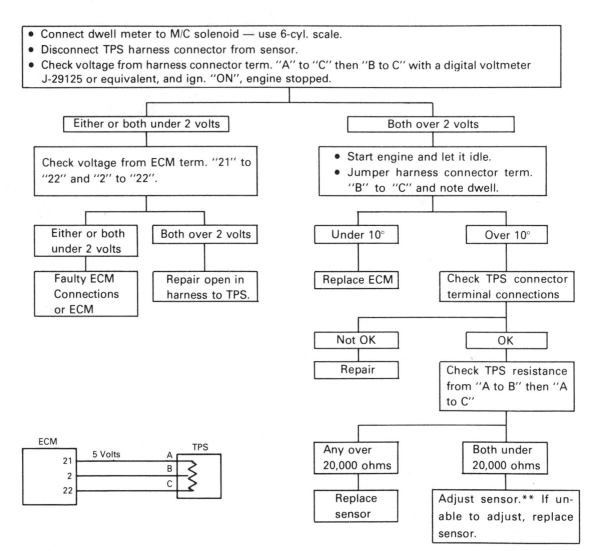

* Use a small blade screw driver on plunger.
** See divisional service manual.

FIGURE 3-60 Typical trouble code chart. *Courtesy of GM Product Service Training.*

ELECTRIC DIVERTER VALVE CHECK (EDV)   Revised: 5-20-82

3.8L — V6 VIN CODE A NON-TURBO, CALIF.
2.8L V-6 "S" TRUCK, A.T., CALIFORNIA
5.0L VIN CODE H

Check for at least 10" of vacuum at valve with engine idling.

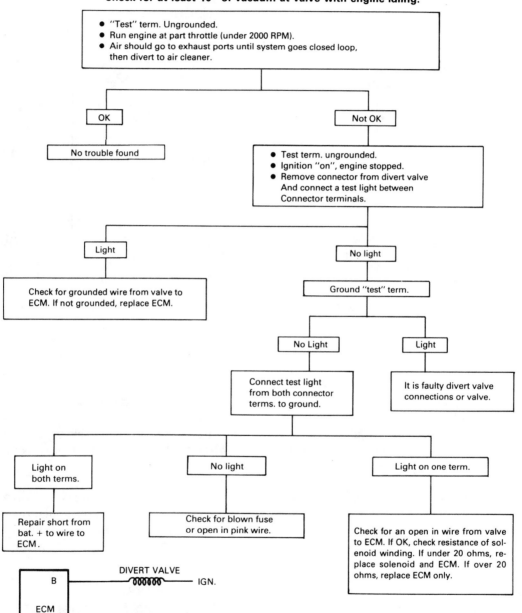

FIGURE 3–61   Typical diagnostic chart for nonmonitored functions. *Courtesy of GM Product Service Training.*

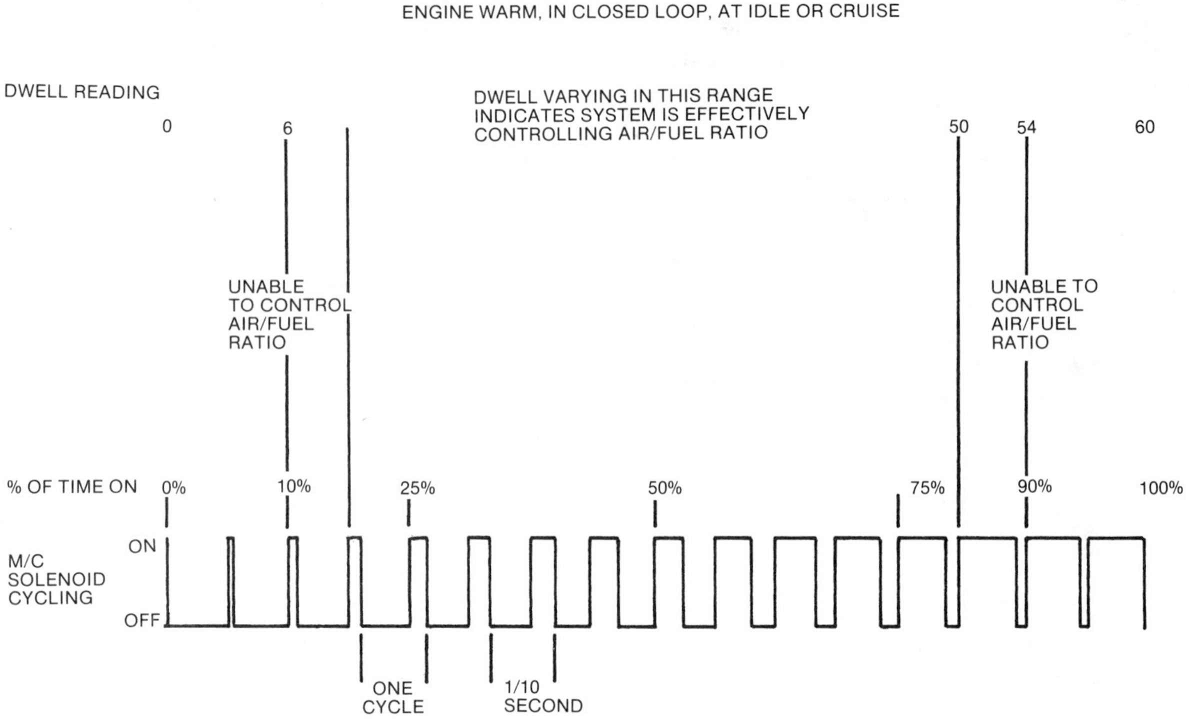

ENGINE WARM, IN CLOSED LOOP, AT IDLE OR CRUISE

DWELL READING

DWELL VARYING IN THIS RANGE
INDICATES SYSTEM IS EFFECTIVELY
CONTROLLING AIR/FUEL RATIO

0   6   50   54   60

UNABLE
TO CONTROL
AIR/FUEL
RATIO

UNABLE TO
CONTROL
AIR/FUEL
RATIO

% OF TIME ON   0%   10%   25%   50%   75%   90%   100%

M/C
SOLENOID
CYCLING

ON

OFF

ONE
CYCLE

1/10
SECOND

FIGURE 3–62   Dwell reading and system performance

desired air/fuel ratio, Figure 3-62. If the dwell is not varying and is between 10° and 50°, it is not in closed loop. If the dwell is stuck below 10° or above 50°, a condition exists that is causing the air/fuel mixture to be either lean or rich and for which the system cannot compensate. It is possible, of course, that the system electronics are malfunctioning and are therefore unable to recognize the existing air/fuel ratio.

Let's assume, for example, that we are getting a steady dwell of 54°, which would indicate a rich condition. To determine if an electronic malfunction is occurring or if the engine is actually running rich, run the engine for two minutes at a fast idle to be sure that it is in closed loop. Next pull a large vacuum hose (just small enough to keep from killing the engine), and observe the dwell. It should begin to go down. If it does not respond, there is an electronic problem; the system is unable to recognize the change in the air/fuel mixture.

Note: Some dwellmeters do not work well on CCC systems. If a particular dwellmeter causes any change in engine performance when it is connected, do not use it.

Carburetor and M/C Solenoid Adjustment

Varying degrees of adjustment capability exist on the four different carburetors used on CCC applications. The 6510-C Holley, used on the T car, has only its idle mixture screw as an external mixture adjustment. Mixture adjustments are made at idle only using a dwellmeter. The E2SE *Varajet* has an idle mixture screw, and down inside its bowl is a lean mixture screw, Figure 3–63. The lean mixture screw is a main metering adjustment and is made at 3,000 RPM using a dwellmeter. The idle mixture screw is adjusted at idle using a dwellmeter. The E4ME *Quadrajet* and the E2ME *Dualjet* carburetors have much more adjustment capability. Their adjustments are identical and include idle mixture screws, an idle air bleed valve, and adjustments on the M/C solenoid.

PERSONAL SAFETY: When performing mixture adjustments at high engine speeds, be certain that all instrument leads are clear of the coolant fan, that you have no loose clothing such as shirt or coat sleeves that can get caught in the coolant fan or drive belts, and that the drive wheels are firmly blocked.

Since the CCC system's introduction, several revisions have been made in adjustment specifications and procedures. A significant procedure revision made in 1983 should be observed when servicing E4ME and E2ME car-

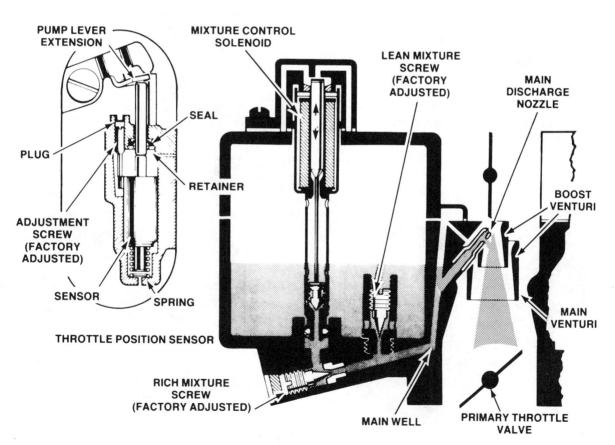

FIGURE 3–63   Rochester Varajet (E2SE). *Courtesy of GM Product Service Training.*

buretors from 1981 through 1984. The E4ME and E2ME carburetors on many 1985 and later models have some slight internal changes that prevent earlier procedures from applying to them. Although such carburetor adjustments should only be performed when clearly indicated, driveability problems can sometimes be eliminated by restoring the proper float level and carefully making updated adjustments.

PERSONAL SAFETY: No attempt should be made to adjust the rich mixture screw on the E2SE carburetor shown in Figure 3–63. Removing and replacing the covering plug produces a potential fuel leak and fire hazard.

ECM and PROM Service

Considerable care should be taken when installing a PROM. A finger coming into contact with the PROM's pins can discharge static electricity into it and damage it, Figure 3–64. To insert the PROM in the ECM, position the PROM in the carrier. The carrier only fits in the

PROM cavity of the ECM in one position; however, the PROM fits in the carrier in either of two ways. If the PROM is installed in the ECM incorrectly, it will be damaged electrically. To avoid this, identify the small half-circle notch on one end of the PROM. Now look in the PROM cavity and identify a similar notch on one end of the PROM seat, Figure 3–65. To avoid bending the pins when installing the PROM, be sure that the PROM is positioned in the carrier with the tips of the pins above the shoulder of the carrier, as shown in Figure 3–66, before placing the PROM and carrier in the PROM cavity. This allows the carrier to guide each pin into place as the PROM is pressed down.

The PROM and ECM are serviced separately; therefore, remove the PROM before exchanging the ECM for a new one.

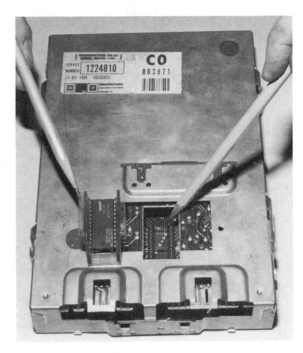

FIGURE 3–65   PROM, PROM carrier, and PROM socket

FIGURE 3–64   PROM

FIGURE 3-66   PROM pins/PROM carrier shoulder

Weather-pack Connectors

All CCC system harness connections under the hood are made with weather-pack connectors, Figure 3-67. They are designed to provide environmental protection for electrical connections so that corrosion and contamination buildup is held to a minimum. Many of the circuits in the CCC system do not carry more than 1 or 2 milliamps at about 100 millivolts (1/10 volt). The resistance caused by oxidation or contamination of an electrical connection can have significant impact on the performance of the system. It is important, then, that the integrity of the weather-pack connectors be maintained. Don't break the seal by inserting a probe into the connector. Voltage readings can be obtained by opening the connector and temporarily installing short jumper wires made for that purpose (many called-for voltage checks require that the circuit be electrically intact). If the weather-pack connector has been damaged, it should be replaced as shown in Figure 3-67 or the connection should be soldered and taped as in Figure 3-68.

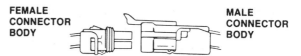

FEMALE CONNECTOR BODY   MALE CONNECTOR BODY

1. OPEN SECONDARY LOCK HINGE ON CONNECTOR

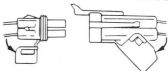

2. REMOVE TERMINALS USING SPECIAL TOOL

J-28742

TERMINAL REMOVAL TOOL

3. CUT WIRE IMMEDIATELY BEHIND CABLE SEAL

4. SLIP NEW CABLE SEAL ONTO WIRE (IN DIRECTION SHOWN) AND STRIP 5.00mm (.2") OF INSULATION FROM WIRE. POSITION CABLE SEAL AS SHOWN.

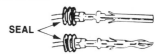

SEAL

FIGURE 3-67   Weather-pack connector. *Courtesy of GM Product Service Training.*

Diagnostic and Service Tips

TPS. The TPS is the one sensor that can be adjusted. Sometimes driveability complaints can be cured by adjusting the TPS to exact specifications, which vary with engine and model year application. Be sure to check the applicable service manual for correct procedures and specifications.

TCC Test Terminal. When using the TCC test terminal, the following points should be remembered:

- Voltage is present when the TCC is not applied and is near zero when the TCC is applied, Figure 3–43, because applying the TCC causes the voltage to be dropped across the apply solenoid.
- The TCC can be applied by grounding the TCC test terminal.

TWISTED/SHIELDED CABLE

1. **Remove outer jacket.**
2. **Unwrap aluminum/mylar tape. Do not remove mylar.**

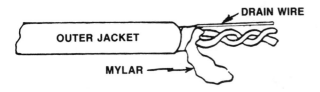

3. **Untwist conductors. Strip insulation as necessary.**

4. **Splice wires using splice clips and rosin core solder. Wrap each splice to insulate.**

5. **Wrap with mylar and drain (uninsulated) wire.**
6. **Tape over whole bundle to secure as before.**

FIGURE 3-68   Wire repair. *Courtesy of GM Product Service Training.*

TWISTED LEADS

1. **Locate damaged wire.**
2. **Remove insulation as required.**

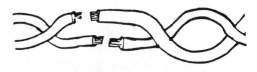

3. **Splice two wires together using splice clips and rosin core solder.**

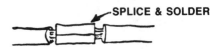

4. **Cover splice with tape to insulate from other wires.**
5. **Retwist as before and tape with electrical tape to hold in place.**

- Applying voltage to the TCC test terminal damages the ECM if the ECM tries to apply the TCC while the voltage is applied. (On earlier models with five-terminal ALCLs, the TCC test terminal is in the fuse panel and thus increases vulnerability to this mistake.)

Surging Complaints. Surging complaints are occasionally heard concerning CCC-equipped vehicles at steady-state speeds around 35 miles per hour or above. This is a fairly common behavior and is often caused by the EGR valve, referred to as *EGR chuggle.* The inert gases introduced in the intake manifold by the EGR valve cause an increased frequency of cylinder misfires. Although this has been occurring since the EGR valve's introduction in the early seventies, it was not noticeable to the driver until a clutch was placed in the torque converter thus making it a solid coupling, compounded by significantly lower axle numerical ratios. This behavior is sometimes interpreted, however, as a malfunction of the TCC. In order to isolate the condition, temporarily disconnect and plug the vacuum line from the EGR valve. Drive the vehicle to see if the behavior still occurs. If it does it is not EGR chuggle.

CAUTION: Do not leave the EGR valve disconnected. To do so is not only against the law but can result in engine damage due to detonation.

CCC System Behavior

Some of the behaviors of the CCC system can be amusing or frustrating, depending on whether or not they are understood. For instance, driving up a long hill in the mountains at near full throttle can set a code and turn on the check engine. Changing altitude rapidly or driving backwards for about a quarter of a mile can have the same effect. There is nothing wrong; the system is just not programmed for those driving conditions.

REVIEW QUESTIONS _____

1. What input values are used by the ECM to calculate fuel mixture in open loop? (See **Blended Enrichment.**)

2. The Hall-effect switch generates what type of signal?

3. What influence does the P/N switch have over engine operation?

4. What two conditions must exist before the TCC is applied?

5. Name at least two ways in which closed-loop operation can be identified.

6. An O_2 sensor voltage of 0.3 causes the ECM to issue a _____ command.

7. What inputs are considered by the ECM when determining the spark timing command?

8. With a dual-bed catalytic converter, where is the air directed under each of the following conditions?
   a. during warm-up
   b. in closed loop
   c. at wide-open throttle

9. What results if the air switching valve fails to switch air away from the exhaust manifold when the system goes into closed loop?

10. How does the ECM know when to control idle speed?

11. How does the ECM control TCC apply?

12. What can happen if the TCC fails to apply during fourth-gear operation with a four-speed automatic?

13. List two likely results if the EGR valve fails to open.

14. List two types of EFE systems.

15. During what operating mode is the canister purge solenoid de-energized, allowing the charcoal canister to be purged?

16. A dwell reading between 10° and 50° and varying suggests what?

17. How is the CCC system put into diagnostics?

18. Why is a high-impedance digital voltmeter required for many CCC system tests?

19. Why must trouble codes be identified as hard codes before the trouble code charts can be used?

ASE-type Questions. (Actual ASE test questions will probably not be this product specific.)

20. Technician A says that the CCC system only responds to the O_2 sensor when the system is in closed loop. Technician B says that the CCC system only uses the O_2 sensor input for fuel control. Who is correct?
    a. A only
    b. B only
    c. both A and B
    d. neither A nor B

21. Technician A says that when the M/C solenoid is down, the mixture is rich and the solenoid is turned off (no power applied to it). Technician B says that when the M/C solenoid is down, power is being applied to it and that the ECM is responding to a lean signal from the O_2 sensor. Who is correct?
    a. A only
    b. B only
    c. both A and B
    d. neither A nor B

22. A car equipped with a CCC system is being tested for dwell. The dwellmeter, connected to the M/C solenoid circuit and set on the six-cylinder scale, indicates a dwell reading of 50°.
    Technician A says that the car is running rich. Technician B says that the car is running lean. Who is correct?
    a. A only
    b. B only
    c. both A and B
    d. neither A nor B

23. A car equipped with TCC is being driven with a test light connected to the TCC test terminal. As the car reaches cruising speed, the light goes out. This indicates that:
    a. either the TCC circuit has failed or the fuse has blown.
    b. the TCC has applied.
    c. the TCC has failed to apply.
    d. the TCC has released.

24. Technician A says that while in diagnostics, the CCC system does not set fault codes. Technician B says that while in diagnostics, the CCC system displays fault codes only if the engine is not running. Who is correct?
    a. A only
    b. B only
    c. both A and B
    d. neither A nor B

GLOSSARY

In addition to the following terms, refer to Figure 3–69 for a list of the abbreviations used in this chapter.

ASPIRATOR   A device using airflow through a restriction in a tube to create a vacuum in another tube.

BLEED   A controlled leak.

DETONATION   Uncontrolled, rapid burning of the fuel charge that results in spark knock.

DIVERT MODE   When the ECM directs the air management system to dump air into the air cleaner or the atmosphere.

DUALJET   A two-venturi carburetor that basically consists of the primary bores and circuitry of a Quadrajet; usually referred to by General Motors as E2ME, the last E meaning it is equipped for CCC system application.

E-CELL   A module containing a cathode and an anode, one of which is gradually sacrificed as current passes through them, causing the circuit to become open.

FAIL SOFT   A feature of system design that provides continued operations (with less than normal performance capability) in the event of certain component or system failures or that activates or deactivates certain systems in order to prevent damage to other components due to component or system failure.

KAM   Keep-alive memory.

A/C	Air-Conditioning
AIR	Air Injection Reaction
ALCL	Assembly Line Communication Link
BARO	Barometric
ECM	Electronic Control Module
EFE	Early Fuel Evaporation
EGR	Exhaust Gas Recirculation
ESC	Electronic Spark Control
EST	Electronic Spark Timing
HEI	High-energy Ignition
ILC	Idle Load Compensator
ISC	Idle Speed Control
MAP	Manifold Absolute Pressure
M/C	Mixture Control
PCV	Positive Crankcase Ventilation
P/N	Park/Neutral
PROM	Programmable Read-only Memory (engine calibration unit)
PWM	Pulse Width Modulated
TCC	Torque Converter Clutch
TERM.	Terminal
TEST LEAD or TERMINAL	Lead or ALCL connector terminal grounded to obtain a trouble code
TPS	Throttle Position Sensor
VAC.	Vacuum
VIN	Vehicle Identification Number
VSS	Vehicle Speed Sensor
WOT	Wide-open Throttle

FIGURE 3–69   Explanation of abbreviations

LIGHT-EMITTING DIODE   A semiconductor device that generates a small amount of light when forward biased (voltage applied with the polarity such that current flows across the PN junction).

MAXIMUM AUTHORITY   A term describing the maximum idle speed that the ISC motor can provide.

MILLIAMP   The most correct form is milliampere, meaning one thousandth of an amp.

MILLIVOLT   One thousandth of a volt.

MINIMUM AUTHORITY   A term describing minimum idle speed, achieved by retracting the ISC plunger until the throttle lever rests on the idle stop screw.

PARAMETER   One of the factors that helps to define operating conditions, such as coolant temperature, engine load, air/fuel mixture, and so forth.

PULSAIR   The name General Motors applies to their pulse injection system.

PURGE VALVE   A vacuum-controlled valve that controls the removal of stored HC vapors from the charcoal canister.

THROTTLE KICKER   A device operated either by vacuum or electrically that when activated during idle bumps the throttle open slightly more.

TIMER CORE   A star-shaped wheel attached near the top of the distributor shaft in an HEI distributor. It has one point per engine cylinder.

VACUUM CONTROL VALVE   A switch that opens and/or closes its ports to supply vacuum in response to coolant temperature change. It is sometimes referred to as a *ported vacuum switch* or a *thermal vacuum switch*. Thermal vacuum switch, however, is more accurately used to refer to a switch located in the air cleaner that opens or closes a port in response to air temperature.

VARAJET   A two-venturi carburetor consisting basically of one primary and one secondary bore and related circuitry from a Quadrajet; usually referred to by General Motors as E2SE, the last E meaning it is equipped for CCC system application.

WOT   Wide-open throttle.

GENERAL MOTORS' ELECTRONIC FUEL INJECTION

Objectives

After studying this chapter, you will be able to:

- list the six most important inputs needed to control engine calibration.
- define the term *pulse width.*
- describe how the computer knows engine temperature, engine load, engine speed, and throttle position.
- name at least two EFI system failures that cause the engine to stop.
- describe the circuit that controls fuel pump operation.
- list the first two steps of EFI diagnostic procedure.
- name several precautions that can prevent damage from occurring to the computer.

Electronic fuel injection (EFI), not to be confused with a fuel injection system used by Cadillac in the late seventies and that was also called EFI, is a single-point injection system. It is often referred to as *TBI* (throttle body injection); that name, however, more accurately describes a major component of the system, Figure 4-1. This system was introduced by General Motors in 1982.

When the EFI system was introduced in 1982, it appeared in two versions: one containing a single TBI unit for four-cylinder engines and a high performance version (Crossfire) with two separate TBI units for eight-cylinder engines. In 1985 and in 1986, a dual TBI (two injectors in one TBI unit) was introduced on some V6 engine applications.

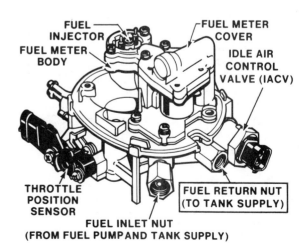

FIGURE 4-1   TBI unit. *Courtesy of GM Product Service Training.*

117

ELECTRONIC CONTROL MODULE _____

The ECM is usually located above or near the glove compartment. On some models, such as the Pontiac Fiero, it is in the console. On some Corvette applications, it is in the battery compartment behind the driver's seat. Its inputs and the functions it controls are shown in Figure 4–2.

ECM for 2.5 liter, EFI Operates Cruise Control

The ECM for selected engines such as the 2.5 liter engine was given more internal circuitry to enable it to operate the cruise control system in 1988.

Throttle Body Backup (TBB)

The *TBB* is a fuel backup circuit within the ECM. It is primarily responsible for providing fuel pulses to the injector solenoid in the event of an ECM failure, which prevents it from running its program. The ECM used on the 2.0 liter engine has a removable *CALPAK* that provides fuel backup. The CALPAK plugs into the ECM in the same way that the PROM does.

Block Learn

The EFI ECM has a learning ability. It accepts and corrects for gradually changing input values that result from sensor wear or driving conditions, such as changing altitude. It even learns to slightly modify output commands to complement a particular driver's driving habits. The aspect of this learning ability of most concern to technicians is covered under **System Diagnosis and Service** in this chapter.

Keep-alive Memory (KAM)

The ECM has a KAM like that discussed in Chapter 3.

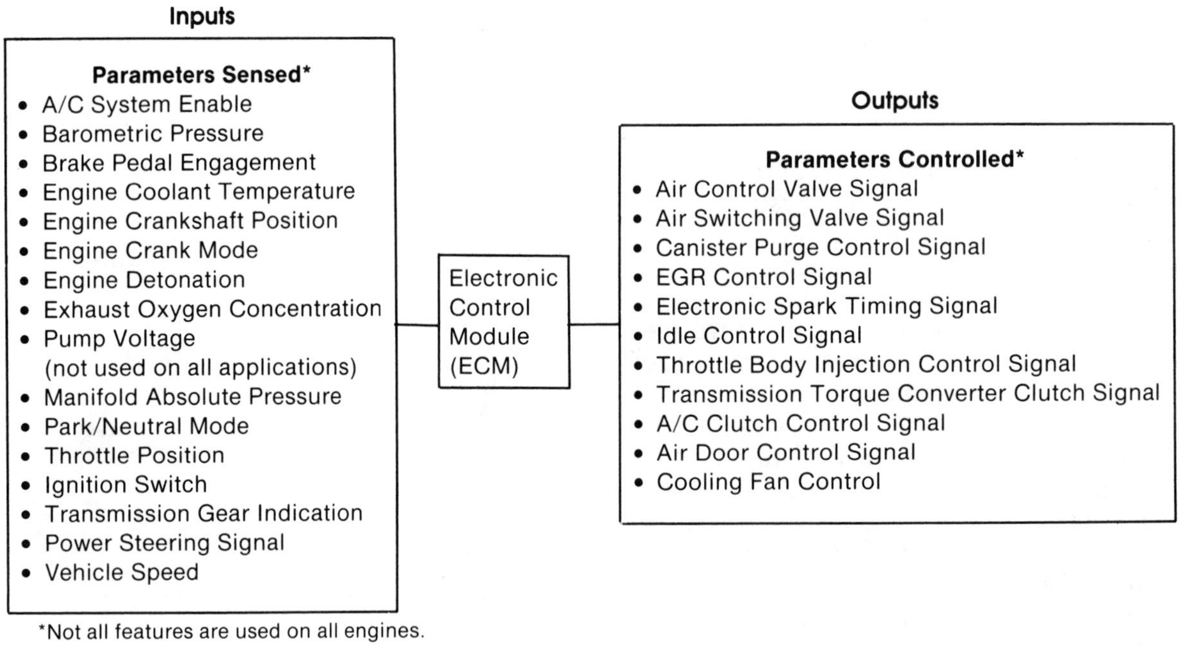

Inputs

Parameters Sensed*
- A/C System Enable
- Barometric Pressure
- Brake Pedal Engagement
- Engine Coolant Temperature
- Engine Crankshaft Position
- Engine Crank Mode
- Engine Detonation
- Exhaust Oxygen Concentration
- Pump Voltage
  (not used on all applications)
- Manifold Absolute Pressure
- Park/Neutral Mode
- Throttle Position
- Ignition Switch
- Transmission Gear Indication
- Power Steering Signal
- Vehicle Speed

Electronic Control Module (ECM)

Outputs

Parameters Controlled*
- Air Control Valve Signal
- Air Switching Valve Signal
- Canister Purge Control Signal
- EGR Control Signal
- Electronic Spark Timing Signal
- Idle Control Signal
- Throttle Body Injection Control Signal
- Transmission Torque Converter Clutch Signal
- A/C Clutch Control Signal
- Air Door Control Signal
- Cooling Fan Control

*Not all features are used on all engines.

FIGURE 4–2   EFI system overview

OPERATING MODES

The EFI system features the typical closed-loop and open-loop modes. During open loop the ECM is programmed to provide an air/fuel ratio most suited to driveability as well as economy and emissions concerns.

Synchronized Mode

In synchronized mode the injector is pulsed once for each reference pulse from the distributor. To put it another way, the injector sprays once for each time a cylinder fires. On dual-TBI units the injectors are pulsed alternately. All closed-loop operation is in synchronized mode as is most open-loop operation.

Nonsynchronized Mode

In nonsynchronized mode the injector is pulsed every 12.5 milliseconds, independent of distributor reference pulses. On dual-TBI systems each injector is pulsed every 12.5 milliseconds; but because they are pulsed alternately, the engine receives a spray every 6.25 milliseconds. Nonsynchronized pulses only occur in response to one or more of the following operating conditions:

- When the injector is on or off (open or closed), time becomes too small for accurate control (about 1.5 milliseconds), as occurs at near full throttle or during deceleration. Even though the injector only opens a few thousandths of an inch, it is a mechanical device opened by a magnetic field and closed by a spring and has a finite speed as contrasted to the speed of current flow. When it is asked to open and close or to close and open at a rate that approaches its response capability limits, the ECM stops making the injector keep up with the reference pulse and allows it to go into a slower mode of operation.
- During prime pulses. On the 1982 Crossfire EFI system, when the ignition was

turned on, if the coolant temperature was low enough, the ECM commanded the injector to deliver a spray or two into the manifold to prime the engine for starting. This action was similar to pumping the throttle before starting a carbureted engine. The Crossfire system was discontinued in 1983.

Cranking Mode

While the engine is cranking, the injector delivers an enriched air/fuel ratio especially suited for starting conditions (high manifold pressure and low airflow velocity). The pulse widths depend on coolant temperature. At −36° C (−32.8° F) the pulse width is calibrated to provide maximum rich air/fuel ratio of 1.5 to 1. At 94° C (201.5° F) the pulse width is calibrated to provide a maximum lean air/fuel ratio of 14.7 to 1. The higher the temperature, the shorter the pulse width or injector on-time, unless the engine is overheated. In case of overheating, the pulse width is widened to provide a slightly richer mixture.

Clear Flood

If the engine floods, depressing the throttle to wide-open throttle (WOT) or within 80% of WOT puts the system into clear flood mode. While cranking in this mode, the ECM issues a pulse command that produces an air/fuel ratio of 20 to 1. It stays in this mode until the engine starts or until the throttle is moved to less than 80%. If the engine is not flooded and the throttle is depressed 80% or more, it is highly unlikely that the engine will start.

Run

When the ECM sees a reference pulse from the distributor that indicates an RPM of 600 or more, it puts the system into open loop. In open loop the ECM does not use the inputs from the oxygen sensor to calculate air/fuel mixture commands. It does, however, watch to see if the oxygen sensor is ready to go into closed

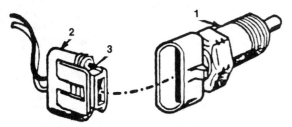

1   **Coolant temperature sensor**
2   **Harness connector to ECM**
3   **Locking tab**

FIGURE 4-3   Engine coolant sensor (new style). *Courtesy of Buick Motor Division.*

loop. Three criteria must be satisfied before the system can go into closed loop:

- The oxygen sensor must produce voltage signals that cross 0.45 volt (450 millivolts). In order to do this, it must reach a temperature of at least 300° C (570° F).
- The coolant temperature must reach a specified temperature, about 65.6° C (150° F).
- A specified amount of time must elapse since the engine was started.

The values for each of these criteria are contained in the PROM. When all of the criteria are met, the ECM puts the system into closed loop. Only the input from the oxygen sensor is now used to make decisions about air/fuel mixture commands.

Semi-Closed Loop

In an effort to achieve greater fuel economy, the ECM for some throttle body injected engines has been programmed to go out of closed loop during cruise conditions at highway speeds. During these periods, the air/fuel mixture may go as lean as 16.5 to 1. The ECM monitors the following parameters and will put the system into this fuel control mode only when they are within predetermined values:

- engine temperature
- spark timing

- canister purge
- constant vehicle speed

The ECM will periodically switch back to closed loop in order to check on engine operating parameters. If parameters are still within required limits, it will go back to the lean calibration, open-loop mode. Operation in the lean calibration, open-loop mode and switching back and forth between closed loop and open loop are smooth and should not be detectable to the driver.

INPUTS

Engine Coolant Temperature (ECT) Sensor

This is the same sensor as that used in the computer command control system and the other General Motors' systems. On later models it had a slightly different appearance also, Figure 4-3.

Manifold Absolute Pressure (MAP) Sensor

The MAP sensor on EFI applications, Figure 4-4, is basically the same as that on CCC applications; but the ECM has been programmed to use the sensor as a barometric pressure sensor also. When the ignition is turned to "run," on its way to "crank," the ECM looks at the MAP sensor reading. The engine is not running (the crankshaft has not moved yet); therefore, manifold pressure is atmospheric pressure. The ECM records this read-

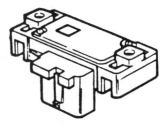

FIGURE 4-4   EFI MAP sensor. *Courtesy of Buick Motor Division.*

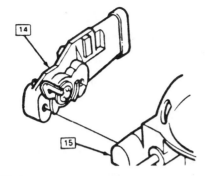

14 THROTTLE POSITION SENSOR
15 TBI UNIT

FIGURE 4–5   Throttle position sensor. *Courtesy of Buick Motor Division.*

ing as a barometric pressure reading and uses it in calculating fuel mixture while in open loop and spark timing. The BARO reading is held and used until the ignition is turned off or until the throttle is pushed to WOT. At this time engine vacuum goes to atmospheric pressure, and an updated BARO reading is taken.

Oxygen (O₂) Sensor

This is the same single-wire sensor as used on other General Motors' systems.

Throttle Position Sensor (TPS)

The TPS is a variable resistor (potentiometer) mounted on the TBI unit, connected to the end of the throttle shaft, Figure 4–5. The TPS on some EFI applications is adjustable; on others TPS adjustment is not provided.

Distributor Reference Pulse (REF)

The REF, sometimes referred to as REF pulse, tells the ECM what engine speed is and what position the crankshaft is in. On most EFI engine applications, the distributor pickup coil provides the REF signal, as described in Chapter 3. It is important to note that if while the engine is running the REF signal fails to arrive at the ECM, fuel injection is stopped.

The earlier 2.5-liter EFI engines used a Hall-effect switch to supply the REF signal, Figure 4–6. The pickup coil was still used for starting, but that was its only function.

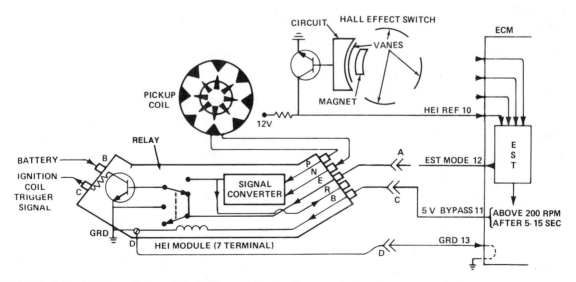

FIGURE 4–6   HEI module with EST and Hall-effect switch. *Courtesy of GM Produ Training.*

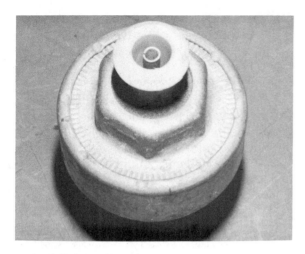

FIGURE 4-7   Knock sensor

Detonation Sensor

A piezoelectric crystal detonation or knock sensor is used on some EFI systems, Figure 4-7.

Vehicle Speed Sensor (VSS)

Two types of VSSs are used on EFI vehicles, as described in Chapter 3. On EFI vehicles the VSS input is also used by the ECM to help identify a deceleration condition. This is discussed further under **Injector Assembly** in the **Outputs** section of this chapter.

Ignition Switch

The ignition switch is one of two power supplies to the ECM. When the ignition is turned on, the ECM initializes (starts its program sequence) and is ready to function. The ignition switch also powers most of the actuators that the ECM controls.

Park/Neutral (P/N) Switch

The park/neutral switch, usually located near the shifter assembly, closes when the vehicle is shifted into park or neutral. This signals the ECM as to whether or not the transmission is in gear. The ECM uses this information to control engine idle speed. If the P/N switch is disconnected or significantly out of adjust-

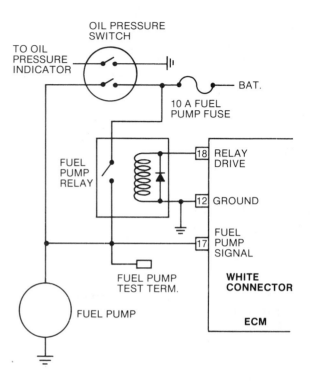

FIGURE 4-8   Fuel pump control circuit

ment, idle quality can suffer while in park or neutral.

Transmission Switches

Most transmissions used in CCC-equipped vehicles have one or more hydraulically operated electric switches screwed into the valve body. These switches provide the ECM with signals that tell it what gear the transmission is in. This information enables the ECM to more effectively control torque converter clutch operation.

Air-conditioning (A/C) Switch

On EFI cars equipped with air-conditioning, the A/C switch is connected by a wire to the ECM. When the A/C is turned on or off, the ECM is informed and uses this information in its control of idle speed.

Ignition Crank Position

The circuit that energizes the starting motor solenoid also signals the ECM that the engine is being cranked. The ECM responds by issuing cranking mode injector commands.

Fuel Pump Voltage Signal

Some EFI systems have a wire from the positive side of the fuel pump power circuit to the ECM. This wire signals the ECM that the pump has been activated, Figure 4-8. Some systems do not use it.

OUTPUTS

Throttle Body Injection (TBI) Unit

The TBI unit is made up of three castings: the throttle body, the fuel meter body, and the fuel meter cover, Figure 4-9. The throttle body contains the throttle bore and valve. It provides the vacuum ports for EGR, canister purge, and so forth, similar to what is found in a carburetor. Mounted to it are the idle air control motor and the TPS. The fuel meter body contains the injector and the fuel pressure regulator, Figure 4-10.

Injector Assembly

The injector is a solenoid-operated assembly. When it is energized by the ECM (pulsed), a spring-loaded metering valve is lifted off its seat a few thousandths of an inch. Fuel under pressure passes through a fine screen filter that fits around the tip of the injector and is sprayed in a conical pattern against the walls of the throttle bore, just above the throttle blade.

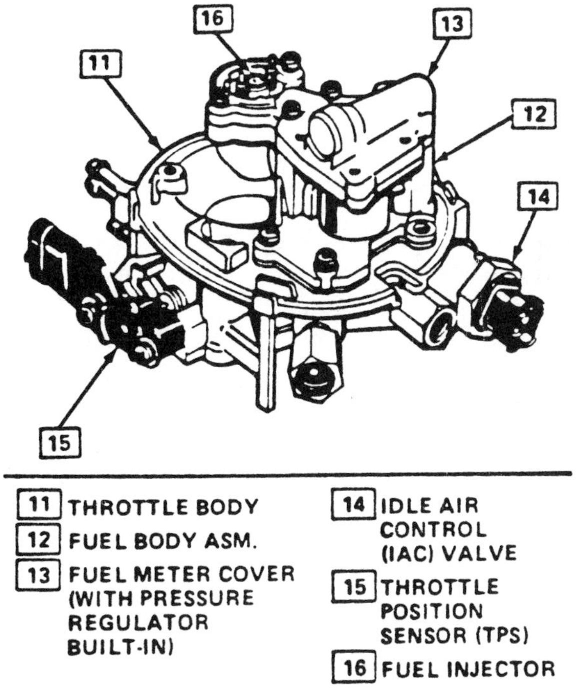

11 THROTTLE BODY	**14** IDLE AIR CONTROL (IAC) VALVE
12 FUEL BODY ASM.	
13 FUEL METER COVER (WITH PRESSURE REGULATOR BUILT-IN)	**15** THROTTLE POSITION SENSOR (TPS)
	16 FUEL INJECTOR

FIGURE 4-9   TBI unit. *Courtesy of Buick Motor Division.*

The TBI

The fascinating thing about the EFI system is that the TBI unit performs every function that a carburetor does and does it with more control. When the TPS is moved rapidly, the injector momentarily sprays more fuel (increased pulse width), similar to an accelerator pump function. On cold starts it sprays additional fuel to duplicate the function of a choke and provides a fast idle. It has no choke, fast idle cam, accelerator pump, and so on; it is all done electronically. On deceleration it leans out the air/fuel ratio (decreased pulse width). This compensates for the fuel that begins to evaporate from the manifold walls as a result of the higher vacuum. On hard deceleration it stops fuel injection completely. Hard deceleration is identified by a closed throttle, a sharp drop in manifold pressure, and a rapid decrease in vehicle speed.

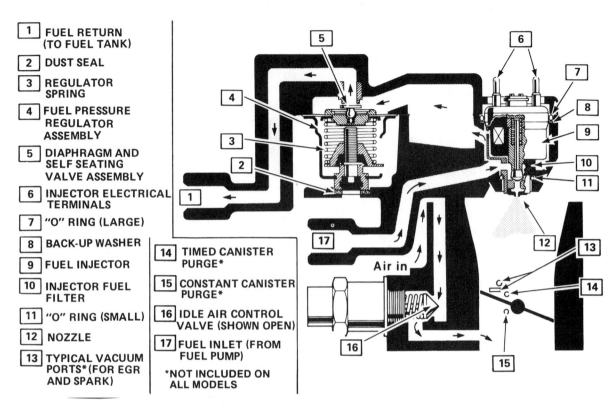

1	FUEL RETURN (TO FUEL TANK)
2	DUST SEAL
3	REGULATOR SPRING
4	FUEL PRESSURE REGULATOR ASSEMBLY
5	DIAPHRAGM AND SELF SEATING VALVE ASSEMBLY
6	INJECTOR ELECTRICAL TERMINALS
7	"O" RING (LARGE)
8	BACK-UP WASHER
9	FUEL INJECTOR
10	INJECTOR FUEL FILTER
11	"O" RING (SMALL)
12	NOZZLE
13	TYPICAL VACUUM PORTS*(FOR EGR AND SPARK)
14	TIMED CANISTER PURGE*
15	CONSTANT CANISTER PURGE*
16	IDLE AIR CONTROL VALVE (SHOWN OPEN)
17	FUEL INLET (FROM FUEL PUMP)

*NOT INCLUDED ON ALL MODELS

FIGURE 4-10   Idle air bypass circuit. *Courtesy of GM Product Service Training.*

Fuel Pressure Regulator

Within the TBI assembly is a fuel pressure regulator, Figure 4-10. The spring side of the diaphragm is exposed to atmospheric pressure; therefore, fuel pressure varies slightly with changes in atmospheric pressure. At lower atmospheric pressures, the fuel pressure is reduced and thus causes slightly less fuel to be injected in a given pulse width. The pressure regulator should provide a pressure ranging from 9 to 13 psi.

On Crossfire systems the rear TBI unit has a pressure regulator, and the front TBI unit has a pressure compensator, Figure 4-11. It works much the same way as the pressure regulator does, except its function is to make up for mo-

mentary fuel pressure drops between the front and rear TBI units to maintain constant operating pressure. Such drops are usually caused by the diaphragm and valve in the pressure regulator moving to a more closed position and thus allowing less fuel to the front TBI unit.

Idle Air Control (IAC)

The IAC assembly controls idle speed. It consists of a small, reversible, electric stepper motor and a pintle valve, Figure 4-12 (number 11). As the motor's armature turns, the pintle valve is extended or retracted depending on the direction the motor is turning. During idle the throttle blade is in a fixed, nearly closed position and thus allows a constant amount of idle air into the intake manifold. The pintle valve allows additional air through the bypass passage. The ECM uses the stepper motor to position the pintle valve for desired idle speed. During warm engine operation, the ECM attempts to maintain a fixed idle speed by adjusting the IAC valve position for load variation (transmission in or out of gear, air conditioner on or off, etc.).

If during idle or at low vehicle speed (below 10 MPH) engine speed drops below a spec-ified RPM, the ECM puts the IAC into antistall mode. The IAC motor retracts the pintle to allow additional air into the intake manifold to raise engine speed. It momentarily raises engine speed above base idle.

The stepper motor is unique; its armature has two separate windings. The direction in which the armature turns depends on which winding is powered. Power is applied in short pulses. Each pulse rotates the armature about 35° and extends or retracts the pintle valve a

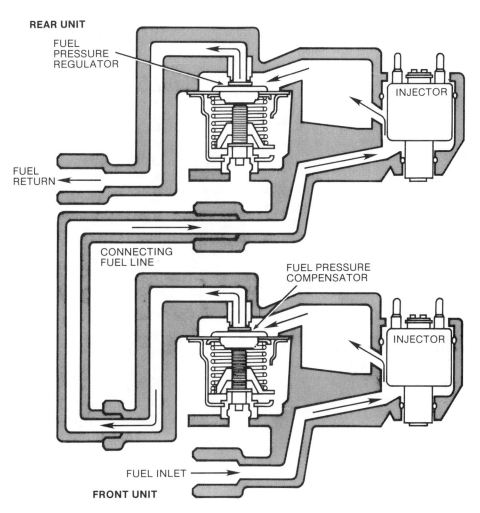

FIGURE 4-11   Crossfire TBI units. *Courtesy of GM Product Service Training.*

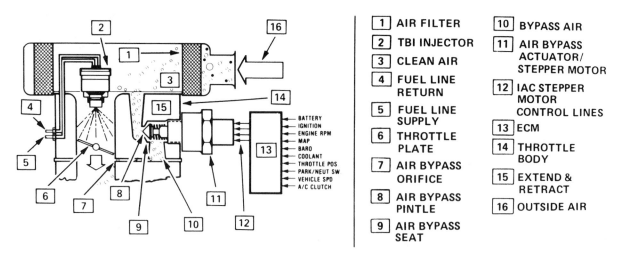

1 AIR FILTER	**10** BYPASS AIR
2 TBI INJECTOR	**11** AIR BYPASS ACTUATOR/ STEPPER MOTOR
3 CLEAN AIR	
4 FUEL LINE RETURN	**12** IAC STEPPER MOTOR CONTROL LINES
5 FUEL LINE SUPPLY	**13** ECM
6 THROTTLE PLATE	**14** THROTTLE BODY
7 AIR BYPASS ORIFICE	**15** EXTEND & RETRACT
8 AIR BYPASS PINTLE	**16** OUTSIDE AIR
9 AIR BYPASS SEAT	

BATTERY
IGNITION
ENGINE RPM
MAP
BARO
COOLANT
THROTTLE POS
PARK/NEUT SW
VEHICLE SPD
A/C CLUTCH

FIGURE 4–12   IAC motor assembly in TBI unit. *Courtesy of GM Product Service Train*

proportionate amount. The ECM applies as many pulses as necessary to achieve its desired idle RPM.

The ECM keeps count of the pulses applied and therefore knows where the pintle valve is at all times. There are 12 pulses (positions) per revolution and 255 total positions. Fully extended (closed bypass passage) is the reference position, zero. Fully retracted (wide-open bypass passage) is step 255.

In order to maintain accurate tracking of the IAC assembly's position, the ECM references itself frequently. When the engine comes off idle and the car starts to move, the ECM begins looking for a reading from the VSS that represents 30 MPH. The first time it sees a 30-MPH signal, it moves the pintle to the park (fully closed) position to reorient the zero reference. It then moves the pintle to a preprogrammed distance from closed. If for any reason the VSS is disabled, the ECM will not be able to reestablish the correct idle speed.

Fuel Pump Control

Fuel and fuel pressure are supplied by an electric pump inside the fuel tank. The fuel

pump assembly contains a check valve that prevents fuel from bleeding from the fuel line back into the tank. If the check valve leaks, the engine must crank long enough before starting to allow the line to be purged of vapor.

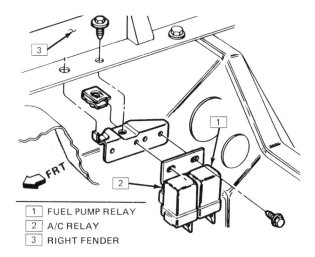

1	FUEL PUMP RELAY
2	A/C RELAY
3	RIGHT FENDER

FIGURE 4–13   Fuel pump and A/C relay. *Courtesy of Buick Motor Division.*

The fuel pump is turned on and off by a fuel pump relay usually mounted in the engine compartment. The fuel pump relay is controlled by the ECM. Be aware that the fuel pump relay is identical in appearance to and usually mounted beside the air-conditioning relay, which is also controlled by the ECM, Figure 4–13. On some models the cooling fan relay is also mounted in the same location and also looks like the other two.

When the ignition is turned on, the ECM activates the fuel pump relay, Figure 4–14. If the ECM does not receive a REF signal within two seconds, it deactivates the relay and reactivates it only when it does receive the REF or when the ignition is turned off and back on again. Once the engine is running, the ECM deactivates the relay when the ignition is turned

off or anytime REF signals stop coming to the ECM. Note also in Figure 4–14 that the oil pressure switch, which operates the oil pressure indicator on the instrument panel, is electrically in parallel with the fuel pump relay. This is done to provide a backup to the fuel pump relay. The engine starts after a fuel pump relay failure if cranked long enough to generate about 4 PSI of oil pressure.

Fuel Pump Test Terminal. A wire is connected to the circuit that powers the fuel pump. The open end of this wire has a terminal to which a jumper lead can be connected to power the fuel pump. A voltmeter connected to the test terminal quickly shows if either the fuel pump relay or oil pressure switch is supplying power to the fuel pump circuit. The terminal is located in the engine compartment or in the

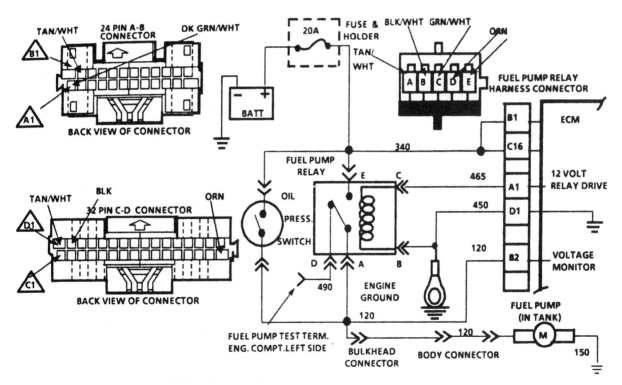

FIGURE 4–14   Fuel pump control circuit. *Courtesy of Buick Motor Division.*

assembly line communication link under the instrument panel, depending on the model and year.

Voltage Correction. During vehicle operation, system voltage can vary considerably as a result of the combination of electrical accessories being used, battery state of charge, and so forth. Variations in system voltage can cause the injector's opening and closing time to vary slightly. This affects the amount of fuel injected during a pulse width. Because of this the ECM monitors system voltage and will multiply the pulse width by a voltage correction factor, Figure 4-15. As system voltage goes down, the pulse width increases; as system voltage goes up, pulse width decreases. If voltage goes down to a criterion value, the ECM increases dwell to ensure good ignition performance and increases idle speed.

Electronic Spark Timing (EST)

EST maintains optimum spark timing under all conditions of engine load, speed, temperature, and barometric pressure. Two basic functions are incorporated within the system. Dwell control is provided to allow sufficient energy in the ignition coil for proper ignition voltage output. Spark is provided at just the right time to start combustion at the moment of peak pressure in the compression stroke. The EST function of the ECM eliminates the vacuum and centrifugal advance mechanisms. The HEI module accepts spark timing commands from the ECM once the engine has started.

During cranking, Figure 4-16, the solid-state switching circuit connects the main switching transistor of the HEI module to the pickup coil. The pickup coil signal turns on and off the main switching transistor, which controls the primary ignition circuit. This is called the *module mode* and provides no spark advance. On applications that do not use a Hall-effect switch, this same signal, after it is converted from an analog to a digital signal by a signal converter circuit in the module, is sent to the ECM as the REF signal. On applications that do use the Hall-effect switch, it supplies the REF signal. Note in Figure 4-17 that terminal R is not used.

When the ECM sees a REF pulse signal that represents an RPM of 600 or more, it decides that the engine is running. It then sends a 5-volt signal through the bypass wire to the solid-state switching circuit. The switching circuit in turn disconnects the pickup coil from the base of the main switching transistor. The EST wire is simultaneously disconnected from ground and connected to the main switching transistor base. The system is now in EST mode, and timing is being controlled by the ECM. The ECM considers barometric pressure, manifold pressure, coolant temperature, engine speed, and crankshaft position and then sends to the HEI module the optimum spark timing command.

Electronic Spark Control (ESC)

The Crossfire EFI system is equipped with ESC. Its function is to retard ignition timing when detonation occurs. The system is essentially the same as the later version of ESC described in the **Outputs** section of Chapter 3.

Torque Converter Clutch (TCC)

The purpose of the TCC is to improve fuel mileage. It does this by eliminating hydraulic

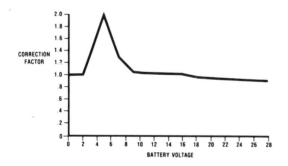

FIGURE 4-15   Voltage correction graph. *Courtesy of GM Product Service Training.*

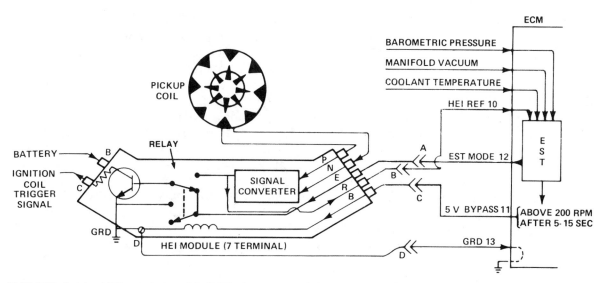

FIGURE 4-16   HEI module with EST. *Courtesy of GM Product Service Training.*

slippage and heat production in the converter once a cruise speed has been achieved. Its operation is essentially the same as the one described in the **Outputs** section of Chapter 3.

Manual Transmissions. On manual transmission applications, the TCC output terminal on the ECM can be used to operate a shift light on the instrument panel. The ECM looks at

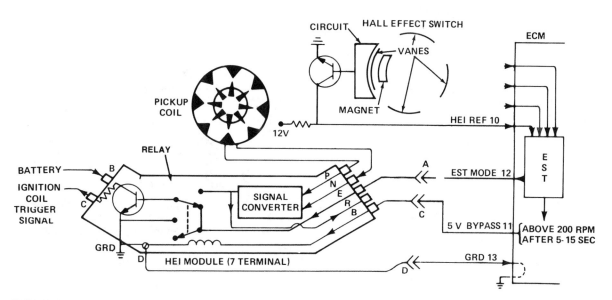

FIGURE 4-17   HEI module with EST and Hall-effect switch. *Courtesy of GM Product Service Training.*

engine speed, engine load, throttle position, and vehicle speed and alerts the driver when to shift into the next gear for optimum fuel economy.

Cooling Fan Control

Transverse engine applications usually use an electrically powered cooling fan to pull air across the radiator fins. On most EFI applications, the ECM has some control over cooling fan operation; the cooling fan can be turned on by the ECM or by a coolant temperature switch. On some applications it has full control. The fan is turned on and off by a relay that the ECM operates, Figure 4–18. On most applications the fan turns on when the coolant temperature is high, when the vehicle is traveling below 30 MPH with the air conditioner on, and in some cases when the air-conditioning high side pressure reaches a certain value.

Hood Louver Control

The Crossfire EFI system features an electrically operated air door that the ECM can open using a relay and solenoid, Figure 4–19. When open the air door allows fresh air from above the engine hood to be directed through a hole in the top of the air cleaner housing to the air cleaner elements. The ECM opens the air door above a specified coolant temperature or at WOT.

Air-conditioning (A/C) Relay

The A/C relay is what actually turns the compressor clutch on and off, Figure 4–20. When the driver turns the A/C switch on, a signal is sent to the ECM and power is available at terminal B of the relay. The ECM waits a half-second and then ground terminal C of the relay. This energizes the relay and engages the compressor clutch (if the engine is at idle, the

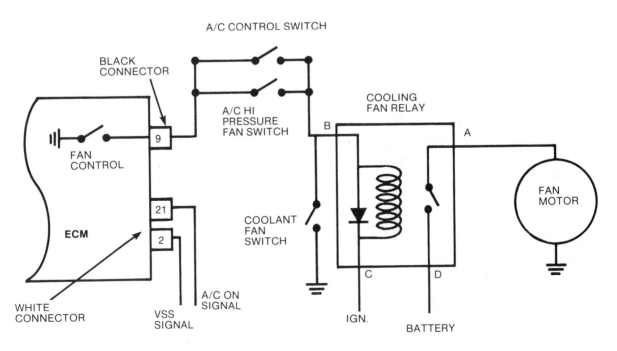

FIGURE 4–18   Cooling fan control circuit (typical)

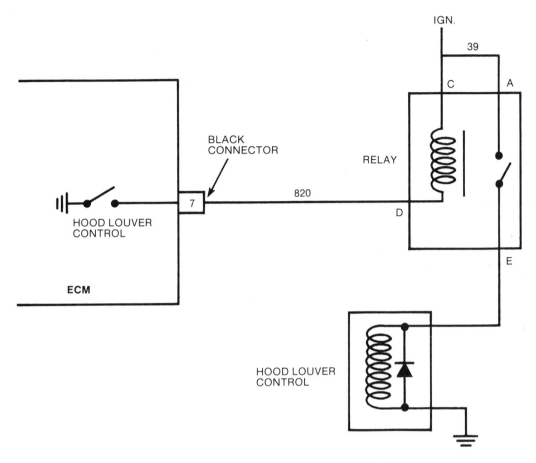

FIGURE 4-19   Hood louver control circuit

half-second allows the ECM time to adjust IAC position before the compressor clutch is engaged). The ECM deactivates the relay at WOT, during heavy engine load, and when IAC is in power steering stall mode. As seen in Figure 4-20, the compressor clutch circuit can be opened by the high pressure switch or the pressure cycling switch, which is a standard A/C control circuit feature.

Related Emission Controls

On all but the Crossfire EFI system, EGR, canister purge, and air management control are not controlled by the ECM. The EGR, canis-

ter purge, and air management controls used on the Crossfire system are essentially the same as those used on V8 engines with computer command control, as discussed in Chapter 3.

Catalytic Converter. Most EFI applications use a single-bed, three-way catalytic converter. No supplemental air is introduced into the converter; therefore, either no air injection is used or a Pulsair system is used with no computer control.

EGR. Many four-cylinder EFI systems use a self-modulating EGR valve with no ECM control (most use a back-pressure EGR, Figure

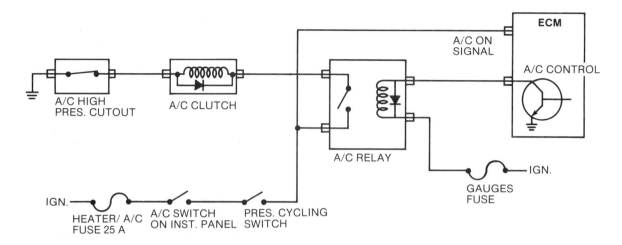

FIGURE 4-20   A/C relay circuit (typical)

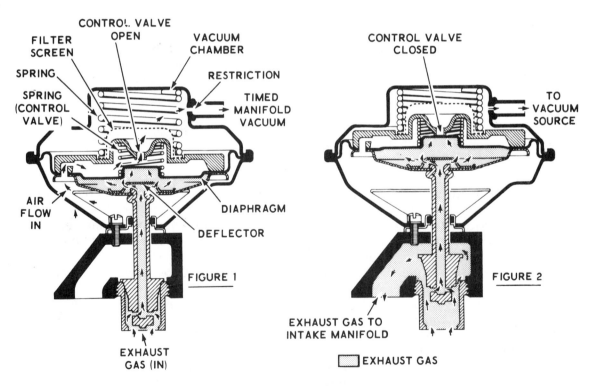

FIGURE 4-21   Back-pressure EGR valve. *Courtesy of GM Product Service Training.*

4-21). Under the main diaphragm found in a conventional EGR valve is a second diaphragm. A small spring pushes downward on the second diaphragm. A passage in the EGR valve stem allows exhaust manifold pressure to enter into a chamber under the second diaphragm. When exhaust pressure is low, the small spring is able to hold the second diaphragm down in spite of the exhaust pressure trying to push it up. Atmospheric pressure flows through a vent in the lower part of the EGR valve housing and through a port in the second diaphragm, into the space between the two diaphragms. If the second diaphragm is in its down position, the atmospheric pressure is free to flow through another port in the main diaphragm into the vacuum chamber. With atmospheric pressure finding its way into the vacuum chamber, enough vacuum is not developed to lift the main diaphragm and open the EGR valve. As the throttle opening is increased, exhaust pressure increases. When exhaust pressure is high enough, the second diaphragm is forced up to close off the port through the main diaphragm into the vacuum chamber. The vacuum chamber is no longer vented to atmospheric pressure, and sufficient vacuum is now developed to open the EGR valve.

Canister Purge. On four-cylinder EFI engines, canister purge control is achieved with a thermal vacuum switch that blocks purge vacuum until the engine is in closed loop. In closed loop any additional fuel from the canister can be compensated for, still maintaining a 14.7 to 1 air/fuel ratio.

Check Engine Light

All EFI systems use a check engine light. If the ECM sees a fault in one of the circuits that it monitors for malfunctions, it turns on the check engine light located on the instrument panel. This warns the driver that a malfunction exists. By grounding the test terminal in the assembly line communication link (ALCL) located under the dash, the ECM flashes the check engine light. These flashes indicate one or more codes that identify the circuit or circuits in which the fault exists. This is explained more fully under **System Diagnosis and Service** in this chapter.

SYSTEM DIAGNOSIS AND SERVICE _____

Self-diagnosis

The self-diagnostic capability and procedure for the EFI system is essentially the same as that of the CCC system described in the **System Diagnosis and Service** section of Chapter 3. An abbreviated discussion that points out the variations from the CCC system is presented here.

Note: Diagnostic procedures for 1987 and later are written to include the use of an ALCL tool, often referred to as a "scanner," where it is applicable. Previously, diagnostic procedures were written around the use of a voltmeter to obtain such measurements as sensor readings.

Approach to Diagnosis

It is important that when any of the diagnostic procedures are being used that the directions be followed carefully. Failing to do so or taking shortcuts usually results in inaccurate conclusions.

Diagnostic Circuit Check. In diagnosing a problem on an EFI system, after verifying that all non-EFI engine support systems are working properly, start with a diagnostic circuit check, Figure 4-22. The diagnostic circuit check either verifies that the system is working properly or refers to another chart for further diagnosis. Be sure that the diagnostic circuit check and trouble code charts come from the

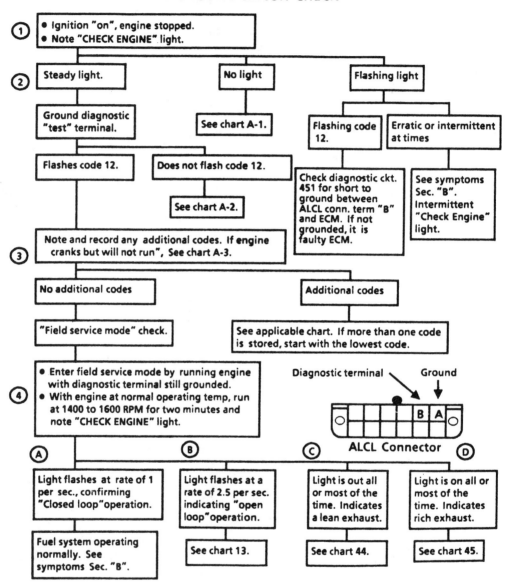

FUEL INJECTION
DIAGNOSTIC CIRCUIT CHECK

①
- Ignition "on", engine stopped.
- Note "CHECK ENGINE" light.

② Steady light. | No light | Flashing light

Ground diagnostic "test" terminal.

See chart A-1.

Flashing code 12. | Erratic or intermittent at times

Flashes code 12. | Does not flash code 12.

See chart A-2.

Check diagnostic ckt. 451 for short to ground between ALCL conn. term "B" and ECM. If not grounded, it is faulty ECM.

See symptoms Sec. "B". Intermittent "Check Engine" light.

③ Note and record any additional codes. If engine cranks but will not run", See chart A-3.

No additional codes | Additional codes

"Field service mode" check. | See applicable chart. If more than one code is stored, start with the lowest code.

④
- Enter field service mode by running engine with diagnostic terminal still grounded.
- With engine at normal operating temp, run at 1400 to 1600 RPM for two minutes and note "CHECK ENGINE" light.

Diagnostic terminal   Ground

B A

ALCL Connector

Ⓐ | **Ⓑ** | **Ⓒ** | **Ⓓ**

Light flashes at rate of 1 per sec., confirming "Closed loop"operation. | Light flashes at a rate of 2.5 per sec. indicating "open loop"operation. | Light is out all or most of the time. Indicates a lean exhaust. | Light is on all or most of the time. Indicates rich exhaust.

Fuel system operating normally. See symptoms Sec. "B". | See chart 13. | See chart 44. | See chart 45.

⑤ Vehicle may be driven in the field service mode and evaluated at any steady speed. This can be helpful in diagnosing driveability problems where the system is rich or lean too long.
- Clear codes and confirm "closed loop "operation and no "light ".
5-14-84   * 5S 1350-6EA '84-85

FIGURE 4-22 Diagnostic circuit check. *Courtesy of Buick Motor Division.*

1983
FUEL INJECTION
FIELD SERVICE MODE CHECK

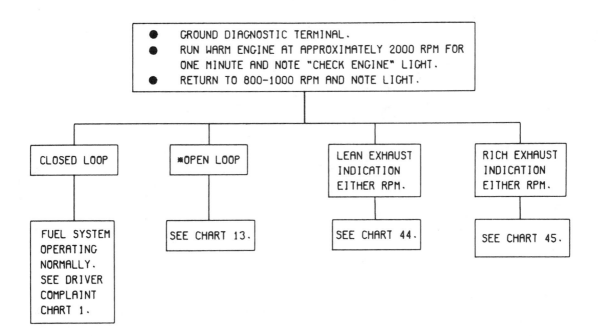

FIELD SERVICE MODE: ENGINE RUNNING, DIAGNOSTIC TERMINAL GROUNDED. "CHECK ENGINE" LIGHT IS "OFF" WHEN EXHAUST IS LEAN AND "ON" WHEN IT IS RICH.

OPEN LOOP: "CHECK ENGINE" LIGHT FLASHES AT A RATE OF 2 TIMES PER SECOND.

CLOSED LOOP: "CHECK ENGINE" LIGHT FLASHES AT A RATE OF 1 TIME PER SECOND.

LEAN EXHAUST INDICATION: "CHECK ENGINE" LIGHT IS OUT ALL THE TIME OR MOST OF THE TIME.

RICH EXHAUST INDICTION: "CHECK ENGINE" LIGHT IS ON ALL THE TIME OR MOST OF THE TIME.

* OPEN LOOP AT IDLE ONLY MAY BE CAUSED BY A COLD OXYGEN SENSOR AFTER A PERIOD OF IDLING. IF THIS IS THE CAUSE, RUNNING ENGINE AT 2000-3000 RPM FOR ONE MINUTE SHOULD WARM IT UP AND GIVE CLOSED LOOP OPERATION AT IDLE.

AFTER ANY REPAIR, CLEAR CODES AND CONFIRM "CLOSED LOOP" OPERATION.

FIGURE 4-23   Field service mode. *Courtesy of GM Product Service Training.*

1982-1983
CHART 1
FUEL INJECTION
DRIVER'S COMPLAINT

● INTERMITTENT "CHECK ENGINE" LIGHT OR STORED CODES.

NOTICE: DO NOT USE DIAGNOSTIC CHARTS FOR INTERMITTENT PROBLEMS. THE FAULT MUST BE PRESENT TO LOCATE THE PROBLEM. IF THE FAULT IS INTERMITTENT, USE OF THE CHARTS MAY RESULT IN THE REPLACEMENT OF NON-DEFECTIVE PARTS.

MOST INTERMITTENT PROBLEMS ARE CAUSED BY FAULTY ELECTRICAL CONNECTORS OR WIRING. DIAGNOSIS MUST INCLUDE A CAREFUL VISUAL AND PHYSICAL INSPECTION OF THE INDICATED CIRCUIT WIRING AND CONNECTORS.

- POOR MATING OF THE CONNECTOR HALVES OR TERMINALS NOT FULLY SEATED IN CONNECTOR BODY (BACKED OUT "TERMINALS").
- IMPROPERLY FORMED OR DAMAGED TERMINALS. ALL CONNECTOR TERMINALS IN PROBLEM CIRCUIT SHOULD BE CAREFULLY REFORMED TO INCREASE CONTACT TENSION.
- HEI DISTRBUTOR EST WIRES SHOULD BE ROUTED AWAY FROM DISTRIBUTOR, IGNITION COIL, SECONDARY WIRING AND GENERATOR.
- CKT 419- "CHECK ENGINE LAMP" TO ECM, SHORT TO GROUND.
- CKT 451- DIAGNOSTIC CONNECTOR TO ECM, SHORT TO GROUND.
- CKT 450 AND 450R- CHECK ECM GROUND AT ENGINE BLOCK ATTACHMENT.
- ELECTRICAL SYSTEM INTERFERENCE CAUSED BY A DEFECTIVE RELAY, ECM DRIVEN SOLENOID, OR A SWITCH CAUSING A SHARP ELECTRICAL SURGE. NORMALLY, THE PROBLEM WILL OCCUR WHEN THE DEFECTIVE COMPONENT IS OPERATED.
- IMPROPER INSTALLATION OF ELECTRICAL OPTIONS, I.E. LIGHTS, 2 WAY RADIO, ETC.
- OPEN AIR CONDITIONING CLUTCH DIODE.

● STALLING, ROUGH OR IMPROPER IDLE SPEED - SEE CHART 11.

● ENGINE CRANKS BUT WILL NOT RUN- SEE CHART 4 OR 4A.

● HARD STARTING, POOR PERFORMANCE, DRIVABILITY, OR FUEL ECONOMY- SEE CHART 7.

● DETONATION (SPARK KNOCK)
- ESC PERFORMANCE CHART 10, IF APPLICABLE.
- EGR CHART 8.

● POOR ENGINE PERFORMANCE WITH AIR CONDITIONING ON.
REFER TO SECTION 1B.

FOLLOWING ANY REPAIRS OR ADJUSTMENTS, ALWAYS CLEAR CODES AND CONFIRM "CLOSED LOOP" OPERATIONS AND NO "CHECK ENGINE" LIGHT.

FIGURE 4-24   Driver's complaint chart. *Courtesy of GM Product Service Training.*

same model year manual as the vehicle being worked on. On earlier systems the diagnostic circuit check is designed to sort out whether or not the fault currently exists or if it is an intermittent fault present at one time but not now. On those systems the trouble code charts are written assuming that a fault exists. If one of them is used to pursue an intermittent code, the chart most often leads to an invalid conclusion because it is beginning with an invalid assumption—that a fault currently exists when in fact none does.

For later systems the diagnostic circuit check and many of the trouble code charts are written to allow the trouble code charts to determine if the code obtained from the diagnostic memory identifies a current fault (hard code) or an intermittent code.

Field Service Mode. Grounding the test terminal with the engine running puts the system into field service mode. In this mode the ECM uses the check engine light to show whether the system is in open or closed loop and whether the system is running rich or lean, Figure 4–23.

Driver's Complaint Chart. The driver's complaint chart, Figure 4–24, acts as a guide to assist in finding the cause of problems not identified by trouble codes. It either suggests things to check for certain problems or refers to other, more specific charts. For later models this chart has been expanded into a complete section in the service manual and is referred to as *Symptoms*. There are other charts in the later service manuals that assist in diagnosing specific EFI component or system malfunctions.

ECM and PROM Service

Considerable care should be taken when installing a PROM or CALPAK (only 2.0-liter engine ECMs are equipped with a *CALPAK*). A finger coming into contact with its pins can discharge static electricity into it and damage it, Figure 4–25. The PROM carrier fits into the

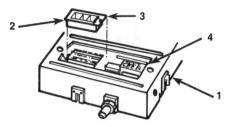

| 1 ECM | 3 PROM CARRIER |
| 2 PROM (ENGINE CALIBRATOR) | 4 CALPAC |

FIGURE 4-25   PROM and CALPAK in ECM. *Courtesy of Buick Motor Division.*

PROM cavity of the ECM in only one position; however, the PROM fits into the carrier in either of two ways. If the PROM is installed in the ECM end for end from its correct position, it will be damaged electrically. Consult the service manual for installation instructions.

The PROM, CALPAK, and ECM are serviced separately; therefore, remove the PROM and CALPAK before exchanging the ECM for a new one.

If the CALPAK is missing or severely damaged, the engine does not start.

Weather-pack Connectors

All EFI system harness connections under the hood are made with weather-pack connectors as discussed in the **System Diagnosis and Service** section of Chapter 3.

Diagnostic and Service Tips

TCC Test Terminal. See **TCC Test Terminal** in the **System Diagnosis and Service** section of Chapter 3.

Vacuum Leaks. A vacuum leak on an EFI system usually results in an increased idle speed while in closed loop.

TBI. The TBI unit has two O-rings, Figure 4–26, between the injector solenoid and the fuel meter body. The O-rings prevent pressurized fuel from leaking past the injector solenoid. A leak at either of these O-rings can result

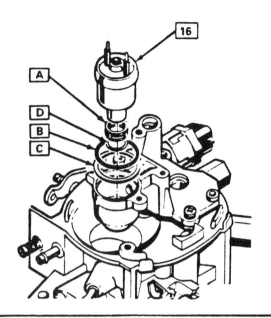

16 FUEL INJECTOR
A FILTER
B LARGE "O" RING
C STEEL BACK-UP WASHER
D SMALL "O" RING

FIGURE 4-26   EFI injector assembly. *Courtesy of Buick Motor Division.*

in dieseling and/or an overly rich mixture. A suspected leak can easily be checked. With the engine off and a jumper lead supplying battery voltage to the fuel pump test terminal, inspect the injector for signs of leakage.

PERSONAL SAFETY: Fuel Pressure Test. Some driveability complaints require a fuel pump pressure test. Before the fuel system is opened, pressure in the line should be relieved; the system is designed to retain operating pressure in the line after the engine is shut off. Opening the line under pressure causes a considerable spray, resulting in a fire and safety hazard. To relieve the pressure, remove the fuel pump fuse and crank the engine for several seconds. See the service manual for test procedures.

Erasing Learned Ability. Anytime battery power is removed from the ECM, its learned memory is lost. This can produce a noticeable change in performance. Learning can be restored by driving the car for usually not more than half an hour at normal operating temperature, part throttle and idle, and at moderate acceleration.

REVIEW QUESTIONS _____

1. What are the six inputs most involved in engine calibration control?
2. What is the second function of the MAP sensor? Under what conditions does it perform this function?
3. Name three functions that the VSS influences.
4. What influence does the A/C switch have on engine operation?
5. What introduces fuel to the engine on EFI applications?
6. How is fuel quantity controlled?
7. Describe the ECM's fuel system control response to each of the following conditions on a four-cylinder EFI application:
   a. the ignition is turned on
   b. no REF signal is sent to the ECM within two seconds
8. What variable control controls the amount of air going into the intake manifold during idle?
9. During what two operating conditions can the antistall mode be activated?
10. What signal alerts the ECM to move the IAC valve back to its parked position?

11. What is the power source for most output devices?

12. What type of device is used by the ECM to activate the fuel pump, cooling fan, and A/C clutch?

13. What are the first two steps when pursuing a driveability complaint on an EFI-equipped vehicle?

14. What is the function of the weather-pack connector?

15. What two functions does the TCC test lead serve?

16. Name six things that should be observed to avoid damage to the ECM.

ASE-type Questions. (Actual ASE test questions will probably not be this product specific.)

17. Technician A says that EFI systems that use a Hall-effect switch do not use a pickup coil in the distributor. Technician B says that the Hall-effect switch, on those EFI systems that use it, provides the signal to the HEI module to control primary circuit switching. Who is correct?
    a. A only
    b. B only
    c. both A and B
    d. neither A nor B

18. The engine of a car equipped with an EFI system dies while being driven. The car is brought into the garage, and it is determined that the ignition system is operating properly, the fuel tank is at least half full, and there appears to be no problem in the fuel lines. The ECM is working properly,

A/C	Air-conditioning
AIR	Air Injection Reaction
ALCL	Assembly Line Communication Link
BARO	Barometric Pressure
CCP	(Charcoal) Canister Purge
DIAGNOSTIC TEST TERMINAL	Lead or ALCL connector terminal grounded to obtain a trouble code
ECM	Electronic Control Module
EFI	Electronic Fuel Injection
EGR	Exhaust Gas Recirculation
ESC	Electronic Spark Control
EST	Electronic Spark Timing
HEI	High-energy Ignition
IAC	Idle Air Control
MAP	Manifold Absolute Pressure
PCV	Positive Crankcase Ventilation
P/N	Park/Neutral Switch
PROM	Programmable Read-only Memory (engine calibration unit)
TBI	Throttle Body Injection
TCC	Torque Converter Clutch
TPS	Throttle Position Sensor
VIN	Vehicle Identification Number
VSS	Vehicle Speed Sensor
WOT	Wide-open Throttle

FIGURE 4-27   Explanation of abbreviations

and no fault codes are stored in it. Technician A says that the problem could be a fault in the reference pulse wire from the distributor. Technician B says that the fault could be a blown fuel pump fuse. Who is correct?

a. A only
b. B only
c. both A and B
d. neither A nor B

19. A car equipped with EFI is brought in with a complaint of poor mileage and black smoke during some driving conditions. A fuel pressure check shows that the fuel pressure is too high. The cause is most likely _____ .

a. a faulty fuel pump
b. a faulty fuel pressure regulator
c. a faulty coolant temperature sensor
d. a faulty ECM

20. Technician A says that an EFI system is put into field service mode by grounding the test terminal with the ignition on and the engine off. Technician B says that the same system is put into field service mode by grounding the test terminal with the engine running. Who is correct?

a. A only
b. B only
c. both A and B
d. neither A nor B

GLOSSARY

In addition to the following terms, refer to Figure 4-27 for a list of the abbreviations used in this chapter.

CALPAK   A removable part of the ECM that plugs in like the PROM and is primarily responsible for fuel calibration.

PICKUP COIL   A small coil of wire in which the signal of a pulse magnetic generator is produced, such as in most electronic distributors.

TBB (Throttle Body Backup)   A backup circuit in the ECM of most EFI systems that operates the injector solenoid in the event of a partial system failure.

GENERAL MOTORS' PORT FUEL INJECTION

Objectives

After studying this chapter, you will be able to:

- name at least two advantages of port fuel injection over other types of fuel induction systems.
- describe two methods used by General Motors' PFI systems to measure a value equivalent to air's mass.
- describe the operation of a cold start valve.
- describe the operation of a turbocharger waste gate.
- describe at least one type of EGR diagnostic switch.
- describe how the idle air control valve controls idle speed.
- describe how the computer becomes aware that spark knock is occurring.
- describe how the EGR valve is modulated on PFI systems.

Port fuel injection (PFI) represents the fourth generation of General Motors' comprehensive computerized engine control systems, excluding the Cadillac digital fuel injection discussed in Chapter 6. It was introduced, with limited application on 3.8-liter and 1.8-liter engines, in 1984. It followed the electronic fuel control (General Motors' first closed-loop fuel control system) introduced in 1978, the CCC system introduced in 1980, and the EFI system introduced in 1982. In 1985 the PFI system was expanded to several engine applications, Figure 5–1. Some General Motors' car divisions have chosen a special name for the PFI system because of some special feature designed for a given engine or body application, such as:

- sequential fuel injection (SFI) on the turbocharged 3.8-liter engine.
- multiport fuel injection (MFI) on most other engine applications.
- tuned port injection (TPI) on the Corvette 5.7-liter engine. The TPI version offers manifold runners matched in shape, length, and cross-sectional area to help maximize *volumetric efficiency* (VE).

Because only air moves through the intake manifold (there is no concern about holding fuel in a vapor state as it passes through the runners), the heated air *(THERMAC)* and the early fuel evaporation systems are eliminated.

PORT FUEL INJECTION
1985

RPO CODE	VIN	TYPE	DISPL.	MANUF.	USER	BODY
LB6	W	60°V6	2.8L	CHEV.	C-P-O B-CA	A-J X
LB8	S	60°V6	2.8	CHEV.	C-P	F
L44	9	60°V6	2.8	CHEV.	P	P
LB9	F	90°V8	5.0L	CHEV.	C-P	F
L98	8	90°V8	5.7L	CHEV.	C	Y
LN7	L	90°V6	3.0L	BUICK	B-O	N
LN3	3	90°V6	3.8L	BUICK	B-O	A-C
LM9	9	90°V6	3.8L	BUICK	B	E-G
LA5	J	L4	1.8L	PONT.	B-P	J

FIGURE 5-1   PFI applications. *Courtesy of GM Product Service Training.*

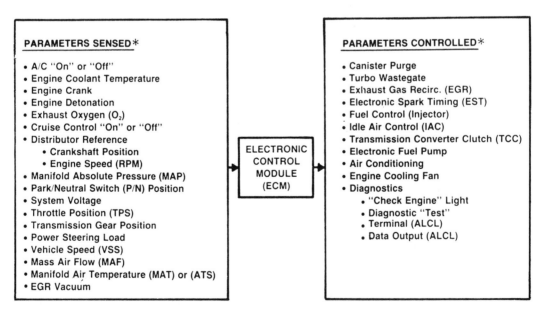

PARAMETERS SENSED*

- A/C "On" or "Off"
- Engine Coolant Temperature
- Engine Crank
- Engine Detonation
- Exhaust Oxygen (O₂)
- Cruise Control "On" or "Off"
- Distributor Reference
  - Crankshaft Position
  - Engine Speed (RPM)
- Manifold Absolute Pressure (MAP)
- Park/Neutral Switch (P/N) Position
- System Voltage
- Throttle Position (TPS)
- Transmission Gear Position
- Power Steering Load
- Vehicle Speed (VSS)
- Mass Air Flow (MAF)
- Manifold Air Temperature (MAT) or (ATS)
- EGR Vacuum

ELECTRONIC CONTROL MODULE (ECM)

PARAMETERS CONTROLLED*

- Canister Purge
- Turbo Wastegate
- Exhaust Gas Recirc. (EGR)
- Electronic Spark Timing (EST)
- Fuel Control (Injector)
- Idle Air Control (IAC)
- Transmission Converter Clutch (TCC)
- Electronic Fuel Pump
- Air Conditioning
- Engine Cooling Fan
- Diagnostics
  - "Check Engine" Light
  - Diagnostic "Test"
  - Terminal (ALCL)
  - Data Output (ALCL)

*NOT ALL SYSTEMS USED ON ALL ENGINES.

FIGURE 5-2   PFI overview. *Courtesy of GM Product Service Training.*

Most of the PFI applications, however, do run engine coolant through a passage in the throttle body to prevent throttle blade icing.

ELECTRONIC CONTROL MODULE

The ECM is usually located under the instrument panel or behind the passenger kick panel. On the P car (Pontiac Fiero) it is in the console between the seats. It contains a removable PROM and CALPAK. The CALPAK provides calibration for cold-start cranking and for fuel backup in the event of an ECM failure. The fuel backup circuit (FBC) provides an operating mode when any of the following conditions occur:

- ECM voltage is below 9 volts (most likely occurs during cranking)
- the PROM is missing or not functioning
- the ECM is unable to provide its normal computer-operated pulses (COP)

The FBC makes use of the throttle position sensor, the coolant temperature sensor, and the engine RPM signal and is powered by the ignition switch. During FBC mode the engine runs erratically and sets a code 52.

Keep Alive Memory (KAM)

The ECM has basically the same self-diagnostic capability and KAM as discussed in Chapter 3. The inputs fed into the ECM and the functions controlled for most systems are shown in Figure 5-2.

OPERATIONAL MODES

The different operating modes of the PFI system control how much fuel to introduce into the intake manifold for various operating conditions.

Starting Mode

When the ignition is turned on, the ECM turns the fuel pump relay on. If it does not see a signal within two seconds telling it that the engine is being cranked, it turns the fuel pump relay off. The fuel pump provides fuel pressure to the injectors. As the system goes into cranking mode, the ECM checks coolant temperature and throttle position to see what the air/fuel ratio should be. The air/fuel ratio ranges from 1.5 to 1 at $-36°$ C ($-33°$ F) to 14.7 to 1 at $94°$ C ($201°$ F). The ECM varies the air/fuel ratio by controlling the time that the injectors are turned on; this is referred to as pulse width.

The 2.8-, 5.0-, and 5.7-liter engine PFI applications also use a cold-start injector. This is an additional injector in the intake manifold that improves cold-start characteristics by spraying additional fuel into the manifold during cranking, Figure 5-3. It is not ECM controlled. Spraying the additional fuel for cold-start enrichment into the manifold gives it more opportunity to evaporate than if it were sprayed into the intake ports. The injector is activated by the cranking system and controlled by a timing mechanism. It is described in more detail in the **Outputs** section of this chapter.

Clear Flood Mode

The clear flood mode works essentially the same as it does on the EFI system. As long as the engine is turning less than 600 RPM and the throttle is held to within 80% of full throttle, the ECM will hold the air/fuel ratio to 20 to 1, except on the 2.8-liter engine, where fuel is cut off completely.

Run Mode

The run mode consists of the open- and closed-loop operating conditions. When the engine is started and the RPM is above 400, the system goes into open loop. In open loop the ECM ignores information from the O_2 sensor and determines air/fuel ratio commands based on input from other sensors (coolant, vacuum

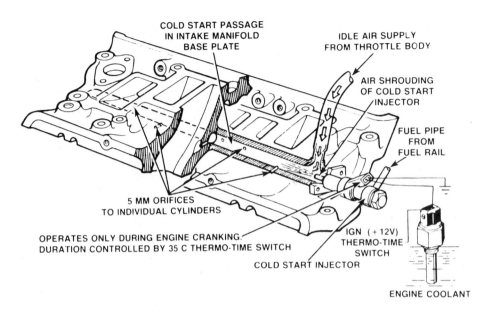

FIGURE 5-3 Cold-start injector (typical). *Courtesy of GM Product Service Training.*

or mass airflow, and throttle position and engine speed if the vacuum sensor is used). The system stays in open loop until:

- the O_2 produces a varying voltage showing that it is hot enough to work properly.
- the coolant is above a specified temperature.
- a specified amount of time has elapsed since the engine was started.

These values vary with engine application and are contained in the PROM. When all three conditions are met, the ECM puts the system into closed loop. In closed loop the ECM uses only O_2 sensor input to calculate air/fuel ratio commands and keeps the air/fuel ratio at a near perfect 14.7 to 1. During heavy acceleration, wide-open throttle, or hard deceleration, the system temporarily drops out of closed loop.

Acceleration Mode

Rapid increases in throttle opening and manifold pressure or airflow signal the ECM to enrich the air/fuel mixture. This compensates for the reduced evaporation rate of the gasoline that results from the higher manifold pressure.

Deceleration Mode

Rapid decreases in throttle opening and manifold pressure or airflow cause the ECM to lean out the air/fuel mixture. If the changes are severe enough, fuel is momentarily cut off.

Battery Voltage Correction Mode

The battery voltage correction feature on the PFI system is different from that on the EFI system. On the PFI system, when battery voltage drops below a specified value, the ECM will:

- enrich the air/fuel mixture
- increase throttle opening, if the engine is idling
- increase ignition dwell

to compensate for a weakened ignition spark.

Fuel Cutoff Mode

When the ignition is turned off, the ECM immediately stops pulsing the injectors; this is done to prevent dieseling. Injection also is stopped anytime the distributor reference pulse stops coming to the ECM.

FUEL SUPPLY SYSTEM

Fuel pressure is supplied by an electric pump in the fuel tank, Figure 5-4. The pump is turned on and off by a fuel pump relay that is controlled by the ECM. A pressure relief valve

in the pump limits maximum pump pressure to between 60 and 90 PSI. This pressure is realized only if flow stops and the pump is working against static pressure. The filter mounted to the bottom of the pump assembly is a 50-micron filter; downstream is a 10- to 20-micron in-line filter. Because of the high operating pressure, an O-ring is used at all threaded connections and all flex hoses have internal steel reinforcement.

Fuel pressure is controlled by the fuel pressure regulator, Figure 5-5, which is usually mounted on the fuel rail, Figures 5-6 and 5-7.

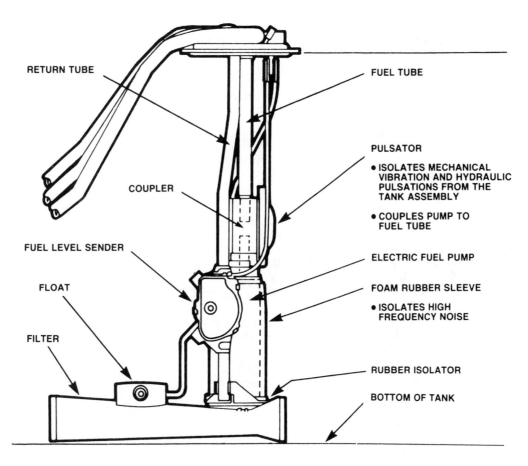

FIGURE 5-4 Fuel pump and sending unit assembly. *Courtesy of GM Product Service Training.*

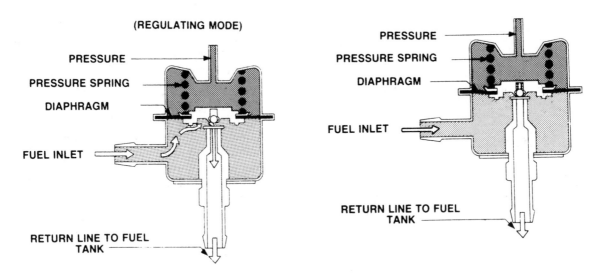

FIGURE 5–5   Fuel pressure regulator. *Courtesy of GM Product Service Training.*

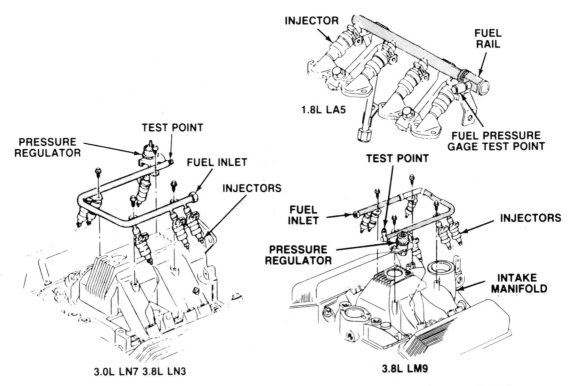

FIGURE 5–6   Fuel rail: 1.8 L, 3.0 L, and 3.8 L. *Courtesy of GM Product Service Training.*

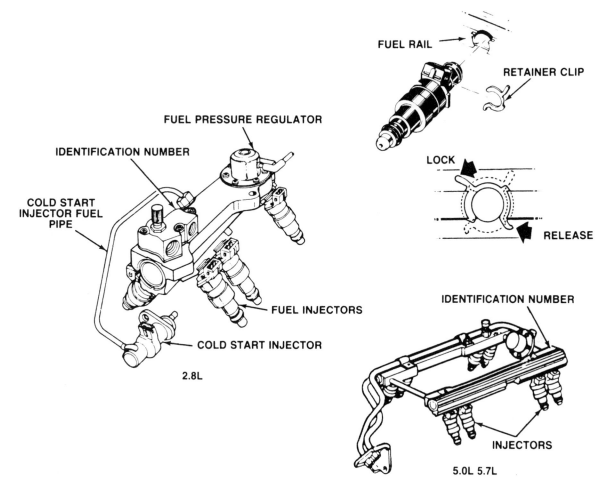

FIGURE 5-7   Fuel rail: 2.8 L, 5.0 L, and 5.7 L. *Courtesy of GM Product Service Training.*

Manifold vacuum is supplied to the spring side of the diaphragm. At light throttle the high vacuum helps to pull the diaphragm up and against the spring. This allows more fuel to return to the tank and thus reduces output pressure. Pressure ranges from 34 PSI at idle to 44 PSI at full throttle. A constant stream of fuel flowing from the tank to the pressure regulator and back to the tank through the return line ensures cool fuel available to the fuel rail and injectors.

This plus the operating pressure eliminate the chances of vapor lock.

INJECTORS

An injector is installed in the intake port of each cylinder with 70 millimeters to 100 millimeters between the tip of the injector and the center of the valve on V-type engines, Figure 5-8. The nozzles spray fuel in a 25° conical pattern. O-rings are used to seal between the nozzle and the fuel rail and between the nozzle and the intake manifold. The O-rings also serve to retard heat transfer from the engine to the

C³I System

For years automotive engineers have generally felt that it did not make much difference when the fuel was sprayed into the intake port as long as precisely the right amount was sprayed. When you consider how fast all of this is happening (the intake valve opens and closes twenty-five times per second at 3,000 RPM), it would not seem to make much difference. The use of the three-coil ignition system (C³I), introduced in 1984 by Buick, required a sequential fuel injection system because the C³I system fires each spark plug each time the plug's respective piston approaches top dead center. As the piston finishes the exhaust stroke and starts the intake stroke, the spark plug fires while both valves (intake and exhaust) are open. This firing serves no intended purpose; it is just a by-product of the ignition system design. If, however, a pressurized, combustible mixture (the first application of the C³I system was on a turbocharged engine) were standing on the other side of the open intake valve, a backfire could occur in the intake manifold. To avoid this, each injector pulse is timed so that the fuel is immediately purged from the manifold by its respective cylinder's intake stroke.

The additional circuitry and related expense seems to have proved to be worthwhile, however, as Buick has rapidly expanded the sequential fuel injection system to its other V6 engines and Ford introduced a sequential fuel injection system in 1986 on some 5-liter applications.

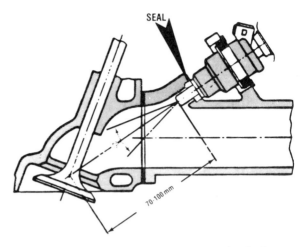

FIGURE 5-8   Port fuel injection (typical). *Courtesy of GM Product Service Training.*

nozzle and to reduce nozzle vibration. The O-rings should be lubricated and/or replaced as necessary when the nozzles are removed. A damaged or hardened O-ring allows a fuel or vacuum leak; PFI systems are very sensitive to vacuum leaks.

On all PFI systems except the sequential fuel injection (SFI) version used on the Buick-built, 3.8-liter turbocharged engine in 1984 and 1985 and on all Buick-built, 3.8-liter engines in 1986, all of the injectors are pulsed simultaneously every crankshaft revolution. Fuel sprayed while the intake valve is closed is simply stored in the intake port until the valve opens. The SFI system pulses each injector one at a time in the firing order.

THROTTLE BODY

The throttle body on a PFI system controls the amount of air that is allowed into the engine's induction system. This is done with a throttle blade and shaft controlled by the accelerator pedal, just as in a carburetor or TBI unit, Figures 5-9 and 5-10. In this case, however, only air passes the throttle blade. An idle

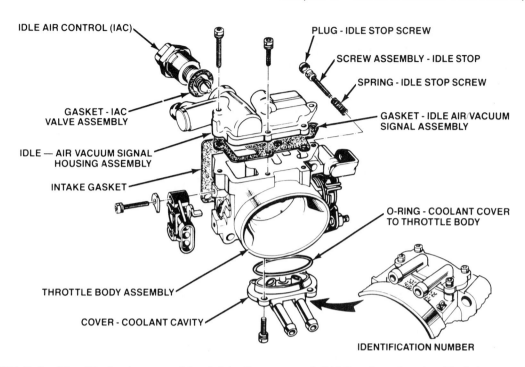

FIGURE 5–9   Throttle body assembly, 2.8 L. *Courtesy of GM Product Service Training.*

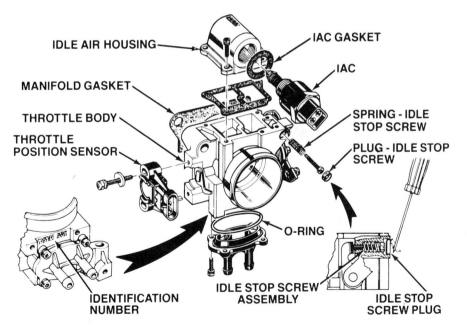

FIGURE 5–10   Throttle body assembly, 3.0 L. *Courtesy of GM Product Service Training.*

stop screw determines how nearly closed the throttle blade is at idle. A throttle position sensor keeps the ECM informed as to the throttle valve's position. Idle speed is controlled by an idle air control motor and valve assembly. It controls the amount of air that is allowed to bypass the throttle blade through a bypass passage and is controlled by the ECM. Most throttle body units have a cavity for engine coolant to flow through to prevent throttle blade icing.

NON-ECM EMISSION CONTROL

A trend seems to be developing whereby as the computerized engine control systems become more efficient, fewer of the traditional emission control devices are needed. This is evidenced by the following list of non-ECM-controlled emission control devices during the first or second year of PFI production.

- 1.8 liter—canister purge, PCV, and internal EGR (no EGR valve)
- 2.8 liter—PCV and deceleration valve (lets additional air in intake manifold during deceleration on manual transmissions only)
- 3.0 and 3.8 liter—PCV
- 5.0 and 5.7 liter—PCV

INPUTS

Coolant Temperature Sensor (CTS)

This is the same sensor used in the computer command control system, the electronic fuel injection system, and other General Motors' systems. The CTS input affects the control of:

- air/fuel mixture (in open loop).
- spark timing.

- spark knock control (on some engine applications).
- engine idle.
- torque converter clutch.
- canister purge (except 1.8 liter as of 1985).
- EGR (except 1.8 liter as of 1985).
- cooling fan (on some applications).

Mass Airflow (MAF) Sensor

Most General Motors' PFI systems use a MAF sensor. Two MAF sensors have been used: one manufactured by *AC,* a division of General Motors, and the other manufactured by Bosch. The Bosch sensor has been used primarily on eight-cylinder engines.

AC. The AC unit, Figure 5–11, contains a screen to break up the airflow. After the air passes through the screen, it flows over an air temperature sensing resistor. A sample tube then directs some of the air to flow over a heated foil sensing element. The power to heat the sensing element comes from a fuse (a relay as on some applications), Figure 5–12. The circuitry of the MAF sensor controls current flow through the foil sensing element to maintain it at 75° C above the incoming air temperature, as measured by the temperature-sensing resistor. The power required to keep the foil sensing element 75° C above incoming air temperature is a measure of mass airflow. This value is sent to the ECM as a digital signal ranging in frequency from 30 cycles per second (30 hertz, or 30 Hz) to 150 Hz (150 Hz represents the highest mass airflow rate). The ECM compares the MAF sensor information to a preprogrammed look-up chart and finds airflow in grams per second. Using this value, engine temperature, and engine RPM, the ECM can determine precisely how much fuel is required to achieve the desired air/fuel ratio. Mass airflow readings are taken and air/fuel mixture calculations are made about 160 times per second.

Bosch. The Bosch MAF sensor works much the same way except that a wire heat

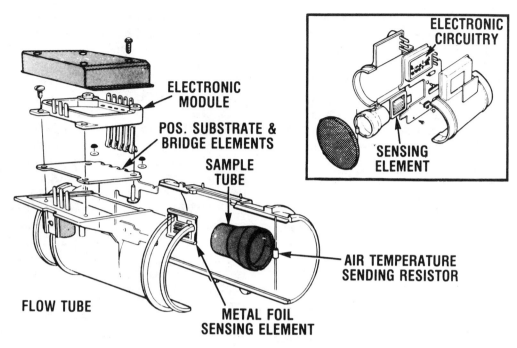

FIGURE 5–11   Mass airflow sensor (A/C type). *Courtesy of GM Product Service Training.*

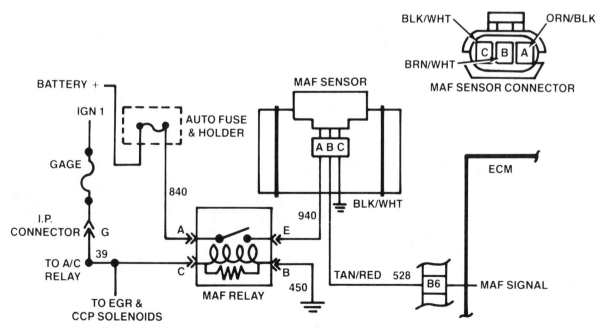

FIGURE 5–12   MAF relay circuit. *Courtesy of GM Product Service Training.*

element is used instead of a foil sensing element. Also, each time the ignition is turned off after the system has been in closed loop, a separate burn-off module, Figure 5-13, momentarily puts enough current through the wire heat element to make it red hot. This burns off any residue accumulation. It is very critical that its surface be clean because an accumulation of deposits will retard its ability to transfer heat.

Speed Density Formula

The General Motors' PFI applications that use the *speed density formula* use a MAP sensor. Using the speed density formula, Figure 5-14, requires knowing the volume, temperature, and flow rate of the air that is being mixed with the fuel. The ECM can determine air volume and flow rate by using data from the MAP sensor, TPS, and the REF signal. The air's temperature is obtained from the manifold air temperature sensor. These three values plus estimates of other engine parameters such as

volumetric efficiency and EGR flow can be used by the ECM to define an approximate equivalent of mass airflow. It is necessary to consider EGR flow because it displaces a percentage of the intake charge with noncombustible materials (CO_2).

Manifold Air Temperature (MAT) Sensor

The MAT sensor looks like and is essentially a coolant temperature sensor with its thermistor sensing element partially exposed instead of being fully enclosed like the CTS. It screws into a hole in the intake manifold, or in some cases into the air cleaner housing, with its nose extending inside. It is sometimes referred to as an air temperature sensor (ATS). It is not used on all applications. Where it is used, it influences air/fuel mixture, spark timing, and idle control.

Throttle Position Sensor (TPS)

The TPS is a potentiometer mounted on the side of the throttle body, Figure 5-10, and

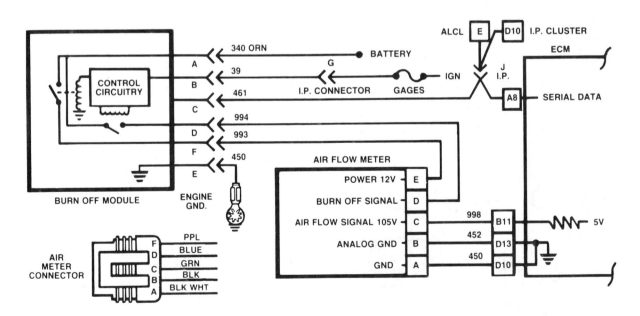

FIGURE 5-13   Bosch MAF circuit and burn-off module. *Courtesy of GM Product Service Training.*

SPEED DENSITY

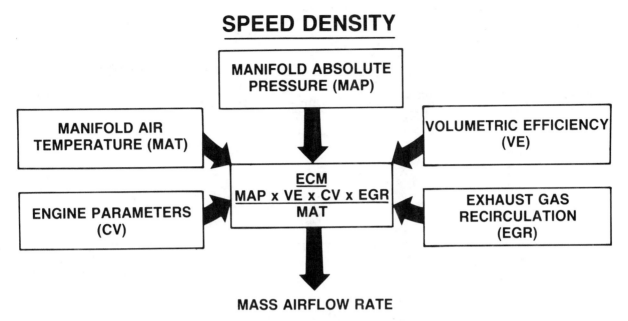

FIGURE 5-14   Speed density formula. *Courtesy of GM Product Service Training.*

attached to the end of the throttle shaft. Its input affects most ECM-controlled functions, including fuel control, spark timing, idle air control, torque converter clutch, and air-conditioning control.

Reference Pulse (REF)

On all but those engines equipped with a distributorless ignition system, the REF signal comes from the distributor pick-up coil. Without this signal the engine does not start or run because the ECM will not pulse the injectors without it. A more complete discussion of the REF signal is found in the **Inputs** section of Chapter 3.

The distributorless ignition systems use sensors to provide information to the ignition module relative to crankshaft and camshaft position and speed (Figure 5–15). The C³I (Computer-Controlled Coil Ignition) systems use Hall-effect switches while both the DIS (Direct Ignition System) and the IDI (Integrated Direct Ignition) systems use a permanent magnetic pulse generator (Figure 5–16).

C³I. There are three different versions of the C³I system: Type I, Type II, and Type III. Each of the three types uses a Hall-effect switch mounted behind the crankshaft balancer (Figure 5–17). The balancer has three vanes, which are positioned to form a circle, that extend back toward the engine. As the crankshaft turns, the vanes alternately pass through the Hall-effect switch, producing a signal that is fed to both the ignition module and the ECM. The Type I system when used on the 3.0 liter engine has a second Hall-effect switch mounted behind the crankshaft balancer with a single vane extending back, nearer the outer circumference of the balancer, that acts as the camshaft sensor. The Type I or Type II systems, when used on the 3.8 liter Turbo engine, feature a Hall-effect switch mounted in what used to be the distributor hole to serve as a camshaft sensor (Figure 5–18). It is driven by the camshaft. These same two systems, when used on the 3.8 liter, non-turbo, SFI engine use a Hall-effect cam sensor mounted in the timing cover and triggered by a vane attached to the

| System | Engine | PM Reluctor | | Hall Effect Switches | | | | |
| | | Crank Pos. Sensors | | CAM/SYNC Sensors | | | | |
		On Crank	Harmonic Balancer	Harmonic Balancer	Dist. Hole	Timing Gear	18x
DIS	2.0	X					
	2.5	X					
	2.8	X					
IDI	2.3	X					
C³I-I	3.0		X	X^C			
I or II^A	3.8 SFI		X			X	
I or II	3.8 TURBO		X		X		
III^B	3.800		X			X	X

A — Type I or II may be used on 3.8 liter engines. Type I has molded one piece coil pack. Type II has independent coils.

B — Introduced in 1988, has independent coils.

C — Now called a "sync" sensor rather than a "cam" sensor.

FIGURE 5-15   General Motors distributorless ignition system applications

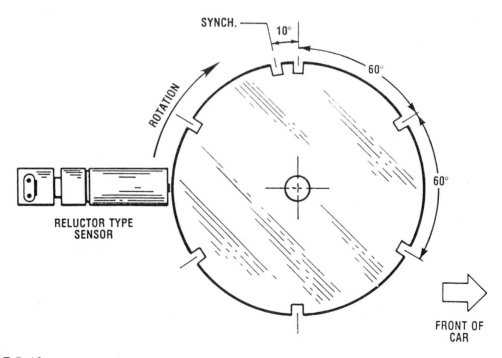

FIGURE 5-16

FIGURE 5-17   Crankshaft position sensor (3.0 L, SFI). *Courtesy of GM Product Service Training.*

DOT

FIGURE 5-18   Camshaft position sensor (3.8L, SFI). *Courtesy of GM Product Service Training.*

camshaft timing gear. The cam sensor signal is used to alert the ECM when the number one cylinder is at TDC so that it can start the firing sequence over again and to time injector pulsing. Loss of this signal will result in a code 41 being set and if the signal is lost during cranking, the engine will not start. In more recent General Motors literature, the camshaft sensor is referred to as a "sync sensor."

On the Type III system, sometimes referred to as the "Fast Start," which also has its sync sensor located in the timing cover, a third sensor is used. It is a Hall-effect switch and is located behind the harmonic balancer along with the crankshaft sensor. This sensor, referred to as the "Crank 18x," is triggered by 18 evenly spaced interruptor vanes protruding rearward from near the outer circumference of the balancer. The Crank 18x sensor provides 18 signals (Figure 5-19) per crankshaft revolution to the ignition module that, in turn, sends it on to the ECM.

On this system, the crank sensor is called the "Crank 3x" sensor, and its three interruptor vanes, nearer the center on the balancer, are not symmetrical as they are on the other C³I systems. The windows between the vanes are 10, 20, and 30 degrees wide as shown in Figure 5-19. The signal produced by the leading and trailing edge of each window constitutes a 3x pulse. The number of 18x signals that occurs during each 3x pulse enables the ECM to determine which 3x pulse it is reading. There will be one 18x signal during the 10° window pulse, two during the 20° window pulse, and three during the 30° window pulse. This enables the system to fire a coil for the appropriate cylinders within the first 120° of crankshaft rotation. This system provides the following advantages:

- faster starts
- more accurate REF signals to the ECM, especially at low engine speed
- increased run reliability as the engine will continue to operate without the cam sensor
- potential for the ECM to read crankshaft acceleration and deceleration rates

DIS and IDI. Both the Integrated Direct Ignition (IDI) and the Direct Ignition System (DIS) systems use a magnetic pulse generator on the crankshaft instead of the Hall-effect switches for crankshaft reference (position and speed). The crankshaft for those engines

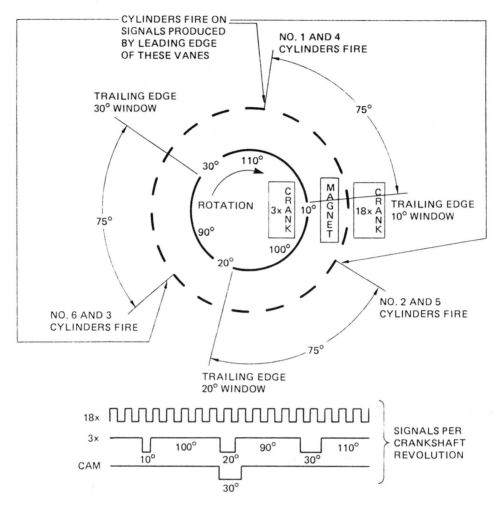

FIGURE 5-19   Combination sensor for C³I-(Type 3)-Fast Start 1

which use either the DIS and the IDI system has a round machined steel disc that is concentric to the crankshaft's centerline for its center counterweight (Figure 5-20). This disc has seven notches cut in it and is called a reluctor. The term "reluctor" refers to any device that changes the reluctance of a material to conduct magnetic lines of force. The same reluctor is used on both four- and six-cylinder engines. The crankshaft sensor is mounted to the engine block and extends through a hole in the crankcase. Its tip, containing a permanent magnet and a wire coil, is spaced .050 inch from the reluctor.

Six of the reluctor's seven notches are spaced evenly around it at 60-degree intervals (Figure 5-16). The seventh notch is called the "sync" notch and is located between notches 6 and 1, 10 degrees from notch 6 and 50 degrees from notch 1. By being placed at an odd position in relation to the other notches, the sync notch provides an irregular signal that the mod-

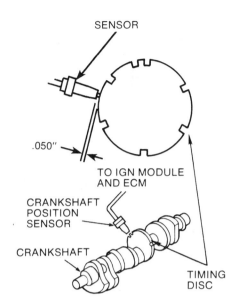

FIGURE 5-20   Crankshaft timing disc

ule can identify. The module uses this signal to synchronize the coil firing sequence to the crankshaft's position in the engine cycle.

On both the DIS and the IDI systems, the notch that produces the signal that results in the appropriate coil being fired during module mode operation is often referred to as the "cylinder event" notch. The crank sensor signal that occurs 60 degrees before the cylinder event notch causes voltage in the REF wire to be low. The cylinder event notch causes voltage in the REF wire to be high. When the engine is operated in the electronic spark timing mode, the ECM uses this signal as an input for calculating:

- ignition timing while operating in EST mode.
- fuel injector pulses.

Oxygen (O_2) Sensor

This is the same single-wire sensor as used on other General Motors' vehicles.

Vehicle Speed Sensor (VSS)

Two types of VSSs are used and are the same as those discussed in Chapter 3.

EGR Diagnostic Switch

On PFI systems that control EGR operation, an EGR diagnostic switch is used to tell the ECM if the EGR is actually being applied when it is commanded to be applied. Two types are used. The four- and six-cylinder engines use the same vacuum-operated switch described in Chapter 3.

The 5.0- and 5.7-liter engines use a thermal switch for exactly the same purpose as the vacuum switch used on the smaller engines. It, however, screws into the base of the EGR valve, Figure 5-21. If the EGR valve is open, the heat of the exhaust gases, flowing past the heat-sensing tip of the thermal switch, cause the switch to close. The ECM wants to see 12 volts on circuit 935 when the EGR is supposed to be off and less than 1 volt when the EGR is supposed to be on.

Detonation Sensor

Most of General Motors' PFI systems incorporate an electronic spark control (ESC) system that uses a piezoelectric knock sensor.

Park/Neutral (P/N) Switch

The park/neutral switch, which can be located near the shifter assembly or in the transmission, is a normally open switch. When the vehicle is shifted into park or neutral, it closes. The ECM monitors the P/N switch and knows if the transmission is in gear or not. This information is used to help control engine idle and EGR control.

Air-conditioning (A/C) Switch

On PFI cars equipped with air-conditioning, the A/C control switch on the instrument panel is connected by a wire to the ECM, Figure 5-22. When the A/C is turned on, the voltage signal travels from the A/C switch to the pres-

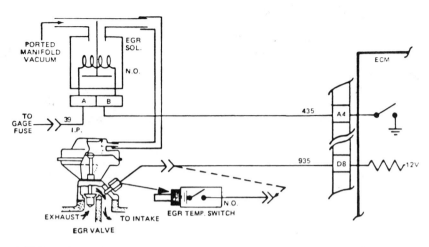

FIGURE 5-21 EGR valve with thermal diagnostic switch. *Courtesy of GM Product Service Training.*

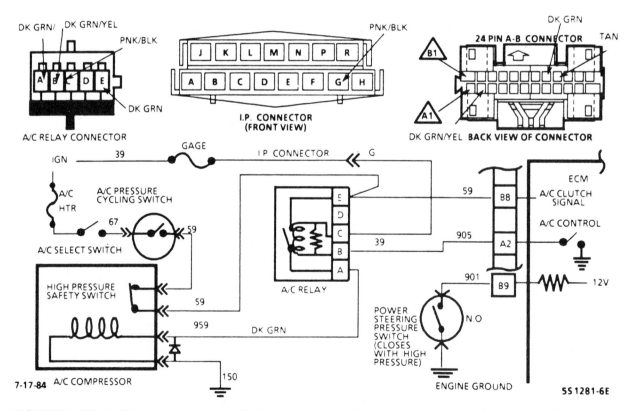

FIGURE 5-22 A/C control circuit with P/S switch (typical). *Courtesy of GM Product Service Training.*

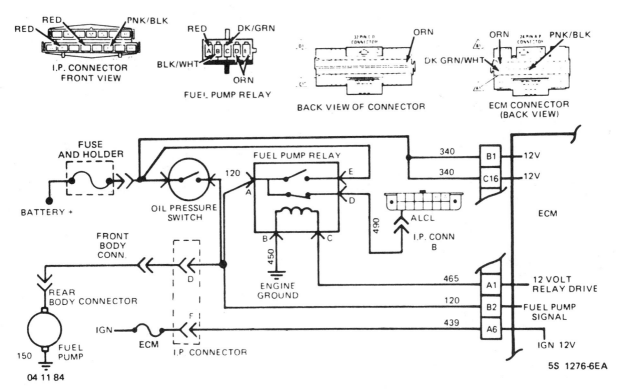

FIGURE 5-23   Fuel pump control circuit (typical). *Courtesy of GM Product Service Training.*

sure cycling switch to the ECM. Some systems also put the high-pressure cut-out switch in series in the same circuit. This signal tells the ECM that the A/C switch is on and the pressure cycling switch is closed. The ECM uses this information to turn on the A/C relay and/or adjust the IAC valve.

Transmission Switches

Transmissions used in PFI-equipped vehicles have one or more hydraulically operated electric switches screwed into the valve body. These switches provide the ECM with signals that tell it what gear the transmission is in. This information enables it to more effectively control torque converter clutch operation.

Cranking Signal

On some engines a wire from the starter solenoid circuit feeds a cranking signal to the ECM to alert it that the engine is being cranked. This puts the system in cranking mode, and an enriched air/fuel mixture is delivered by the injectors for easier starting. On other engines the cranking signal is fed to a cold-start injector in the intake manifold and provides necessary starting fuel enrichment. The cold-start injector is not ECM controlled.

System Voltage/Fuel Pump Voltage

The ECM monitors system voltage through one of its battery voltage inputs. If voltage drops below a preprogrammed value, the system goes into battery voltage correction mode. Some PFI systems have a wire from the positive side of the fuel pump power circuit to the ECM, Figure 5-23. This signal is used to make fuel delivery compensations based on system voltage and causes a code 54 to be stored if pump voltage is lost while the engine is running.

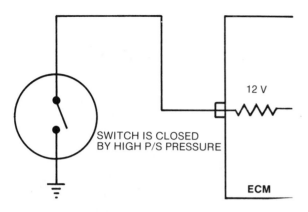

FIGURE 5-24   Power steering switch (typical)

Power Steering (P/S) Switch

A power steering switch is used in PFI systems to alert the ECM when power steering pressure is high enough to significantly affect engine idle performance, Figure 5-24. The P/S switch is normally open and is closed by high P/S pressure. When the switch is open, feedback voltage to the ECM is about 12 volts; when the switch is closed, voltage is less than 1 volt. When the switch closes, the ECM increases idle air bypass and retards ignition timing. On some systems the A/C compressor clutch is also disengaged.

Cruise Control

When cruise control is engaged, the ECM is notified by the cruise control module. The ECM uses this information to modify its control of the torque converter clutch.

OUTPUTS

Injectors

The fuel injector is a solenoid-operated nozzle controlled by the ECM, Figure 5-25. The ECM pulses it by grounding the return side (low-voltage side) of the injector circuit, Figure 5-27. Fuel under nearly constant controlled pressure is sprayed past the open needle and seat assembly into the intake port of each cyl-

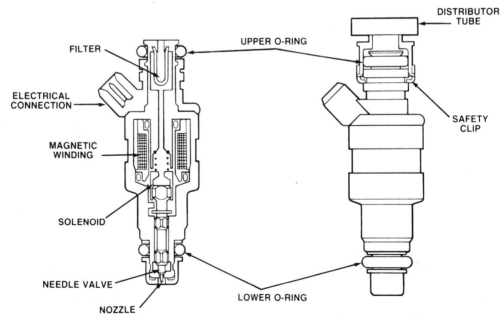

FIGURE 5-25   Injector (cutaway). *Courtesy of GM Product Service Training.*

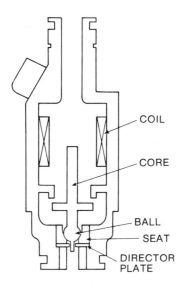

FIGURE 5-26   Multec injector

inder. On all except SFI applications, all injectors are simultaneously pulsed once, in each crankshaft revolution. The SFI system pulses each injector individually, once per engine cycle, in the firing order, Figure 5-28. During each pulse the injectors are held open for about one or two milliseconds at idle to perhaps six or seven milliseconds at WOT (1 millisecond = 1/1000 of a second).

Multec Injector. Rochester Products, a division of General Motors, has developed a new style injector for multi-point injection systems (Figure 5-26). It is referred to as a "Multec," which is short for "Multiple Technology Injector." This injector features a lower opening voltage requirement (important during cold-weather cranking), fast response time, better fuel atomization, improved spray pattern con-

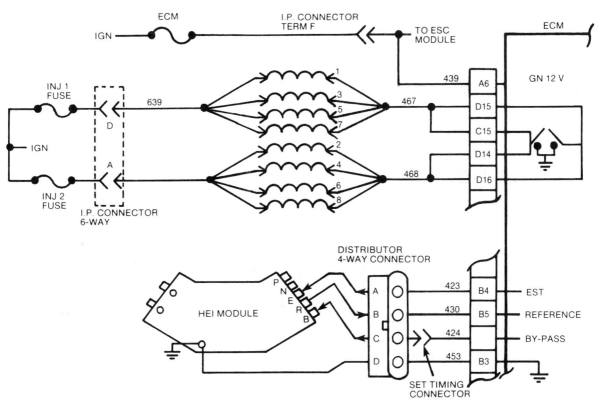

FIGURE 5-27   Injector control circuit.

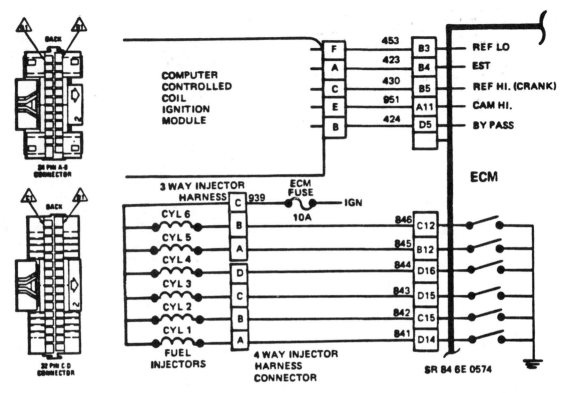

FIGURE 5-28   Injector control circuit for SFI system. *Courtesy of GM Product Service Training.*

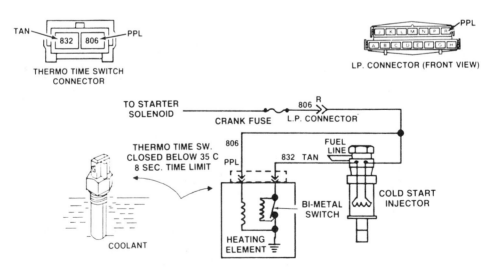

FIGURE 5-29   Cold-start injector control circuit (typical). *Courtesy of GM Product Service Training.*

trol, and less susceptibility to fouling due to fuel blends or contamination. The traditional pintle, as shown in Figure 5-25, is replaced with a ball shape on the solenoid core. When the ball is lifted off its seat, fuel sprays through the seat opening and past the director plate into the intake port.

Cold-start Injector. The cold-start injector is controlled by a thermal vacuum switch screwed into the engine water jacket, Figure 5-29. When the starter solenoid is engaged, voltage is applied to circuit 806 by way of the crank fuse. Circuit 806 powers a heat element in the thermo-time switch. Branching off circuit 806 is circuit 832, which supplies voltage to the cold-start injector. Ground for the cold-start injector is provided by a bimetal switch in the thermo-time switch. If the bimetal switch is cool and closed when voltage is applied to the injector, the injector is activated. As soon as voltage is applied to the injector, however, it is also applied to the heat element in the thermo-time switch. If the coolant is at 20° C (68° F) or below, the bimetal switch remains closed for a maximum of eight seconds before the heat from the heat element causes it to open. Therefore, the cold-start injector is turned on for a maximum of eight seconds or less, depending on coolant temperature, while and only while the engine is being cranked. If the engine is started while hot, the injector does not come on.

Electric Fuel Pump

Other than providing a higher fuel pressure, the fuel pump for the PFI system operates and is controlled just like the pump for the EFI system discussed in the **Outputs** section of Chapter 4.

Electronic Spark Timing (EST)/ Distributorless Ignition Systems

EST maintains optimum spark timing under all conditions of engine load, speed, temperature, and air density or mass. The distribu-

tor type ignition systems used on PFI engines operate the same as those discussed in the **Outputs** section of Chapter 4, except that none of them use Hall-effect switches.

Given the developments that we have seen in computerized engine controls, there is no need to continue using a distributor with its tendencies toward wear and mechanical failure because:

- reference signals can be taken directly from the crankshaft and camshaft
- an ignition module can be mounted almost anywhere
- spark advance can be more effectively managed by a microprocessor
- secondary spark distribution can be managed by separate coils and a module

The term "distributorless ignition system" has come to be used as a generic label for an ignition system that does not use a distributor and may be abbreviated in some literature as "DIS." General Motors, however, most often uses the same abbreviation as an abbreviation for "Direct Ignition System," one of their specific distributorless ignition systems. General Motors has introduced several different distributorless ignition systems on various engine applications:

- the C^{3}I (Computer-Controlled Coil Ignition) of which there are three types
- the DIS (Direct Ignition System)
- the IDI (Integrated Direct Ignition)

Each system has its own, unique module that receives information about engine speed, and crankshaft and camshaft position from its sensors. The module monitors the information and passes it on to the ECM. A coil pack, with one ignition coil for each pair of cylinders, is mounted on and controlled by the module (Figure 5-30). The module performs the same function as an HEI module in a distributor type ignition system with the same REF, bypass, EST, and ground circuits connecting the mod-

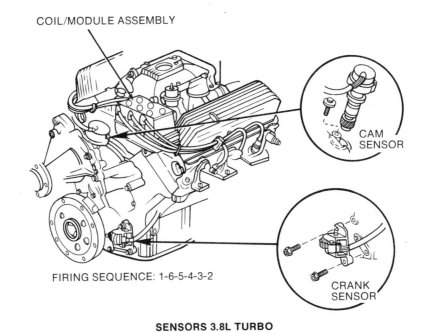

COIL/MODULE ASSEMBLY

CAM SENSOR

CRANK SENSOR

FIRING SEQUENCE: 1-6-5-4-3-2

SENSORS 3.8L TURBO

FIGURE 5–30   C³I control circuit. *Courtesy of GM Product Service Training.*

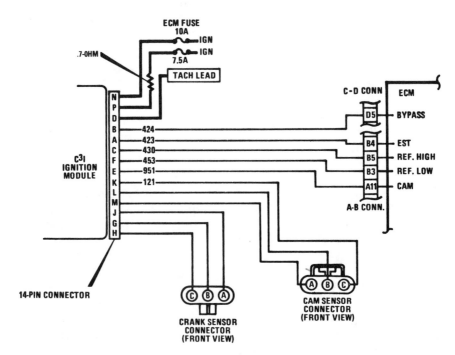

FIGURE 5–31   C³I control circuit. *Courtesy of GM Product Service Training.*

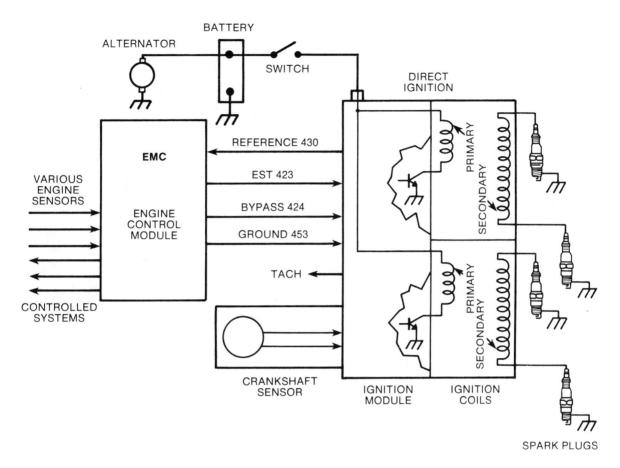

FIGURE 5-32   Four-cylinder DIS systems schematic. *Courtesy of GM Product Service Training.*

ule to the ECM (Figure 5–31). In this case the module must select and fire each coil in the correct sequence. The module and coil pack may be mounted anywhere on the engine. The end of each coil's secondary winding is connected to a spark plug by way of a spark plug wire (Figure 5–32). Each time a coil's primary circuit is opened that coil fires both of its spark plugs simultaneously.

The spark plugs that the coil fires are in companion cylinders; those cylinders arrive at TDC at the same time and are opposite each other in the firing order. For instance, if number 1 cylinder is being fired at the end of its com-

pression stroke (assume the firing order is 1-6-5-4-3-2), number 4 will be at the top of its exhaust stroke and will be fired also. It only takes about 2kV to 3kV, (k stands for kilo, which means 1000, and V stands for volts) to fire a spark plug during the exhaust stroke (about the same as having to fire the rotor gap in a distributor-type ignition system).

The C³I system was introduced in 1984 on the 3.8 liter, turbo, Buick engine. The C³I II system soon followed and C³I III was introduced in 1988 on the 3800 (a new 3.8-liter V-6 also introduced in 1988). The ignition module is connected to the ECM by a 14-pin connector,

Figure 5–31. Ignition switch voltage at terminal "N" powers the module; ignition switch voltage at terminal "P" powers the ignition coils. The .7 ohm resistance wire in the circuit to "P" is to prevent coil overheating during high ambient temperatures. If the ignition is turned on without a cranking signal appearing within 1 to 2 seconds, the module will shut off primary circuit current to prevent coil overheating. The ECM has a spark timing range capability of 0° to 70° and will provide dwell times between 3 milliseconds at high RPM and 15 milliseconds at low RPM.

On Type I and Type II C³I systems and on DIS and IDI systems, the module has to see a signal from the cam or sync sensor during cranking before it knows what position the crank is in and which coil to fire. Therefore, no spark occurs until the module sees the sync signal. Once that signal has occurred, the module will know what position the crankshaft is in and will fire a coil on the next appropriate cylinder event signal, which will be the second cylinder in the firing order. Depending on what position the crankshaft is in when cranking starts, the crankshaft may spin more than a full revolution before a spark occurs. On these systems, the first cylinder to fire will always be the second cylinder in the firing order. The C³I Type III (Fast Start) system provides faster starting due to its improved sensor input.

IDI. The newest system is called "Integrated Direct Ignition" (IDI), and is used on the new 2.3-liter Oldsmobile Quad 4 engine. The IDI system is similar to the Direct Ignition System. It uses the same crankshaft-mounted, seven-notch reluctor and crankcase-mounted magnetic sensor (Figure 5–16), but uses a different module and harness connectors. No part of the IDI system is visible from the exterior of the engine. The coil, module, spark plugs, and wires are mounted under the top most part on the engine (Figure 5–33).

DIS. During module mode operation (see Chapter 3 under "Electronic Spark Timing"),

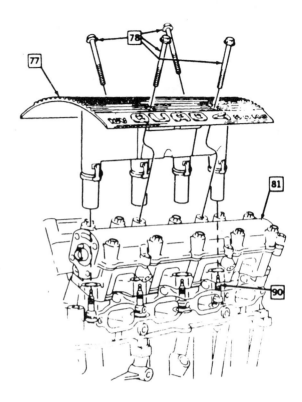

77. IGNITION COIL AND MODULE ASM.
78. BOLTS, IGNITION COIL AND MODULE ASM. TO CAMSHAFT HOUSINGS
81. COVER, CAMSHAFT HOUSING (INTAKE SHOWN)
90. SPARK PLUG

FIGURE 5–33

on four-cylinder applications, the signal produced by the sync notch tells the ignition module to skip the next signal (produced by the number 1 notch) and to fire the 2–3 coil on the signal produced by the number 2 notch (Figure 5–34). The module will then skip the signals produced by notches 3 and 4 and will fire the 1–4 coil in response to the signal produced by notch number 5. Notches number 6 and 7 pass the sensor without being responded to and the process begins again. Notice that the module is not concerned with cylinder firing order; it is concerned with coil firing order.

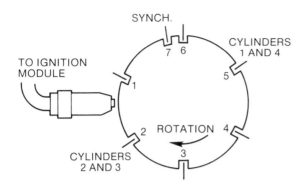

FIGURE 5-34 Four-cylinder coil firing sequence

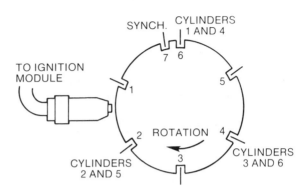

FIGURE 5-35 Six-cylinder coil firing sequence

On the six-cylinder engine, the signal produced by notch number 7 references the ignition module to skip the signal from notch 1 and respond to the signal from notch number 2 to fire the 2–5 coil (Figure 5–35). The module then skips the signal from notch number 3 and fires the 3–6 coil on the signal from notch number 4. The signal from notch number 5 is skipped and the 1–4 coil is fired on the signal from notch number 6.

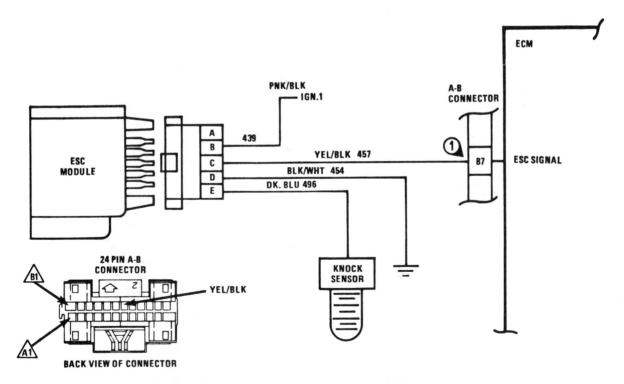

FIGURE 5-36   ESC circuit. *Courtesy of GM Product Service Training.*

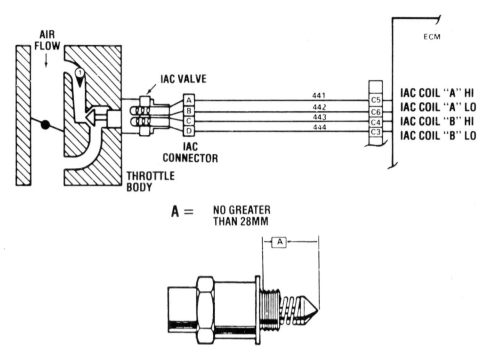

FIGURE 5-37   IAC motor control circuit. *Courtesy of GM Product Service Training.*

Electronic Spark Control (ESC)

All PFI systems except the 2.8-liter applications are equipped with ESC. Its function is to retard ignition timing when detonation occurs. It has its own hybrid ESC module that works in conjunction with the ECM, Figure 5-36. Its operation is essentially the same as the later style ESC system discussed in the **Outputs** section of Chapter 3. The conditions under which ignition timing are retarded in response to spark knock are:

- engine speed is above 1,100 RPM.
- ignition voltage to the ECM is above 9 volts.
- no code 43 is present.

Idle Air Control (IAC)

The IAC assembly is a stepper motor and pintle valve just like it is on the EFI system. It is screwed into the throttle body and controls idle speed, Figure 5-37. With one exception it works like the IAC used on the EFI system (**Outputs** section of Chapter 4). The exception is that the ECM rereferences its count of IAC's pintle valve position by moving the valve to its zero position (closed) each time the ignition is turned off instead of when the ECM gets a 30-MPH signal from the VSS, as the EFI system does. The ECM then issues a preprogrammed number of pulses to open the valve to a ready position for the next start-up. The following inputs can affect idle speed performance:

- battery voltage
- coolant temperature
- throttle position
- vehicle speed
- mass airflow
- engine speed
- A/C clutch signal
- power steering signal
- P/N switch signal

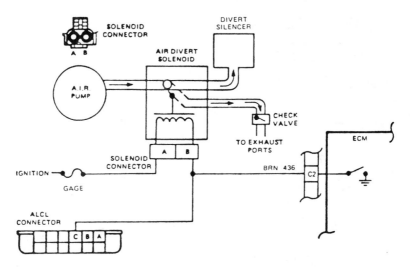

FIGURE 5-38   Divert valve control circuit (2.8 L). *Courtesy of GM Product Service Training.*

Air Injection Reaction (AIR) System

Only some of the PFI systems use an AIR system. They are much the same as those systems discussed in the **Outputs** section of Chapter 3 and will be discussed here only briefly.

2.8-liter Engines. The 1985 2.8-liter engine equipped with manual transmission uses an AIR system with a single-bed, three-way converter. During warm-up the ECM energizes the divert valve solenoid, Figure 5-38, which then allows vacuum to be applied to the divert valve. The pump's air is directed to the exhaust ports. As the engine approaches normal temperature (just before going into closed loop), the ECM turns off the solenoid, which in turn blocks vacuum from the divert valve. The divert valve then directs the air to the air cleaner (divert mode).

5.0- and 5.7-liter Engines. These engines use dual-bed, three-way catalytic converters. The air pump pumps air to the system control valve; the control valve, which contains two valves in one housing, can direct air to any one of three places, Figure 5-39:

- to the exhaust ports during engine warm-up.

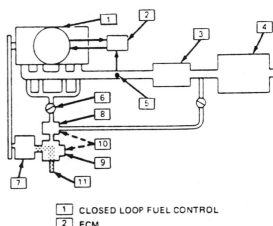

1	CLOSED LOOP FUEL CONTROL
2	ECM
3	REDUCING CATALYST
4	OXIDIZING CATALYST
5	O₂ SENSOR
6	CHECK VALVE
7	AIR PUMP
8	PORT SOLENOID
9	CONVERTER SOLENOID
10	ELECTRICAL SIGNALS FROM ECM
11	BY PASS AIR TO ATMOSPHERE

FIGURE 5-39   Air management system (5.0 L and 5.7 L). *Courtesy of GM Product Service Training.*

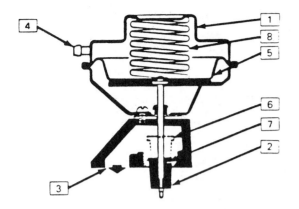

1	EGR VALVE	5	DIAPHRAGM
2	EXHAUST GAS	6	VALVE OPEN
3	INTAKE AIR	7	VALVE CLOSED
4	VACUUM PORT	8	SPRING

4S 0536-6E

FIGURE 5-40   Standard EGR valve. *Courtesy of GM Product Service Training.*

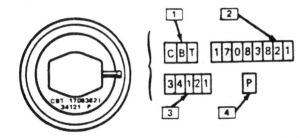

1	ASSEMBLY PLANT CODE
2	PART NUMBER
3	DATE BUILT
4	LOOK HERE FOR LETTER

P = POSITIVE BACK PRESSURE
N = NEGATIVE BACK PRESSURE
BLANK = PORTED VALVE

4S 0538-6E

FIGURE 5-41   EGR valve identification. *Courtesy of Buick Motor Division.*

- to the oxidizing chamber of the catalytic converter during closed-loop operation.
- to atmosphere (divert mode—during WOT, deceleration, when any failure oc-

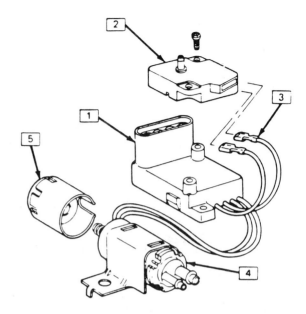

1	EGR VACUUM CONTROL ASSEMBLY BASE
2	EGR VACUUM DIAGNOSTIC CONTROL SWITCH
3	DIAGNOSTIC SWITCH CONNECTORS
4	EGR SOLENOID
5	FILTER

4S 1182-6EA

FIGURE 5-42   EGR vacuum control assembly. *Courtesy of GM Product Service Training.*

curs that causes the ECM to turn on the check engine light, or when RPM is high and causing the pressure relief valve inside the control valve to open, although pressure relief is not really a divert mode). Divert mode is intended to prevent converter overheating.

Exhaust Gas Recirculation (EGR)

The EGR valve lets small amounts of exhaust gas into the intake manifold to lower combustion temperature, thereby reducing the production or oxides of nitrogen (NO_x). The

EGR valve is opened by ported vacuum (manifold vacuum on some applications), which pulls the diaphragm up against the diaphragm spring, Figure 5-40. Three types of EGR valves are used: the standard ported valve as shown in Figure 5-40, a positive back-pressure valve, and a negative back-pressure valve. The other two valves are very similar in design to the standard valve except that they are sensitive to and modulated by exhaust back pressure or a combination of exhaust back pressure and manifold vacuum, respectively. They can be identified by:

- a P stamped on top of the positive back-pressure valve, Figure 5-41
- an N stamped on top of the negative back-pressure valve
- a blank space on top of the port valve

Regardless of the valve used, vacuum to it is controlled by a vacuum control assembly, Figure 5-42. The control assembly solenoid is pulsed many times per second by the ECM. When the solenoid is turned on, vacuum to the EGR port is blocked. The strength of the vacuum signal, which controls how far the EGR valve opens, is controlled by varying the pulse width. The solenoid is fitted with a vent filter, which must be replaced periodically. The ECM controls the solenoid using input concerning coolant temperature, throttle position, and mass airflow. The EGR valve is not opened unless the engine is warm and above idle.

Digital EGR Valve. A new concept in EGR valves is featured on the 3800, V-6 engine for 1988. The digital EGR valve provides the following advantages:

- vacuum is not required for its operation
- increased control of EGR flow
- faster response to ECM commands
- greater ECM diagnostic capability

The digital EGR valve is directly operated by solenoids rather than a vacuum diaphragm. Instead of trying to regulate EGR flow by controlling how far the valve opens or by duty cycling it, three different valves are used to open or close three different sized orifices. Each valve can be opened or closed independently of the others, allowing any of seven different increments of exhaust gas flow to be admitted into the induction system (Figure 5-43). Orifice number one, when open, will flow 14 percent of the maximum EGR. Orifice number two, when open, will flow 29 percent of the maximum EGR, and number three will flow 57 percent of maximum when it is open. By manipulating which valve or combination of valves are open at any given time, EGR flow can be easily and accurately controlled.

Increment	Orifice #1 (14%)	Orifice #2 (29%)	Orifice #3 (57%)	EGR Flow
0	closed	closed	closed	0%
1	open	closed	closed	14%
2	closed	open	closed	29%
3	open	open	closed	43%
4	closed	closed	open	57%
5	open	closed	open	71%
6	closed	open	open	86%
7	open	open	open	100%

FIGURE 5-43 Increments of EGR flow with digital EGR

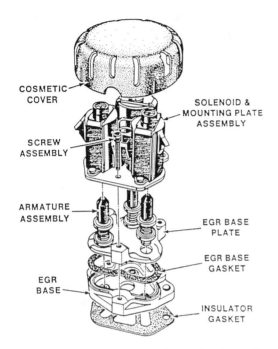

FIGURE 5-44   *Courtesy of GM Products Service Training.*

The digital EGR assembly consists of the EGR base, the EGR base plate and the solenoid and mounting plate assembly (Figure 5-44). When the base and base plate are fitted together, there is a sealed cavity between them with the base forming the floor of the cavity and the base plate forming the cavity ceiling. There are four holes in the base (floor). The center hole is always open and allows exhaust gas to enter the cavity. The other three holes are closed by the three EGR valve pintles (Figure 5-45). The pintles are attached to their pintle shafts by a ball joint-type connection so that they can more readily align with the seats they rest on and maintain a more positive seal.

As shown in Figure 5-45, the pintle and its shaft are part of the armature assembly. Each pintle-shaft-armature assembly is held down by its armature return spring, which is attached to the lower portion of the shaft. When one of the solenoid windings, in the solenoid assembly, is turned on by the ECM, the resulting magnetic field lifts the corresponding armature

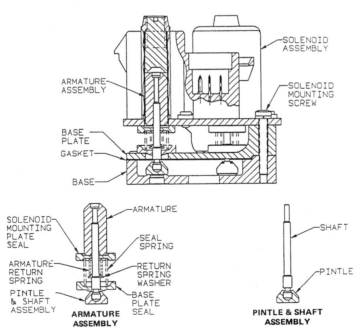

FIGURE 5-45   Digital EGR assembly. *Courtesy of GM Products Service Training.*

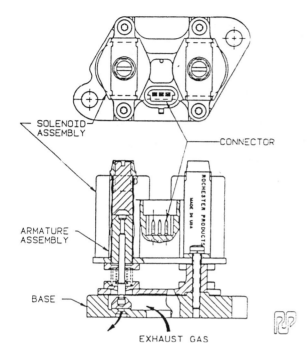

SOLENOID ASSEMBLY

CONNECTOR

ARMATURE ASSEMBLY

BASE

EXHAUST GAS

FIGURE 5-46   Digital EGR valve. *Courtesy of GM Products Service Training.*

assembly. This opens the valve and compresses the return spring. When the ECM turns the solenoid off, the spring drives the armature assembly back down and closes the valve.

There are two seals on the pintle shaft. The lower seal is pushed down against the top of the base plate. It prevents exhaust gases in the base cavity from leaking out between the stem and the base plate. On this EGR valve design, ambient air has much less tendency to leak into the induction system by way of the EGR valve stem seal than it does on the traditional EGR valve design. This is because exhaust pressure in the base cavity prevents a low pressure from developing under the seal.

The upper seal is pressed up against the bottom of the solenoid assembly. It keeps dirt and dust from entering the armature cavity of the solenoid assembly to reduce wear and provide greater reliability. A spring fitted between

the seals holds them in place. The return spring also helps hold the upper seal in place.

Some engine applications use a two-solenoid digital EGR valve, which works the same way except that it only has three different EGR flow rates (Figure 5-46). When the first valve is open, about 33% of total EGR will flow. When the second valve is open, about 66% of total EGR will flow. When both valves are open, 100% of EGR is allowed to flow.

By not relying on vacuum to operate the EGR valve assembly, the ECM can be made to monitor EGR system operation for fault diagnosis without any additional external hardware. If a short or open occurs in any one of the solenoid circuits, a code will be set in the ECM's keep alive memory. Each solenoid has its own assigned code number.

Charcoal Canister Purge

The ECM operates a solenoid that controls vacuum to the purge valve on the charcoal canister. The solenoid is turned on and blocks vacuum to the purge valve when the engine is cold or at idle, Figure 5-47. The ECM turns the solenoid off and thus allows any stored hydrocarbons to be purged through the purge valve when:

- the engine is warm.
- the engine has run for a preprogrammed amount of time since it was started.
- a preprogrammed road speed is reached.
- a preprogrammed throttle opening is achieved.

In addition to the purge valve, some engine application canisters also have a non-ECM-controlled control valve, Figure 5-48. It is controlled by ported vacuum. When vacuum is applied to it, it allows canister purging to occur through another port connected to manifold vacuum.

For 1987, Chevrolet 5.0 liter and 5.7 liter engines have a revised canister purge strategy. When the engine is operating in the closed loop

173

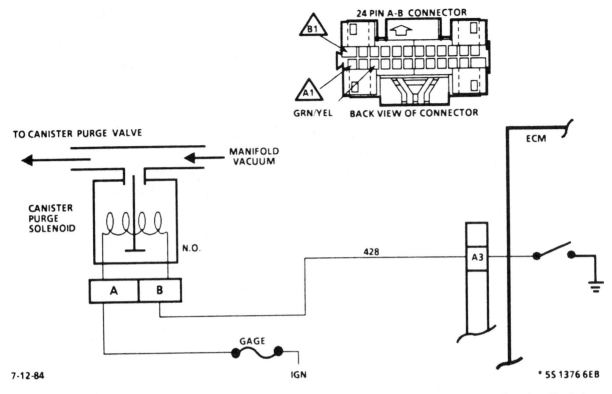

FIGURE 5-47   Canister purge control solenoid circuit. *Courtesy of GM Product Service Training.*

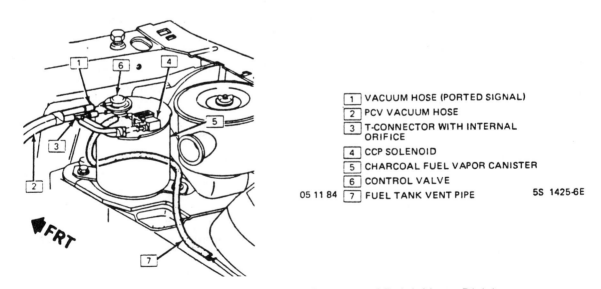

1	VACUUM HOSE (PORTED SIGNAL)
2	PCV VACUUM HOSE
3	T-CONNECTOR WITH INTERNAL ORIFICE
4	CCP SOLENOID
5	CHARCOAL FUEL VAPOR CANISTER
6	CONTROL VALVE
7	FUEL TANK VENT PIPE

05 11 84   5S 1425-6E

FIGURE 5-48   Canister purge and control valves. *Courtesy of Buick Motor Division.*

mode, the ECM will duty cycle the purge solenoid to control the amount of vapors that are
admitted into the engine's induction system
rather than just turning the purge solenoid on
to stop purging and off to allow purging. The
ECM will use input from the O_2 sensor to determine the volume of vapors to be purged from
the canister. If a rich condition is indicated by
the O_2 sensor, purge volume is reduced. This
strategy is intended to provide improved driveability.

Turbocharger

One of the major factors that controls
power is how much air and fuel are put into the
cylinders. Putting more fuel in is a fairly simple
matter; putting more air in is not so simple (the
proper air/fuel ratio must be maintained). The
amount of air that can be put into the cylinder is
limited by atmospheric pressure unless some
means is used to force air in at greater than
atmospheric pressure. This is accomplished
with some type of supercharger of which the
turbocharger is the most popular and generally
the most efficient.

The turbocharger is a centrifugal, variable-
displacement air pump driven by otherwise
wasted heat energy in the exhaust stream. It
pumps air into the intake manifold at pressures
that are limited only by the pressure that can be
developed without significantly increasing the
air's temperature.

Most normally aspirated engines (those
that rely on atmospheric pressure to fill the
cylinder) only fill the cylinder to about 85% of
atmospheric pressure (they achieve 85% volumetric efficiency, or VE, at full throttle with the
engine at its maximum torque speed). Turbo-
charging produces VE values in excess of
100%.

Turbo Operation. Upon leaving the manifold, the hot exhaust gases flow through the
vanes of the exhaust turbine wheel and thus
cause it to spin, Figure 5–49. The more exhaust
coming out of the engine, the faster the turbine

Turbochargers

Because a turbocharger puts more
air into the cylinder, it raises the engine's
compression pressure. Because it does
not produce significant boost pressure at
low exhaust flow rates, it allows an engine
the benefits of low compression during
light throttle operation and high compression during heavy throttle operation.
Low compression offers lower combustion chamber temperatures, which result
in less wear and lower NO_x emissions.
High compression provides higher combustion chamber temperatures, which result in more complete combustion and
more power for the amount of fuel consumed.

Another way to look at it is in terms of
displacement. If an engine is operating at
an atmospheric pressure of 14 PSI and the
turbocharger is producing a boost of 14
PSI, the cylinders are being charged with
a pressure of 28 PSI. The engine is con-
suming approximately double the amount
of air that it would without the turbocharger; we have effectively doubled the
engine's displacement. The engine cer-
tainly consumes more fuel but not as
much as it would if it had twice the displacement. This is because the higher
compression pressure and temperature
produces more complete combustion.

spins. It can achieve speeds in excess of
130,000 RPM. The exhaust turbine drives a
short shaft that drives a compressor turbine.
The high speed produces centrifugal force,
which moves the air from between the vanes
and thus causes it to flow out radially. The air is
then channeled into the intake manifold. As air
is thrown from the turbine vanes, a low pressure develops in its place. Atmospheric pressure pushes more air through the air cleaner,

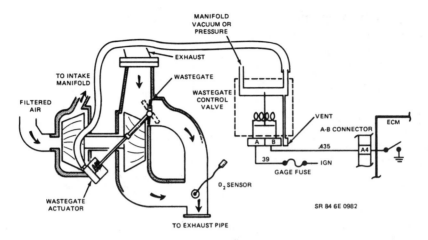

FIGURE 5-49   Turbocharger and wastegate control (typical). *Courtesy of GM Product Service Training.*

the mass airflow sensor (if used), and the throttle body.

A criterion turbine speed must be reached before a pressure boost is realized; that speed varies with turbine size and turbocharger design. If turbine speed goes to high, boost pressure (and therefore charge temperature) will go too high and will cause preignition and engine damage. To limit boost pressure, a wastegate is used to divert exhaust gases away from the exhaust turbine and thus limit its speed.

Wastegate.   The wastegate is opened by a wastegate actuator. A spring in the actuator holds the wastegate closed. When manifold pressure (turbo boost) reaches approximately 8 PSI, it overcomes the actuator spring and opens the wastegate. If, however, engine operating parameters are favorable (coolant temperature, incoming air temperature, etc.), the ECM pulses a wastegate solenoid, which in turn bleeds off some of the pressure acting on the actuator. When this occurs boost pressure is allowed to rise to 10 PSI before the wastegate opens.

A code 31 is set if:

- an overboost is sensed by the MAP sensor on the 1.8 liter.

- the ECM, which monitors wastegate solenoid circuit operation, sees a malfunction in the circuit while the solenoid is being operated between a 5% and a 95% duty cycle on the 3.8 liter.

Transmission Converter Clutch (TCC)

Notice that the name has changed slightly from this unit's application on CCC and EFI systems, where TCC meant torque converter clutch; it is still the same part. The purpose of the TCC is to increase fuel economy. It does this by eliminating hydraulic slippage during cruise conditions and by eliminating heat production in the torque converter, especially during overdrive operation. For a more detailed explanation on TCC operation, see **Torque Converter Clutch** in the **Outputs** section of Chapter 3.

Electric Cooling Fan

All General Motors' vehicles with *transverse* mounted engines and some with *longitudinal* engines are equipped with an electric cooling fan to pull air through the radiator and A/C condensor. Control of the fan varies somewhat with engine application. In all cases, however, the fan is turned on when:

- coolant temperature exceeds a specified value. This can be done by either the ECM or a coolant temperature override switch on some applications and only by the ECM on others.
- when A/C compressor output pressure (head pressure) exceeds a specified val-

ue. This is done by a switch in the high-pressure side of the A/C system on some engine applications. It is done by the ECM on other applications (on those applications an A/C head pressure switch feeds head pressure information to the ECM).

On most applications the ECM turns on the fan anytime the A/C is on and vehicle speed is

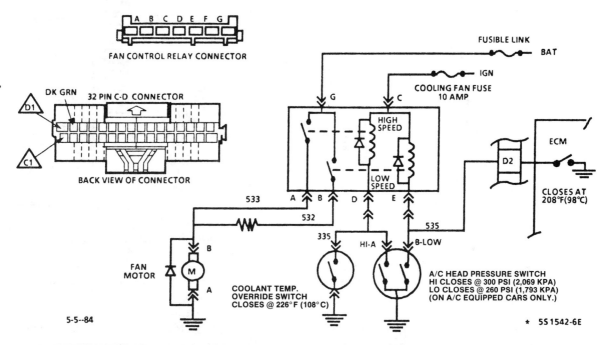

COOLANT FAN CONTROL OPERATING CONDITIONS (WITH A/C)					
A/C SW.	HEAD PRESS.	ROAD SPEED	ENG. TEMP.	FAN SPEED	FAN ON BECAUSE:
OFF/ON	UNDER 260 PSI	UNDER 45 MPH	OVER 98°C	LOW	ECM TURNED ON
OFF/ON	UNDER 260 PSI	UNDER 45 MPH	UNDER 95°C	OFF	
OFF/ON	UNDER 260 PSI	OVER 45 MPH	UNDER 106°C	OFF	
ON	OVER 260 PSI	N/A	N/A	LOW	A/C HEAD PRESS. SW. (LOW) TURNED ON
ON	OVER 300 PSI	N/A	N/A	HIGH	A/C HEAD PRESS. SW. (HIGH) TURNED ON
OFF/ON	N/A	N/A	OVER 106°C	HIGH	COOLANT TEMP. OVERRIDE SW. TURNED ON
			WITHOUT A/C		
IGN. SW.		ROAD SPEED	ENG. TEMP.	FAN SPEED	FAN ON BECAUSE:
ON		N/A	OVER 98°C	LOW	ECM TURNED ON
ON		OVER 45 MPH	UNDER 108°C	OFF	
ON		OVER 45 MPH	OVER 108°C	HIGH	COOLANT TEMP. OVERRIDE SW. TURNED ON

FIGURE 5-50   Coolant fan control circuit (typical). *Courtesy of Buick Motor Division.*

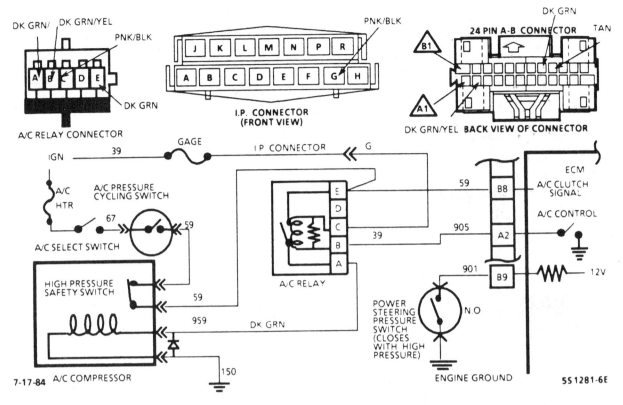

FIGURE 5–51   A/C control circuit (typical). *Courtesy of GM Product Service Training.*

less than a specified value. Others turn it on under a specified speed whether the A/C is on or off. Once the vehicle reaches a criterion speed, enough air is pushed through the radiator without the aid of the fan, unless overheating or high A/C head pressure conditions exist. This is a fuel economy feature. Study Figure 5–50 as a typical example.

Air-conditioning (A/C) Control

The ECM controls the relay that turns on and off the A/C clutch for two and in some cases three reasons, Figure 5–51.

- When the A/C control switch (on the instrument panel) is turned on, the ECM delays A/C clutch engagement for 0.4 second to allow time to adjust the IAC valve.

- The ECM disengages the A/C clutch during WOT operation.
- On some applications the ECM turns off the A/C clutch if power steering pressure exceeds a specified value during idle. On others the power steering switch is in series with either the A/C clutch or the control winding of the A/C relay and disengages the A/C clutch without relying on the ECM.

SYSTEM DIAGNOSIS AND SERVICE _____

Self-diagnosis

The self-diagnostic capabilities and procedures of the PFI systems are essentially the same as those of the CCC system discussed in the **System Diagnosis and Service** section of Chapter 3. In 1988 General Motors made a significant increase in the self-diagnostic capability of most PFI engine applications. The ECM was designed to monitor more circuits, both input and output, and new diagnostic code numbers were assigned. It is important to note that from year to year and in some cases from engine to engine, code numbers may have different meanings. Be sure that the service literature you are using is applicable to the engine being serviced and to the model year.

TCC Test Lead. The TCC test lead is in cavity F of the ALCL. It can be used to monitor TCC circuit operation with a voltmeter or test light, or it can be used to ground the TCC test lead overriding the ECM.

Fuel Pump Test Lead. On some vehicles the fuel pump test lead is found in cavity G of the ALCL. On others it is found on the left side of the engine compartment. A voltmeter or test light can be connected to this lead to determine if the fuel pump relay or oil pressure switch has supplied power to the fuel pump power lead. Or a jumper wire can be connected from the test lead to a 12-volt source. This powers the fuel pump for fuel pressure or injector tests.

CAUTION: When powering the fuel pump by the above method, be sure that there are no fuel leaks and avoid causing any sparks to occur.

Note: Diagnostic procedures for 1987 and later are written to include the use of an ALCL tool, often referred to as a "Scanner," where it is applicable. Previously, diagnostic procedures were written around the use of a voltmeter to obtain such measurements as sensor readings.

Diagnostic Procedure

Diagnostic Circuit Check. After making a thorough visual inspection, paying particular attention to possible causes of vacuum leaks, and correcting any problems found, perform the diagnostic circuit check. This procedure is outlined in the service manual and is designed to identify the type of problem causing the complaint and to refer to the next step or chart to use.

Fuel Pressure Test. Some diagnostic charts call for a fuel pressure test. Fuel pressure is critical to the performance of a fuel injection system. The fuel rail has a fuel pressure test fitting that contains a schrader valve like that used in an A/C pressure test fitting or in the valve stem of a tire.

PERSONAL SAFETY: While the fuel pressure gauge hose is screwed to the test fitting, a shop towel should be wrapped around the fitting to prevent gasoline from being sprayed on the engine. Remember, the fuel is under high pressure and the residual check valve in the fuel pump can hold pressure for some time after the engine is shut off.

Injector Balance Test

In order to realize some of the advantages of PFI, the injectors must all deliver the same amount of fuel to each cylinder. The injector balance test tests injector performance and uniform fuel delivery. Check the applicable service manual for specific directions.

Integrator and Block Learn

The integrator serves General Motors' fuel-injected engines in much the same way that dwell, a measure of duty cycle, serves CCC applications. Let's assume that while in closed loop a particular fuel-injected engine is operating with a slight vacuum leak. The oxygen sensor reading indicates a lean mixture until the ECM increases injector pulse width enough to compensate for the vacuum leak and get the air/fuel mixture back to 14.7 to 1. If an ALCL

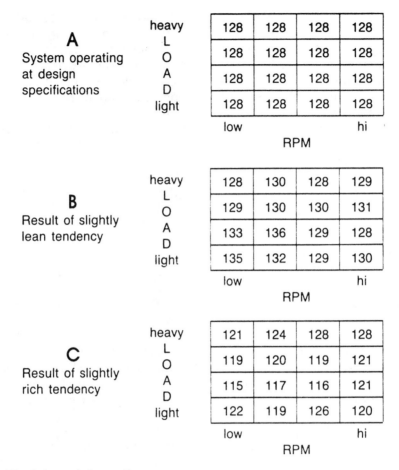

A

System operating
at design
specifications

heavy

L
O
A
D

light

128	128	128	128
128	128	128	128
128	128	128	128
128	128	128	128

low   hi

RPM

B

Result of slightly
lean tendency

heavy

L
O
A
D

light

128	130	128	129
129	130	130	131
133	136	129	128
135	132	129	130

low   hi

RPM

C

Result of slightly
rich tendency

heavy

L
O
A
D

light

121	124	128	128
119	120	119	121
115	117	116	121
122	119	126	120

low   hi

RPM

FIGURE 5-52   Block learn information

tool were connected to the ALCL with "integrator" selected, a number would be shown on the ALCL tool representing the injector pulse width currently being used. In a normal situation where no compensation is required, that number would be 128. If, as in our example, pulse width has been increased to compensate for a tendency to be lean, the number would be above 128. If the pulse width has been decreased to compensate for a tendency to be rich, the number would be below 128.

Block learn functions much like the integrator, except it represents:

- pulse width adjustments that have become a trend over a longer period of time
- learned behavior on the part of the ECM that has caused it to modify its original pulse width programming.

Once the engine has been running in closed loop long enough for block learn information to be stored in memory (as a number), that information is retained in memory until it is modified by new information or until the ignition has been turned off. Some earlier Buick and Cadillac systems would hold block learn numbers in memory until battery power was disconnected from the ECM.

Block learn numbers are stored in sixteen different cells within a square block, as shown in Figure 5-52. Each cell represents a different combination of engine speed and load. Remember that block learn only functions during closed loop operation. If the pulse width that was originally programmed for each of the sixteen combinations of engine speed and load requires no adjustment in order to maintain a 14.7 to 1 air/fuel ratio, the block learn number in each cell will be 128 as in Figure 4-52A. If, for some reason, the engine tends toward a slightly lean condition, the block learn numbers may look similar to those in Figure 5-52B. If, due to some fault, the engine has a tendency toward running slightly rich, the block learn numbers might look similar to those in Figure 5-52C.

If block learn numbers are above 128 but the integrator is at 128, the system has a lean tendency that is being corrected by block learn with wider pulse widths, and the integrator is not having to make any additional corrections. In other words, block learn has modified "normal" to a wider pulse width and the integrator is not having to correct.

If block learn numbers are 128 but the integrator is above 128, the system is trying to run lean but has not been doing so long enough for block learn to have made any correction.

If block learn is well above 128 and the integrator is the same or higher, the system is running lean and the ECM is not able to compensate enough to correct it. The same number relationships would occur if the system were trying to run rich, except that the numbers would be below 128.

ECM, PROM, and CALPAK Service

If correctly used diagnostic procedures call for an ECM to be replaced, check the ECM and PROM service and part numbers to be sure that they are the right units for the vehicle. If a

PROM is being replaced, check that it has not been superceded by a replacement PROM. This can be done with the aid of the dealer's parts department or with service bulletins. Some aftermarket service manuals contain service bulletins. Carefully remove the PROM and CALPAK from the ECM to be replaced. The service (replacement) ECM does not contain a PROM or CALPAK. Be careful not to touch the pins of either unit with your fingers because static electricity applied to the pins can damage them. Store them in a safe place while they are out of the ECM.

When the PROM is reinstalled, be sure to orient it properly to the PROM carrier, Figure 5-53. If it is installed backwards, it will be destroyed when the ignition is turned on.

Weather-pack Connectors

All PFI system harness connections under the hood are made with weather-pack connectors as discussed in the **System Diagnosis and Service** section of Chapter 3.

Service Tips

IAC Assembly. If the engine is run with the IAC harness disconnected or if the IAC assembly is removed and reinstalled, idle speed will probably be incorrect when it is reconnected.

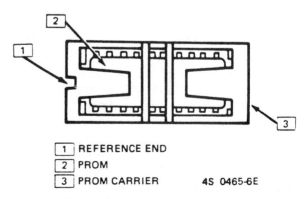

| 1 | REFERENCE END |
| 2 | PROM |
| 3 | PROM CARRIER | 4S 0465-6E

FIGURE 5-53   PROM and PROM carrier. *Courtesy of GM Product Service Training.*

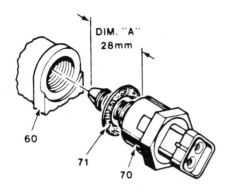

60 IDLE AIR/VACUUM SIGNAL HOUSING

70 IDLE AIR CONTROL VALVE (IAC)

71 IAC GASKET

05 24 84   5S 1455-6E

FIGURE 5-54 IAC motor. *Courtesy of GM Product Service Training.*

To rereference idle speed, simply start the engine with the IAC harness connected. Allow the engine to warm up and then turn the ignition off. The 3.0- and 3.8-liter engines rereference simply by cycling the ignition on and off.

If the IAC assembly is removed, be sure that the distance from the tip of the pintle valve to the motor housing does not exceed 28 millimeters (1-1/8 inch), Figure 5-54; otherwise, it will be damaged when it is installed. See service manual for directions. When it is installed, use correct torque specifications; overtorquing can distort the IAC housing and cause the motor to stick. Complaints of engine stalling as the car comes to a stop can be a result of this condition. Also, if the IAC assembly is replaced, be sure to get the right part. There are three different pintle shapes, and the correct one must be used.

Minimum Idle Speed Adjustment. Minimum idle speed is the speed achieved when the IAC valve is in its closed position; the only air getting into the intake manifold is what goes by the throttle blade. An idle stop screw is used to position the throttle blade. The head of the idle stop screw is recessed and covered with a metal plug. The plug can be removed and the idle stop screw adjusted; however, this should only be done if the throttle body is being replaced or if a service procedure calls for it. It is not a routine adjustment.

In order to make the minimum idle speed adjustment, the engine should be warm and the IAC valve fully extended (closed). Consult the appropriate service manual for specific directions and specifications.

TPS Adjustment. The TPS on PFI systems is adjustable. Adjustment is checked by connecting a voltmeter between terminals A and B at the sensor connector with the ignition on. Gain voltmeter access to the terminals by disconnecting the three-wire weather-pack connector and inserting three short jumper leads to temporarily reconnect the circuit. Performing a minimum idle speed adjustment will probably cause the TPS to need adjustment also.

Injector O-rings. The O-ring that seals between the bottom of the injector and the manifold is a potential source of vacuum leaks. Carefully examine the O-rings and/or replace them when the injectors are serviced.

3.8-liter Turbo (SFI) Injector Leads. Because the injectors are pulsed individually and in fir-

CYLINDER	WIRE COLOR
1	Black
2	Dk. Green
3	Pink
4	Red
5	White
6	Lt. Green

FIGURE 5-55 SFI injector wire color codes. *Courtesy of GM Product Service Training.*

ing order on the 3.8-liter turbocharged engine, the wires that power the injectors must be attached in the firing order. These wires are color-coded as shown in Figure 5-55.

Fuel Line O-rings. Threaded connections between the fuel pump and the fuel rail use O-rings to reduce the possibility of leaks. When replaced they should only be replaced with O-rings designed to tolerate exposure to gasoline.

Fuel Flex Hose. The fuel flex hose used on PFI systems is internally reinforced with steel mesh. No attempt should be made to repair it. It should be replaced if it becomes unusable or questionable.

PERSONAL SAFETY: The factory-recommended procedures should be carefully adhered to when servicing fuel lines or fuel line connections. The fuel is under high pressure, and failure to comply with factory-recommended procedures can result in a fuel leak and, consequently, fire.

REVIEW QUESTIONS _____

1. Name at least two systems typically found on engines with a carburetor or a TBI unit that are not used on a PFI system.
2. What is the major weakness of PFI?
3. During what two driving conditions are the above weaknesses most pronounced?
4. How are the fuel lines used on PFI systems different from those used on other systems, such as carburetion or TBI?
5. What causes the fuel pressure regulator to vary the fuel pressure from 34 PSI to 44 PSI?
6. Name at least two ECM responses to the power steering switch opening during idle or speeds below 10 MPH?
7. What circuit powers the cold-start injector?

8. List two unique features of the C³I system?
9. If the wire from terminal C of the ESC module to terminal B7 of the ECM became open, what would the result be?
10. How is vacuum to the EGR valve controlled?
11. What opens the wastegate on PFI applications?
12. Name at least two conditions that cause the cooling fan to come on.
13. What does the check engine light do if the ECM sees a fault in a circuit it monitors and then the fault clears up?
14. What is the primary function or functions of the diagnostic circuit check?
15. Name at least one precaution that should be taken before connecting or disconnecting the fuel pressure test gauge.
16. What removable part is found in the ECM besides the PROM?
17. What is a clamping diode or resistor?
18. What can be the result of overtorquing the IAC motor assembly when installing it in the throttle body?
19. Name at least one major concern relevant to the O-ring that seals between the injector and the manifold.
20. What is the combustion air source during the minimum air adjustment?

ASE-type Questions. (Actual ASE test questions will probably not be this product specific.)

21. Technician A says that on most General Motors' PFI systems, the injectors are all pulsed at the same time. Technician B says that they are divided into two groups, with each group being pulsed alternately. Who is correct?
    a. A only
    b. B only
    c. both A and B
    d. neither A nor B

22. A 5-liter tuned port injection engine is hard to start cold. Inspection shows that the cold-start injector is not working. This could be the fault of _____ .
    a. the ECM
    b. the thermal time switch
    c. either A or B
    d. neither A nor B

23. A car with a PFI system is hard to start, especially after sitting for an hour or more. A fuel pressure test shows that fuel pressure is normal but drops to zero quickly after the engine is turned off. The hard-starting problem is most likely the result of _____ .
    a. a defective fuel pump
    b. a plugged injector
    c. a defective check valve in the fuel pump
    d. none of the above

24. Technician A says that the PFI system re-references the IAC motor each time the VSS produces a signal that equals 30 miles per hour. Technician B says that the PFI system rereferences the IAC motor when the ignition is cycled. Who is correct?
    a. A only
    b. B only
    c. both A and B
    d. neither A nor B

25. Technician A says that if the wastegate fails to open, detonation will probably occur. Technician B says that if the wastegate fails to open, piston damage can occur. Who is correct?
    a. A only
    b. B only
    c. both A and B
    d. neither A nor B

GLOSSARY

In addition to the following terms, refer to Figure 5–56 for a list of the abbreviations used in this chapter.

AC A parts manufacturing and distributing division of General Motors.

C³I A name used by General Motors to describe their three-coil ignition system that uses a module instead of a distributor.

HIGH-PRESSURE CUT-OUT SWITCH A pressure-operated switch located in the high-pressure side of an air-conditioning system that opens when freon pressure goes too high. When the switch opens, it disables the A/C compressor clutch.

LONGITUDINAL Parallel to the length of the car.

MICRON A unit of measurement that describes the size of a small particle; a 1-micron particle is one millionth of an inch in diameter.

NORMALLY ASPIRATED Using atmospheric pressure to force air into the engine.

PRESSURE CYCLING SWITCH A pressure-operated switch in the low-pressure side of the air-conditioning system that opens as freon pressure drops below a specific pressure. When the switch opens, the A/C compressor clutch is disabled.

RHEOSTAT A variable resistor that usually has two leads and that serves as a voltage or current limiter in series with a load.

SPEED DENSITY A term that refers to a method of calculating the flow rate and density of air. The result reveals the air's ability to evaporate fuel.

THERMAC The term used by General Motors to identify the heated air inlet system. The word is a shortened form of thermostatic air cleaner.

THERMO-TIME SWITCH An electrical switch, usually applied to a bimetal strip, that uses heat to open or close a set of electrical contacts. The amount of heat is controlled so that the time it takes to open or close the contacts is always the same.

A/C	Air-conditioning
AIR	Air Injection Reaction
ALCL	Assembly Line Communication Link
ATS	Air Temperature Sensor
BARO	Barometric Pressure
C^3I	Computer-Controlled Coil Ignition
CO	Carbon Monoxide
DIAGNOSTIC "TEST" TERM.	Lead or ALCL connector that is grounded to get a trouble code
DVM (10 Meg.)	Digital Voltmeter with 10 million ohms resistance
ECM	Electronic Control Module (controller)
EECS	Evaporative Emissions Control System
EFE	Early Fuel Evaporation
EFI	Electronic Fuel Injection
EGR	Exhaust Gas Recirculation
ESC	Electronic Spark Control
EST	Electronic Spark Timing
HC	Hydrocarbons
HEI	High Energy Ignition
IAC	Idle Air Control
IP	Instrument Panel
IGN	Ignition
MAF	Mass Airflow
MAP	Manifold Absolute Pressure
MAT	Manifold Air Temperature
MFI	Multiport Fuel Injection. Individual injectors for each cylinder are mounted in the intake manifold. The injectors are fired in groups rather than individually.
MPH	Miles Per Hour
N.C.	Normally Closed
N.O.	Normally Open
NO_x	Oxides of Nitrogen
O_2	Oxygen (sensor)
PCV	Positive Crankcase Ventilation
PFI	Port Fuel Injection
P/N	Park/Neutral
PROM	Programmable Read-only Memory (calibrator)
RPM	Revolutions per Minute
SFI	Sequential Fuel Injection. Used on 3.8 L Turbo. Each injector is individually fired.
TACH	Tachometer
TBI	Throttle Body Injector (unit)
TCC	Transmission Converter Clutch
THERMAC	Thermostatic Air Cleaner
TPI	Tuned Port Injection
TPS	Throttle Position Sensor
TVS	Thermal Vacuum Switch
V	Volt
VIN	Vehicle Identification Number
VSS	Vehicle Speed Sensor
WOT	Wide-open Throttle

FIGURE 5-56   Explanation of abbreviations

TRANSVERSE   At right angle (perpendicular) to the center line of the vehicle.

VOLUMETRIC EFFICIENCY (VE)   A measure of cylinder filling efficiency. VE is expressed as the percentage of atmospheric pressure to which the cylinder is filled at the completion of compression stroke.

WASTEGATE   A valve or door that when opened allows the exhaust gas to bypass the exhaust turbine of a turbocharger. It is used to limit turbocharger boost pressure.

6 CADILLAC'S DIGITAL FUEL INJECTION

Objectives

After studying this chapter, you will be able to:

- list the functions that the ECM controls.
- describe the sensors that feed information to the ECM.
- list the functions that the BCM controls.
- describe the relationship between the ECM and the BCM.
- describe the features of the self-diagnostic system.
- put the system into diagnostics and get any stored codes displayed.
- put the system into any of its self-diagnostic features (switch test, ECM data, BCM data, cooling fans override, or ECM outputs cycling).

Digital fuel injection (DFI) was introduced in mid-1980. By 1981 it had replaced all other fuel induction systems on eight-cylinder Cadillac engines except diesels. Prior to DFI Cadillac used a system they called electronic fuel injection (a Bendix-patented, single-function, analog computer–controlled, fuel injection system). The Cadillac EFI system was used from 1975 through 1980. In 1979 and 1980, some California 6.0-liter engine applications used an early version of the computer command control system, referred to then as computer-controlled catalytic converter. For clarification any use of the EFI abbreviation in this chapter refers to the EFI system in current production. Since 1981 all four- and six-cylinder-equipped Cadillacs have used either CCC or EFI systems.

Cadillac's DFI system is very similar in operation to the EFI system; the EFI system probably came directly from DFI, minus a few features. The DFI system uses a dual throttle body injection (TBI) unit with two injectors. Do not confuse dual throttle body with the Crossfire EFI dual TBIs. The Crossfire system used two separate TBI units; the DFI's dual TBI is in a configuration similar to that of a conventional two-barrel carburetor, Figure 6–1.

Fuel pressure is supplied by an electric fuel pump located in the fuel tank. The pressure is controlled by a pressure regulator located in the TBI unit. Each injector is pulsed alternately to the other, with a pulse occurring each time a cylinder fires during normal driving.

The two most unique features of the DFI system are, first, the type of functions it controls and monitors in addition to engine parameters such as air-conditioning and heating, fuel mileage information, and cruise control.

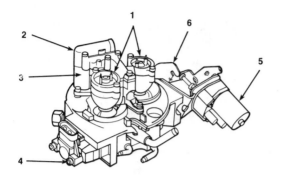

1 FUEL INJECTORS   4 THROTTLE POSITION SENSOR
2 FUEL METER COVER   5 IDLE SPEED CONTROL (ISC)
3 FUEL METER BODY   6 THROTTLE LEVER

FIGURE 6-1   TBI unit. *Courtesy of Cadillac Motor Car Division.*

Secondly, the DFI system has an elaborate self-diagnostic capability that has expanded since its introduction.

The 1981 DFI system controlled the *modulated displacement,* or 8-6-4, system in which selected cylinders were disabled (by preventing the rocker arms from opening the valves) when power requirements were low. Under acceleration all eight cylinders were used. During steady-state driving, when minimal power was required, only six cylinders were used; and during very low power demand or deceleration, only four cylinders were used. That system was used on only the 6-liter engine, which was discontinued in all but limousine and commercial chassis vehicles in 1982, when the 4.1-liter V8 was introduced.

Cadillac Allante

Cadillac's new two-seater sports car, introduced in 1987, features the 4.1-liter engine with a sequential, port fuel injection system. This system is similar to the non-turbo, sequential, PFI system on Buick V-6, except that it uses an O_2 sensor in each exhaust manifold. This allows the system to control the air/fuel ratio more closely. It can adjust the pulse width

on one side of the engine independently of the other side.

In 1985 Cadillac introduced a body computer (body computer module, BCM) on the new front-wheel drive C body (C designates a particular size of car body) to work in conjunction with the ECM. It took over all A/C and related functions, data display, and some diesel engine controls for those vehicles equipped with diesel engines. Although this text is aimed at computerized engine controls, a brief unit on the BCM is included in this chapter as an introduction to what is in store.

ELECTRONIC CONTROL MODULE (ECM)

The ECM for the DFI system has more capability than do those used with other General Motors' systems. It is located on the right side of the instrument panel and contains a removable PROM. All inputs fed into it and functions it controls are shown in Figure 6-2.

ECM Learning Ability

As driving conditions change, sensor input values change. Not all of them, however, change in proportion to each other. For instance, if a car that is normally driven at sea level starts on a trip that takes it to a significantly higher altitude, the ECM will see:

- that engine speed and road speed still match up as they normally do, provided the transmission converter clutch is still applied but the throttle position that normally accompanies that road and engine speed is now different.
- that engine vacuum is less than it normally is at the present throttle position.

If this trip is of long enough duration so that the new combination of values are seen consistently, the ECM will learn to accept them as normal

1985/86 FWD "C" DIGITAL FUEL INJECTION

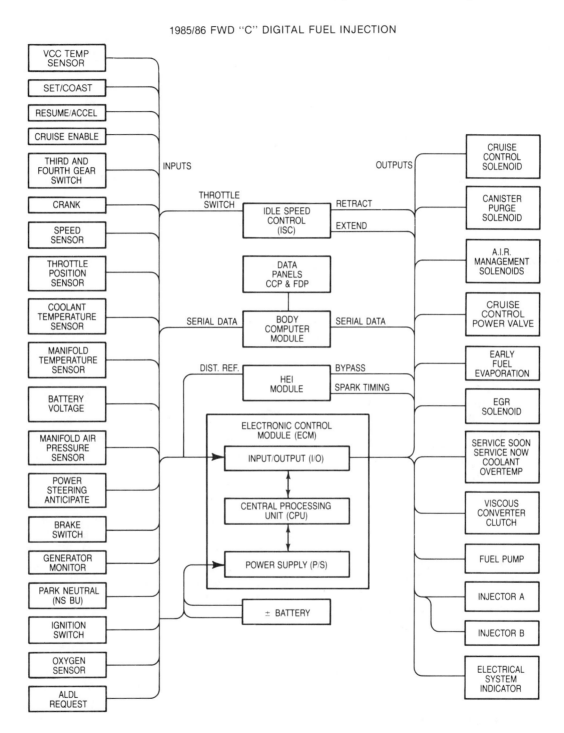

FIGURE 6–2   DFI overview. *Courtesy of Cadillac Motor Car Division.*

and will issue spark timing and other commands that provide optimum driveability. If battery power is removed from the ECM, it will forget what it has learned and performance can be noticeably affected until it relearns. To help the ECM relearn, the car should be driven at normal operating temperature, part throttle, idle, and moderate acceleration until normal performance returns.

Keep-alive Memory (KAM)

The ECM has the same KAM as discussed in Chapter 3.

OPERATING MODES _____

The ECM operates the DFI system in one of several operating modes, depending on prevailing conditions. Although some of the DFI modes are the same as EFI operating modes, others are slightly different.

Starting Mode

When the ignition is turned on, the ECM turns on the fuel pump relay and checks coolant temperature and throttle position. When the engine starts cranking, the injectors are pulsed so as to deliver an air/fuel ratio that ranges from 1.5 to 1 at –36° C (33° F) coolant temperature to 14.7 to 1 at 94° C (201° F) coolant temperature.

Clear Flood Mode

If for some reason the engine becomes flooded, pressing the throttle to the floor (or within 80% of full throttle) and cranking it will cause the injectors to be pulsed so that an air/fuel ratio of 25.5 to 1 is produced. This is leaner than the EFI clear flood mode. This ratio is maintained until the engine starts or until the throttle is allowed to close to less than 80%.

Run Mode

This mode is used during most of the driving time and presents one of two conditions:

open loop or closed loop. When the engine is first started and exceeds 400 RPM, the system goes into open loop. The ECM calculates air/fuel requirements based on coolant temperature, manifold pressure, throttle position, and engine RPM (in that order of influence). Oxygen sensor inputs are not used.

The ECM puts the system into closed loop when it sees:

- a varying voltage from the oxygen sensor that crosses 0.45 volt. This means it is hot enough to work properly.
- that coolant temperature is above 70° C (158° F).
- that sixty seconds have elapsed since the engine was started. This allows all sensors to stabilize.

The ECM now uses only oxygen sensor information to determine injector pulse width. The injector is turned on for only a few milliseconds during maximum fuel requirements.

Acceleration Mode

If the ECM sees throttle position change rapidly toward a more open position and this change is confirmed by a like change in manifold pressure, the ECM will take the system out of closed loop and put it into acceleration mode. The higher manifold pressure produced by the greater throttle opening reduces the fuel evaporation rate. More fuel is now required than is provided for a 14.7 to 1 air/fuel ratio. If the throttle position sensor indicates wide-open throttle (WOT) operation, the system will stay in this enrichment mode until throttle opening is reduced.

Deceleration Mode

When the ECM sees rapid changes in throttle position and manifold pressure values that indicate that the throttle is closing, it reduces the amount of fuel being injected. If deceleration is severe, fuel will be cut off to compensate for the fuel that will begin to evaporate from the manifold floor and walls due to the sudden drop in pressure.

Battery Voltage Correction

If battery voltage drops significantly, the ECM will make the following compensations in order to maintain the best possible operating efficiency:

- increase injector pulse width to maintain the proper air/fuel ratio
- increase idle RPM
- increase ignition dwell time to maintain sufficient ignition performance

INPUTS

All of the sensors used in the DFI system are listed here, even though many of them are the same as those used in other General Motors' systems. Some vary slightly as indicated.

- coolant temperature sensor (CTS)
- throttle position sensor (TPS)
- oxygen (O_2) sensor
- vehicle speed sensor (VSS)
- ignition switch
- crank signal

Manifold Absolute Pressure (MAP) Sensor

The earliest DFI systems used Bendix-built MAP and BARO sensors located inside the car.

Distributor Reference (REF)

The DFI system has consistently used the pickup coil in the distributor as the source of the reference pulse.

Manifold Air Temperature (MAT) Sensor

The MAT sensor is the same as that discussed in Chapter 5. It measures air/fuel mixture temperature in the intake manifold.

Throttle Switch

This is the same as the ISC switch discussed in the **Inputs** section of Chapter 3 and can be studied there. The DFI system uses an idle speed control motor like the one used on CCC systems instead of the idle air control motor that EFI and PFI systems use.

Battery Voltage

The ECM monitors battery voltage to determine when it is necessary to go into voltage correction mode.

Park/Neutral (P/N) Switch

The park/neutral switch indicates to the ECM when the transmission is in park or neutral as opposed to being in gear. This information is used to control idle speed, cruise control, and the torque converter clutch or viscous converter clutch system. The front-wheel drive C body, introduced in 1985, uses the 440-T4 transaxle, which uses a viscous converter clutch. This transaxle and converter clutch were also introduced in 1985 (see **Viscous Converter Clutch** in the **Outputs** section of this chapter).

Third- and Fourth-gear Switches

These are hydraulically operated switches that are screwed into passages of the valve body. When the transmission shifts into third gear, the third-gear switch opens; when it shifts into fourth gear, the fourth-gear switch opens, Figure 6-3. By monitoring these switches, the ECM knows if the transmission is in third, fourth, or one of the lower gears. The ECM uses this information to control the viscous converter clutch.

Viscous Converter Clutch (VCC) Temperature Sensor

A thermistor is located in the transmission to monitor transmission fluid temperature, Figure 6-3. This information influences the ECM's control of the viscous converter clutch.

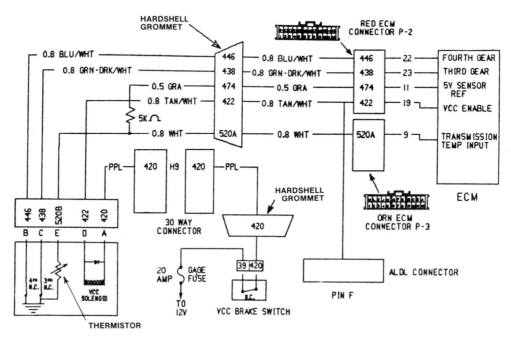

FIGURE 6-3   VCC control circuit. *Courtesy of Cadillac Motor Car Division.*

Generator Monitor

On some applications the ECM monitors generator voltage by means of a wire that taps into the fuel pump power circuit (ECM terminal 17, red connector P2 in Figure 6-4). Although this circuit does not connect directly to the generator (alternator), it does reflect system voltage, which is a function of generator output. If the ECM sees less than 10.5 volts or more than 15.5 volts while the engine is running, it will turn on the service now light and set a code 16 in its diagnostic memory. On cars equipped with a body computer, all engine function codes are preceded by the letter E.

Power Steering (P/S) Anticipate Switch

If power steering pressure reaches a criterion pressure, it will open a hydraulically operated switch. When the ignition is on, battery voltage is passed through the switch to the ECM. If the switch opens and thus causes signal voltage at the ECM terminal to drop to zero during idle, the ECM will increase throttle blade opening to support the additional engine load. If the switch opens above 40 MPH, the ECM will turn on the service engine soon light and record a code 40. This switch is not found on DFI systems prior to 1985.

Cruise Control Enable (ON/OFF Switch)

When the cruise control switch is turned on (upper left corner of Figure 6-5), it supplies a 12-volt signal to the ECM. This alerts the ECM to operate the cruise control system. On 1986 Eldorado and Seville models, the cruise control operation was taken over by the body computer.

Cruise Set/Coast Switch

When the set/coast switch is pressed, a 12-volt signal is delivered to the ECM, Figure 6-5. If the cruise control system is in a coast mode, the ECM will engage and maintain the current vehicle speed. If it is in an engaged mode when the switch is pressed, the ECM will allow the vehicle speed to coast down.

Cruise Resume/Accelerate Switch

When the resume/accelerate switch is moved toward resume, a 12-volt signal is delivered to the ECM, Figure 6-5. The ECM resumes a controlled preset speed that was disengaged as a result of brake application. If the switch is moved toward accelerate while the system is engaged at a particular speed, the ECM will open the throttle and accelerate the vehicle as long as the switch is held. When the switch is released, the ECM maintains the speed at which it was released.

Cruise Control Brake Switch

When the ignition is on, it supplies battery voltage to the brake switch assembly, Figure 6-5. When the switch is closed (brake not applied), voltage is fed through the switch to the ECM. When the brake is applied, the switch is opened. The voltage signal at the ECM disappears and the ECM knows that the brake is applied. The ECM uses this information to disengage the cruise control if it is engaged.

ALDL Request

When the service technician grounds the test terminal of the ALDL (assembly line data link, most often referred to as assembly line communication link) under the dash, the ECM responds by putting the electronic spark timing in a fixed timing mode. This mode is used to check ignition timing.

Serial Data

Information fed to the ECM from either of the display panels or from the body computer module comes via a data link and is correctly referred to as serial data. This information can be concerning driver-selected temperature or air-conditioning mode or fuel mileage, or it can be requested information from the BCM.

OUTPUTS

Injectors

Each of the two injectors contains an electric solenoid, Figure 6-6. When the solenoid is

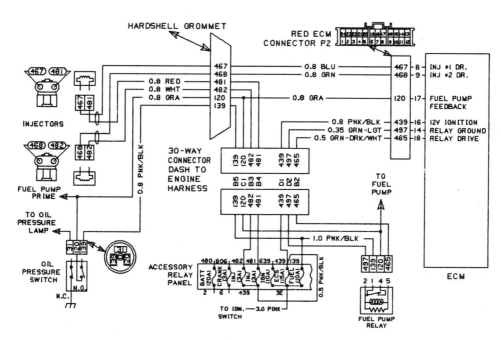

FIGURE 6-4   Fuel pump control circuit. *Courtesy of Cadillac Motor Car Division.*

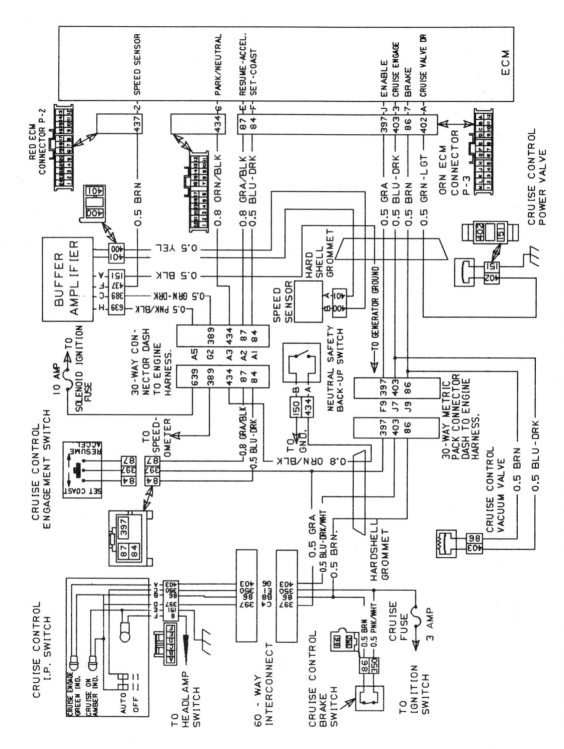

FIGURE 6-5   Cruise control circuit. *Courtesy of Cadillac Motor Car Division.*

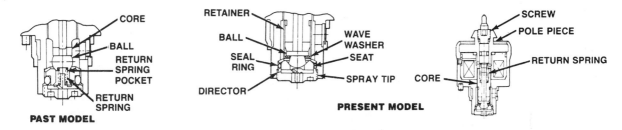

FIGURE 6-6   Injector comparison. *Courtesy of GM Product Service Training.*

pulsed by the ECM, the injector valve is lifted off of its seat a few thousandths of an inch. Fuel is sprayed out under a controlled pressure of 9 to 12 PSI. The amount of fuel sprayed is a function of how long the valve is held open (pulse width).

Each injector has its own power supply circuit and fuse. Voltage is supplied by the ignition switch. The ECM grounds each injector circuit individually to pulse the injectors.

Fuel Pressure Regulator. The pressure regulator provides an operating pressure ranging from 9 to 12 PSI to the injectors. Its operation is the same as that of EFI, which is discussed in the **Outputs** section of Chapter 4.

Fuel Pump Operation

Fuel and fuel pressure are supplied by a twin-turbine electric pump inside the fuel tank. The pump is turned on and off by a fuel pump relay mounted under the instrument panel on the relay panel. Its operation and control is like that of the EFI system pump discussed in the **Outputs** section of Chapter 4. The 1980 DFI system, however, did not have the oil pressure switch as a backup feature.

Electronic Spark Timing (EST)

The EST operation is the same as that of the EFI system discussed in the **Outputs** section of Chapter 4 with the exception that DFI does not use a Hall-effect switch.

Idle Speed Control (ISC)

The DFI system uses the same ISC motor to control idle speed that the CCC system

does. Although the motor's operation and adjustment provisions are the same as those discussed in the **Outputs** section of Chapter 3, the ECM's strategy for its use varies a little from CCC. The ECM uses information from the following sensors and switches to control throttle blade position in any of three operational modes (some of these inputs are used to identify the operational mode):

- VSS
- TPS
- throttle switch
- REF
- A/C clutch
- CTS
- MAP
- P/N switch
- ignition switch

During Cranking and Initial Warm-up. During cranking the throttle is held in a preprogrammed position. When the engine has started, the throttle is held in a preprogrammed fast idle position to allow intake manifold conditions to stabilize. If the engine is started hot, the fast idle period will be short. If the engine is started cold, the fast idle period will be longer.

Curb Idle. When the engine is idling at normal operating temperature, curb idle is maintained at a preprogrammed speed. As engine loads change as a result of the A/C clutch cycling on and off, the transmission being put into gear or in neutral, or momentary high P/S pressure occurring, the ECM adjusts throttle blade opening to maintain the desired speed.

Deceleration. During deceleration manifold pressure goes very low, (high vacuum) and thus causes liquid fuel on the floor and walls of the manifold to evaporate. This results in high HC and CO production unless air is let into the manifold to mix with the additional fuel going into the combustion chambers. The ECM looks at vehicle speed and manifold pressure to determine severity of deceleration and how much throttle opening to provide. During hard deceleration it also stops fuel injection.

AIR Management

The air injection reaction (AIR) system used on DFI is essentially the same as the AIR system described in the **Outputs** section of Chapter 3 for a vehicle with a dual-bed catalytic converter. However, the DFI ECM's strategy for its use varies slightly. The ECM de-energizes the divert solenoid and thus puts the system into divert mode under any of the following conditions:

- acceleration or WOT operation. (Some DFI systems go into divert mode when the vehicle exceeds 60 MPH.)
- deceleration.
- extremely cold weather starting.
- certain failures that bring on either of the service engine lights.
- high RPM, which causes pump pressure to exceed the calibration of a pressure relief valve built into the control valve. This is not a divert valve function although it achieves the same results.

Note that any control valve solenoid or related electrical circuit failure results in a fail-safe condition (the failure leaves the valve in the safest operating mode). A mechanical failure within the valve may not work out so well.

Torque Converter Clutch (TCC) or Viscous Converter Clutch (VCC)

The TCC is a clutch within the torque converter that when applied eliminates the characteristic hydraulic slippage of a torque converter. When it is released, the converter acts like a conventional torque converter. Its apply and release are controlled primarily by the ECM. Its function is to improve fuel economy. For an in-depth description of the TCC and its operation, refer to **Torque Converter Clutch** in the **Outputs** section of Chapter 3. The parameters by which it is controlled are slightly different in the DFI system, and those are presented here.

The VCC, introduced in 1985 on the Cadillac front-wheel drive C body, works the same way that the TCC does except that when applied it does not provide 100% lock-up between the engine and the transmission input shaft as the TCC does. The VCC clutch assembly is made up of a rotor sandwiched between two parts called a cover and a body, Figure 6-7. The rotor and body surfaces that face each other have a series of fins that mesh together and allow a small space between each fin, Figure 6-8. This space is filled with silicone fluid, which drives the rotor. When the VCC is applied, it allows just enough constant slippage (about 40 RPM at 60 MPH) to eliminate some of the roughness that occurs with a 100% lock-up between the engine and transmission. It is controlled in the same way that a TCC is controlled except that one additional parameter is included in its control—transmission fluid temperature.

TCC/VCC Control Parameters. The ECM grounds the TCC/VCC apply solenoid circuit when:

- the criterion speed is reached as reported by the VSS. On some models this can be as low as 24 to 36 MPH depending on throttle position and transmission oil temperature.
- the engine is warm. If the coolant temperature is 18° C (64° F) or above when the engine is started, the ECM will apply the clutch when the coolant reaches 60° C (140° F) on VCC applications. If the coolant is colder than 18° C at start-up, the ECM will use a fixed-time delay be-

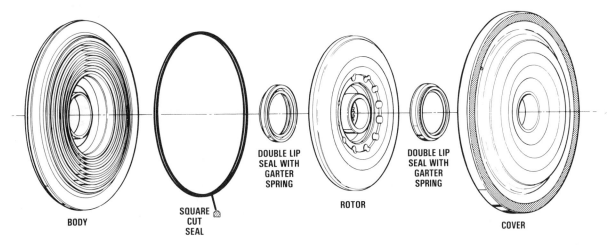

FIGURE 6–7   Viscous converter clutch (exploded view). *Courtesy of GM Product Service Training.*

fore allowing the clutch to be applied. This is to improve low-temperature driveability.

- TPS is within a preprogrammed window. (The window varies depending on what

APPLY OIL
VISCOUS FLUID
SEALS
CLUTCH MATERIAL

FIGURE 6–8   Viscous converter clutch. *Courtesy of GM Product Service Training.*

gear the transmission is in.) TPS is compared to VSS, and the ECM selects different apply speeds for different combinations of TPS and VSS.

Two other switches can cause the clutch to be disengaged on VCC applications. They are the brake switch and the over-temperature protection switch, Figure 6–9. TCC applications do not use the over-temperature switch. These switches are not controlled by the ECM but are in the apply solenoid circuit and must be closed before the circuit can be completed to the ECM. They are both normally closed. The brake switch is opened when the brake is depressed. The over-temperature protection switch opens when transmission oil temperature exceeds 157° C (315° F).

The ECM on the 1987 Allante also controls the transmission 2–3 and 3–4 shifts. This is achieved by way of two ECM-controlled solenoids in the transmission's valve body. The solenoids block or open passages that allow oil to move the 2–3 and 3–4 shift valves.

Exhaust Gas Recirculation (EGR)

When open the EGR valve allows a small amount of exhaust gas (mostly carbon dioxide and nitrogen, neither of which burn) into the

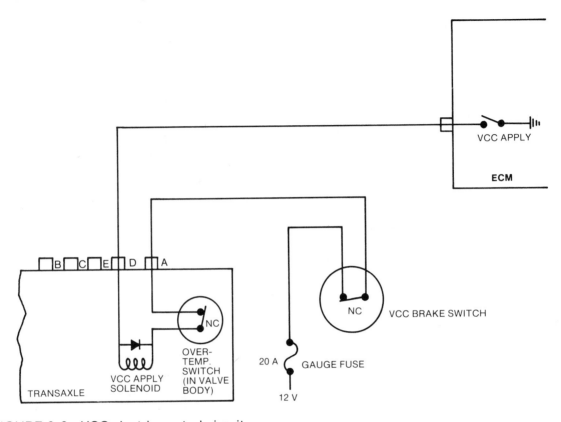

FIGURE 6-9   VCC electric control circuit

intake manifold. The nonburnable exhaust gas displaces its equivalent amount of burnable material and slows down flame *propagation.* This reduces combustion temperature and thereby reduces NO_x production.

The DFI system uses a negative back-pressure EGR valve, Figure 6-10. It has some self-modulation capability. Exhaust pressure from the exhaust crossover (early fuel evaporation) passage in the intake manifold is sensed through a small hole in the lower portion of the valve itself. The hollow valve stem allows the exhaust pressure to get into a chamber beneath a second diaphragm below the main diaphragm. This pressure helps a small spring push the lower (second) diaphragm up so that it blocks a bleed hole in the main diaphragm. When the bleed hole is blocked, ambient air

vented into the space between the two diaphrams is no longer allowed to bleed into the vacuum chamber above the main diaphragm. This allows sufficient vacuum to be developed in the vacuum chamber (using manifold vacuum as the source) to pull the diaphragm up to open the EGR valve. When the valve is open, the same hole that admitted exhaust pressure also admits manifold vacuum. If the net result of the engine vacuum versus exhaust pressure is a positive or near positive pressure, the lower diaphragm will be pushed up and the valve will be opened. If, however, the net result is a negative pressure, the lower diaphragm will be held down and the bleed valve will be open. This will cause the EGR valve to stay closed.

EGR Modulation. In addition to the EGR control provided by the negative back-pressure

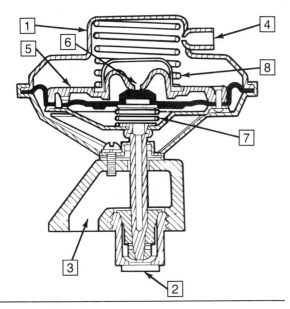

1	EGR VALVE	6	VACUUM BLEED HOLE
2	EXHAUST GAS		
3	INTAKE AIR	7	SMALL SPRING
4	VACUUM PORT	8	LARGE SPRING
5	DIAPHRAGM		

FIGURE 6-10 Negative back-pressure EGR valve. *Courtesy of Cadillac Motor Car Division.*

EGR valve, the DFI system on the 4.1-liter engine also controls vacuum to the EGR vacuum port. This control is achieved by means of a duty-cycled solenoid that while turned on bleeds atmospheric pressure into the vacuum supply hose to the EGR valve and thus causes the valve to close, Figure 6-11. When the solenoid is not turned on, it allows manifold vacuum to get to the EGR valve and thus causes it to open. By duty-cycling the solenoid, the ECM can control the amount of EGR opening and thus the EGR flow rate. The duty cycle varies for different driving conditions to provide the best possible driveability characteristics while still maintaining control of NO_x. A 10% duty cycle provides full vacuum to the EGR and a

90% duty cycle provides no vacuum to it. The ECM uses information from the following sensors to define prevailing driving conditions and the amount of EGR needed:

- CTS (no EGR when coolant temperature is low)
- TPS
- MAP
- ISC throttle switch (no EGR at idle)
- REF

Charcoal Canister Purge

Like most systems the DFI system uses an evaporative emission control system that consists primarily of a charcoal canister. The fuel tank is vented through the canister, and any evaporated fuel that tries to escape from the tank is stored in the activated charcoal within the canister. When the engine is running in

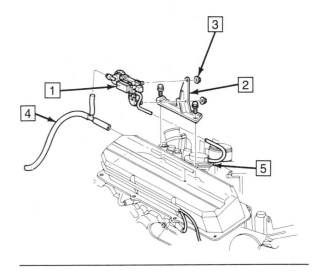

1	EGR SOLENOID
2	BRACKET
3	NUT
4	HOSE ASM.
5	EGR VALVE

FIGURE 6-11   EGR solenoid. *Courtesy of Cadillac Motor Car Division.*

closed loop, manifold vacuum draws air through the canister and into the manifold, Figure 6-12. Any fuel vapors stored in the canister are purged by this airflow. Because the ECM has no way of knowing how much, if any, fuel is stored in the canister, it waits until the system is in closed loop so that it can compensate for any fuel introduced by the canister before it allows purging to occur.

Purge is controlled by two valves: the control valve and a solenoid-operated valve in the vacuum line that goes to the control valve, Figure 6-13. The control valve contains a diaphragm and spring. The spring pushes the diaphragm down and thus closes the passage that

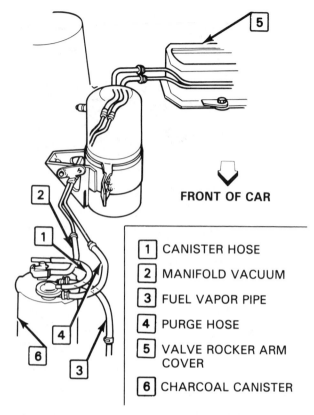

FRONT OF CAR

1	CANISTER HOSE
2	MANIFOLD VACUUM
3	FUEL VAPOR PIPE
4	PURGE HOSE
5	VALVE ROCKER ARM COVER
6	CHARCOAL CANISTER

FIGURE 6-13   Charcoal canister hose connections. *Courtesy of Cadillac Motor Car Division.*

otherwise allows air and fuel vapors to flow to the purge port. When manifold vacuum is applied to the top of the diaphragm, it is lifted and it opens the passage to the purge port. By turning on or off the solenoid, the ECM controls vacuum to the control valve. When the solenoid is energized, vacuum to the control valve is blocked. The ECM turns the solenoid off and allows purge when:

- the system is in closed loop and off idle.
- faults exist that trigger codes 13, 16, 44, or 45.

An electrical failure in the solenoid or its circuit causes purging to occur anytime the engine is running. This can cause:

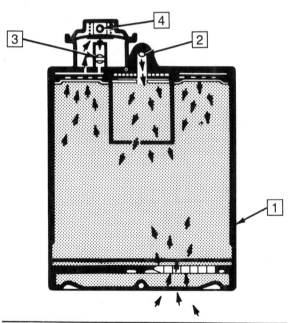

1	CHARCOAL CANISTER
2	TO FUEL TANK
3	PURGE VALVE (TO MANIFOLD)
4	CONTROL VALVE (TO MANIFOLD VACUUM)

FIGURE 6-12   Charcoal canister. *Courtesy of Cadillac Motor Car Division.*

- an overrich condition during warm-up or idle.
- an overlean condition during warm-up or idle (purging an empty canister).
- dieseling on shutdown.

Early Fuel Evaporation (EFE)

The EFE system applies heat to the area just beneath the throttle valves to prevent throttle blade icing and to enhance fuel evaporation during engine warm-up. During the warm-up period, this helps to:

- reduce the richness requirement of the air/fuel ratio.
- improve driveability.
- reduce emissions.

Rear-wheel drive DFI vehicles have used exhaust heat for this purpose without ECM control. A valve (EFE valve similar to a throttle or choke valve) is placed in the exhaust pipe on one side of the engine where it attaches to the exhaust manifold. A vacuum actuator containing a spring and a diaphragm opens and closes the valve. A thermal vacuum switch controls vacuum to the actuator. When coolant temperature is below 49° (120° F), the thermal vacuum switch routes vacuum to the EFE actuator. Vacuum applied to the actuator pulls the diaphragm against the spring and closes the EFE valve. This forces exhaust gases from that side of the engine to flow through a passage in the head to the intake manifold EFE passage. The exhaust crosses beneath the *plenum* to the other head, where it finds its way through another passage to the exhaust pipe on that side. When the coolant temperature goes above 49° C, a temperature-actuated vacuum control valve blocks vacuum to the actuator. The spring pushes the diaphragm back down and opens the EFE valve. Exhaust gas is no longer forced through the EFE passage.

Front-wheel drive DFI systems use a ceramic-covered electric heat element under the TBI unit for its EFE heat source, Figure 6–14. The heat element is powered by a relay (EFE relay) that is ECM controlled. The ECM activates the relay when all of the following conditions exist:

- MAT is less than 75° C (167° F).
- CTS indicates less than 106° C (223° F).
- battery voltage is more than 10 volts.

When any of the above conditions are no longer met, the EFE relay is turned off.

If EFE is off while the vehicle is being driven, the ECM will turn it back on if all of the following conditions occur:

- MAT is less than 38° C (100° F).
- CTS indicates less than 106° C.
- battery voltage is greater than 12 volts.

When any of the above conditions are no longer met, the EFE relay is turned off.

One other set of conditions causes the ECM to bring on EFE heat. If throttle position is open more than 30° (0° being closed and 90° being fully open) and MAT is less than 60° C (140° F), the EFE will be turned on for at least fifteen seconds to prevent throttle blade icing.

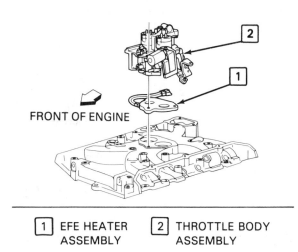

FRONT OF ENGINE

| 1 | EFE HEATER ASSEMBLY | 2 | THROTTLE BODY ASSEMBLY |

FIGURE 6–14   EFE grid. *Courtesy of Cadillac Motor Car Division.*

Cruise Control Vacuum Solenoid and Cruise Control Power Valve

When the ignition is turned on, power is fed to the cruise control on/off switch and to the brake switch, Figure 6–15. The cruise control on/off feeds power to:

- the resume/acceleration and set/coast switches.
- the cruise on indicator light.

The brake switch feeds power to:

- the ECM (as a signal that the brake is not applied).
- the cruise control vacuum solenoid.
- the cruise engaged indicator light.

When the set/coast switch is depressed (at speeds above 25 MPH), the selected speed is recorded in the ECM's memory, and the ECM grounds the cruise vacuum solenoid. The solenoid routes vacuum to the cruise control servo. By means of a vacuum diaphragm and spring, the servo controls throttle blade position and therefore vehicle speed. As vacuum is applied, the throttle is opened; as vacuum is removed, the spring closes the throttle. The amount of vacuum applied to the servo is controlled by the cruise control power valve, which is controlled by the ECM. The ECM supplies a pulsed voltage signal to the power valve; this is one of the few instances in which the ECM supplies the power instead of the ground. As the power valve is turned on and off, the servo is alternately exposed to vacuum or atmospheric pressure. The ECM pulses the power valve at whatever rate is necessary to make the indicated speed from the VSS match the speed it has recorded in its memory. As the ECM begins controlling vehicle speed, it also turns on the cruise control engaged light.

When the cruise control system is engaged (holding a speed), it can be disengaged by any one of three ways:

- The cruise control on/off switch can be turned off, in which case the selected speed is erased from the ECM's memory.
- The brake pedal can be depressed and thus causes the ECM to disengage the cruise control vacuum solenoid and thus vent the servo to atmospheric pressure. Moving the resume/accelerate switch to the resume position and releasing it after the brake has been released causes the system to reengage.
- Many faults that cause the ECM to turn on one of the service engine lights also cause it to disengage the cruise control system.

If while the system is engaged the resume/accelerate switch is moved and held, the system will accelerate the vehicle until the switch is released. The system will then engage on whatever speed the vehicle was traveling when the switch was released. If the set/coast switch is depressed and held while the system is engaged, it will disengage and the throttle will close. The system will reengage on whatever speed exists when the switch is released.

Air-conditioning (A/C) Cut-out

When the ECM sees the TPS give a WOT indication, it signals the electronic climate control (ECC) module or the body computer module (BCM), depending on year and model, to momentarily turn off the A/C compressor. This allows more power for acceleration and reduces the heat load on the engine. DFI systems do not use an A/C relay. The A/C compressor clutch is energized and controlled by the ECC module or the BCM. Refer to **Compressor Clutch Control** and **Programmer and Power Module** in the **Body Computer Module** section of this chapter for more information on compressor clutch control.

Service and Coolant Over-temperature Lights

If either of the two modules, the ECM or the BCM, sees a fault in any of the circuits they monitor, they will turn on one of the diagnostic indicator lights, Figure 6–16. The light selected depends on the nature of the problem.

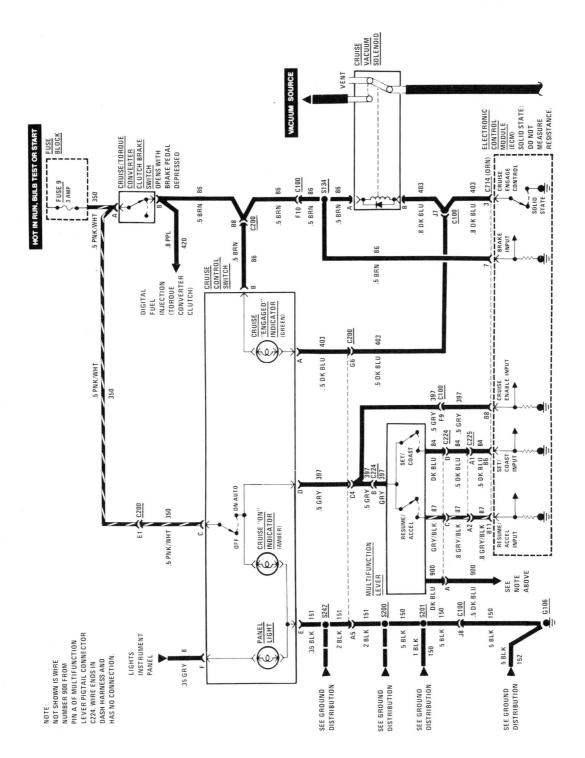

FIGURE 6-15   Cruise control circuit. *Courtesy of Cadillac Motor Car Division.* **(Continued on page 204)**

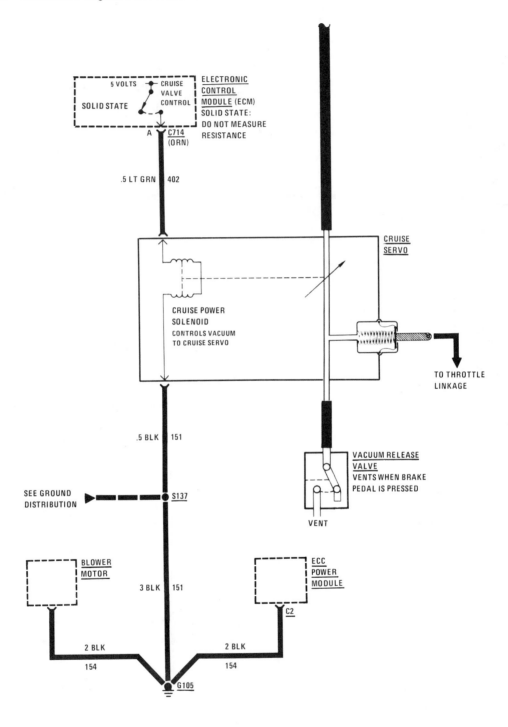

FIGURE 6-15   (Continued)

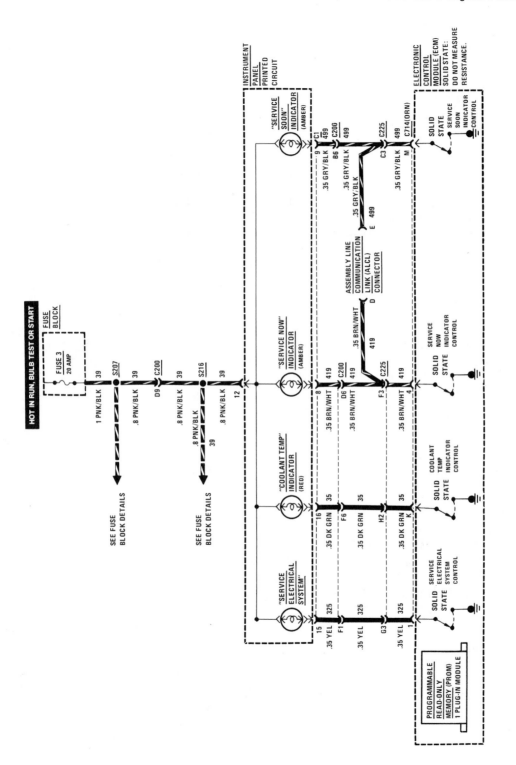

FIGURE 6–16   Indicator lights circuit. *Courtesy of Cadillac Motor Car Division.*

BODY COMPUTER MODULE (BCM)

The BCM is a microcomputer just as is the ECM. It has the basic components (ROM, RAM, PROM, input/output interface, etc.) and about the same capability as the ECM; however, it is in charge of different functions within the vehicle. On the 1985 C body Cadillac, where it was introduced, it performs the following functions:

- controls the electronic climate control (air-conditioning system)
- controls the electric cooling fans, which draw air through the A/C condensor and radiator
- controls power to the power windows, trunk release, and *astroroof* (an electrically opened and closed window in the roof)
- provides and displays information for the driver
- controls information display panel dimming for visual clarity and comfort
- monitors for BCM system faults, stores codes to identify the faults, and in some cases provides a fail-safe operating condition to compensate for a failure

In addition to having sensors and switches for input information and actuators to send commands to, the BCM and the ECM interact with each other. In some cases information contained within the ECM is sent by way of a data link to the BCM as input information, and an output can be sent to the ECM to serve as ECM input. Figure 6–17 shows the kind of information that the two computers can exchange.

Electronic Climate Control (ECC)

In order to maintain the driver-selected temperature inside the vehicle, the BCM must select the most appropriate air inlet and outlet modes and blower speed. It also monitors the condition of ECC components.

BCM/ECM DATA TRANSFER

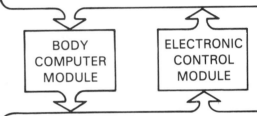

CIRCUIT 459-ECM TO BCM DATA

- REQUESTED DIAGNOSTIC DATA
- FUEL ECONOMY DATA
- VEHICLE SPEED
- COOLANT TEMPERATURE
- ENGINE RUN STATUS
- WIDE OPEN THROTTLE STATUS

BODY COMPUTER MODULE

ELECTRONIC CONTROL MODULE

CIRCUIT 491-BCM TO ECM DATA

- DIAGNOSTIC ACTION REQUEST
- OUTSIDE AIR TEMPERATURE
- A/C HIGH SIDE TEMPERATURE
- A/C CLUTCH STATUS
- REAR DEFOG STATUS
- HIGH BLOWER STATUS
- HIGH COOLING FANS STATUS

FIGURE 6–17   BCM/ECM interaction. *Courtesy of Cadillac Motor Car Division.*

ECC Inputs

Input information that the BCM needs to control the ECC system comes from several sources, Figure 6–18.

Climate Control Panel (CCP). The driver selects desired temperature and operating mode (automatic, hi fan, lo fan, economy, front window defog, rear window defog, etc.) on the CCP. The CCP communicates this information in digital form via a data link to the BCM.

ECM. The ECM informs the BCM when the throttle is fully opened or when coolant tem-

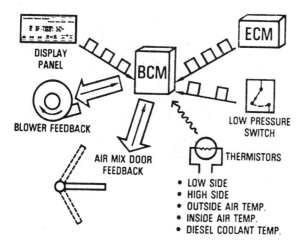

DISPLAY PANEL

ECM

BCM

BLOWER FEEDBACK

LOW PRESSURE SWITCH

AIR MIX DOOR FEEDBACK

THERMISTORS

- LOW SIDE
- HIGH SIDE
- OUTSIDE AIR TEMP.
- INSIDE AIR TEMP.
- DIESEL COOLANT TEMP.

FIGURE 6-18   BCM information sources. *Courtesy of GM Product Service Training.*

perature goes beyond a criterion value. In either case the BCM commands the compressor clutch to be disengaged.

Temperature Sensors. Strategically located thermistors provide information about outside air temperature, inside air temperature, and refrigerant temperature on both the low- and high-pressure sides of the A/C system. A switch located in the low-pressure side of the A/C system notifies the BCM if A/C pressure goes too low.

Air Mix Door Position. A potentiometer attached to the air mix door in the air distribution system housing provides feedback information to the BCM about the door's position. The air mix door blends heated and refrigerated air to produce the desired air temperature, Figure 6-19. By monitoring voltage at the blower motor, Figure 6-20, the BCM can determine blower speed.

ECC Outputs

Program Number. When the car is first started, the BCM compares inside air temperature to driver-selected temperature. It calculates what operational mode the ECC should

be in to achieve the selected air temperature. In its calculation it considers the influence of outside air temperature. This calculation produces a program number between 0 and 100. A low number means inside air temperature is higher than selected temperature and calls for cooling. A high number means just the opposite. The program number dictates blower speed and air distribution system door positions, Figure 6-19. As inside air temperature comes closer to selected temperature, the BCM updates the program number, the blower speed decreases, and air distribution system doors are adjusted. When the temperatures are the same, the program number reaches an equilibrium point. The BCM tries to maintain the program number at equilibrium. The program number is displayed on the CCP only when the system is in diagnostics.

Air Mix Door. The air mix door is moved by an electric motor (inside the programmer) and connecting linkage. All of the other doors are moved by vacuum motors controlled by vacuum solenoids. To determine what the air mix door position should be, the BCM checks:

- coolant temperature (requested from the ECM) to determine what heated air temperature will be.
- refrigerant temperature on the low-pressure side to determine what cold air temperature will be. If the A/C is not operating, outside air temperature is used instead.
- the program number.

The BCM uses the air mix door feedback to see if the door is in the right position. Because the linkage between the air mix door motor and the door is adjustable, it must be carefully adjusted to be sure that its position coincides with the BCM-commanded position.

Compressor Clutch Control. The BCM controls the compressor clutch. Before applying the clutch, it checks to see that:

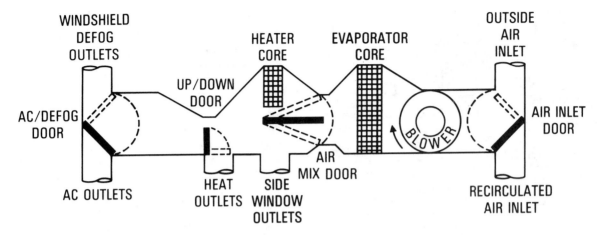

FIGURE 6-19   ECC mode doors. *Courtesy of GM Product Service Training.*

- outside temperature is above 0° C (32° F).
- engine coolant and/or A/C system high-

pressure side temperatures are not too high.

- refrigerant pressure (indicates refrigerant charge) is not too low.
- the throttle is not wide open.

When the compressor is operating, the BCM continuously rechecks these parameters and disengages the clutch if any of them are not maintained to preprogrammed values.

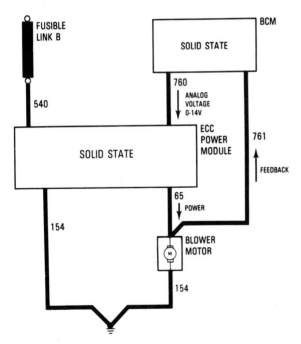

FIGURE 6-20   EEC blower motor control circuit. *Courtesy of GM Product Service Training.*

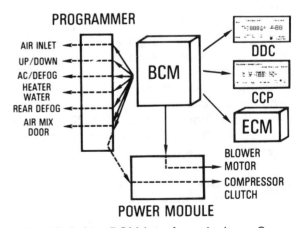

FIGURE 6-21   BCM interface devices. *Courtesy of GM Product Service Training.*

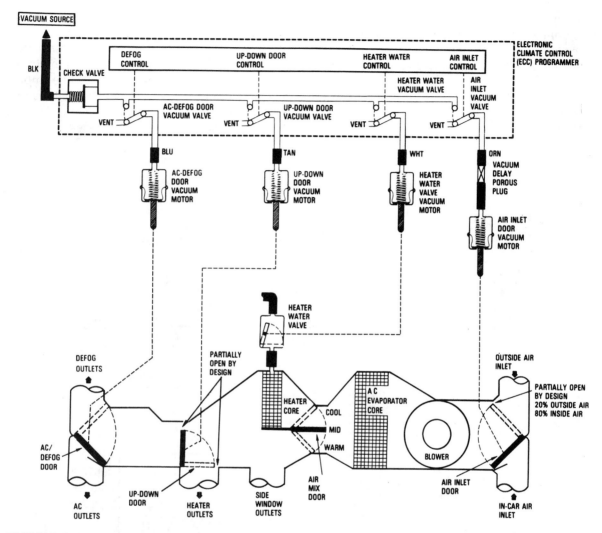

FIGURE 6-22   EEC programmer control of EEC mode. *Courtesy of GM Product Service Training.*

To cycle the clutch, the BCM looks at outside air temperature and low-pressure side temperatures. The compressor clutch is activated when the low side temperature exceeds 10° C (50° F). The clutch is turned off again when low side refrigerant temperature drops to –1° C (30° F).

Programmer and Power Module. The BCM uses two interface devices for many ECC functions: the ECC programmer and the ECC power module, Figure 6-21. The BCM sends instructions for air door positions, heater water control, rear window defog, and compressor clutch through a data link to the programmer. The rear window defogger is a heat element in the rear window glass powered by a relay. The programmer commands the actuators for each of these functions, Figure 6-22, except the compressor clutch. The compressor clutch on/off signal is transmitted from the programmer to the power module, and the power module actually turns the clutch on and off. The compressor clutch on/off instructions, which the BCM sends to the programmer, are also sent to the ECM so that it can anticipate the change in engine load and adjust engine speed.

The BCM sends the blower motor instructions straight to the power module in the form of a variable voltage. The power module amplifies this signal and keeps it proportional to volt-

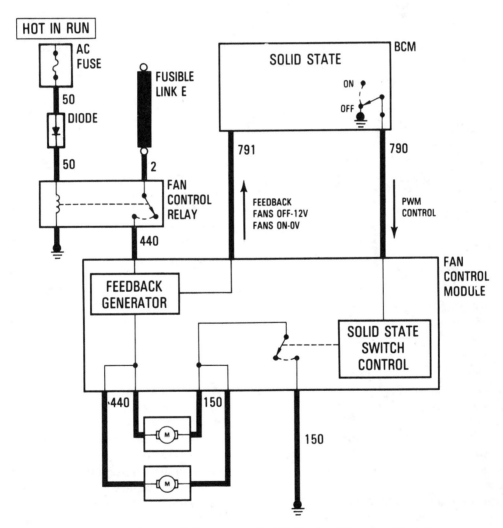

FIGURE 6-23   Cooling fans' control. *Courtesy of GM Product Service Training.*

age received from the BCM and issues it as a command to the blower motor, Figure 6-20.

Electric Cooling Fans

Inputs. The BCM receives inputs concerning cooling fan operation from two sources: the A/C high-pressure side thermistor, located at the condensor outlet, and coolant temperature information from the ECM.

Outputs. The speed of the two coolant fans is controlled by a pulse width–modulated (PWM) signal produced by the BCM and sent to the fan control module, Figure 6-23. The module completes or opens the ground side of the fans' power circuit in response to the BCM signal. The PWM signal is cycled once for each time period (32/1,000 of a second). If the signal voltage is high during only a short portion of the period, Figure 6-24, the fan motors run at low speed because their power is supplied in short pulses and they spend most of each revolution coasting, although their momentum keeps their speed constant. The greater the portion of each period during which the pulsed voltage is high, the faster the fan motors turn.

The fans are turned on at 40% full speed when either the coolant temperature reaches 106° C (223° F) or the A/C high-pressure side temperature reaches 61° C (142° F). They are turned on at full speed when either the coolant temperature reaches 116° C (241° F) or the A/C high side temperature reaches 72° C (162° F). They are turned off when either the coolant temperature drops to 102° C (216° F) or the A/C high side temperature drops to 57° C (135° F). The lower turn-off temperatures prevent the fans from indecisively turning on and off if one of the temperatures hovers at the on value.

The feedback signal sent back to the BCM from the fan control module allows the BCM to monitor fan motor performance. This signal is near 12 volts when the fans are off or stalled and near 0 volts when they are on or disconnected. The BCM compares feedback voltage to the commands it sent; if they do not match, it will set a fault code (F41).

Retained Accessory Power

After the ignition has been turned off (turned to lock) or to accessory, the BCM provides power to the power windows, astroroof, and the trunk release for ten minutes or until either a door is opened or the courtesy lights are turned on, Figure 6-25. When the ignition is turned off, the BCM grounds the retained power accessory relay for ten minutes. The relay then connects circuit 300 to battery voltage from an always hot source. Circuit 300 powers the trunk release, the power windows, and astroroof. If a door jam switch, the courtesy light switch, or the ignition switch is turned on, the BCM will unground the relay and the relay will reconnect circuit 300 to the ignition run terminal. Circuit 300 will now be powered only when the ignition switch is turned on.

Driver Information and Display

The BCM displays information about fuel and temperature on either the fuel data center (FDC) or the CCP. The driver selects the information to be displayed by pressing the appropriate button on one of the display panels. On vehicles equipped with digital display instrument clusters, the driver can choose to have the information displayed in either English or metric units by moving the English/metric switch on the digital cluster. Vehicles not equipped with the digital cluster display in either English or metric, depending on whether or not a wire (circuit 811) leading from terminal J2-13 of the BCM is connected to ground. If the wire is grounded, information will be expressed in metric units; if it is not grounded, information will be expressed in English units.

Fuel Data Inputs. For fuel data information, the BCM gets information from the fuel gauge sending unit (a potentiometer), the gauge fuse, and the ECM. The potentiometer is used to measure fuel level just as fuel gauges have done for years. An arm with a float on it moves the movable contact of the potentiometer as the fuel level changes. By comparing

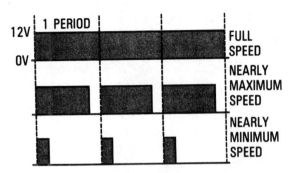

FIGURE 6-24   PWM command to cooling fan motors. *Courtesy of GM Product Service Training.*

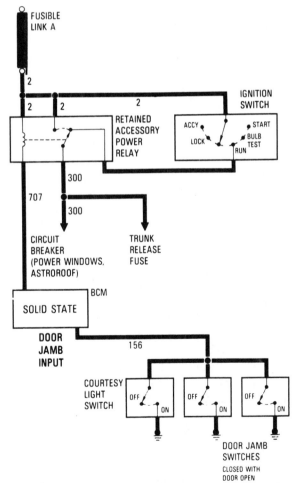

FIGURE 6-25   Retained accessory power circuit. *Courtesy of GM Product Service Training.*

return voltage from the potentiometer to a voltage reading taken at the gauge fuse, the BCM can calculate fuel level, Figure 6-26.

Vehicle speed and distance traveled (distance traveled can easily be calculated from VSS information because 4,004 VSS pulses equal 1 mile using the pulse generator–type VSS) plus injector pulse width information are sent from the ECM to the BCM via the data link.

Fuel Data Outputs. Normally the BCM displays fuel level by way of the FDC as discussed above. Upon request it displays:

- *instantaneous fuel mileage.* If the INST button on the FDC panel is pressed, the BCM looks at fuel being consumed (from injector pulse width information) and vehicle speed; from this it calculates the fuel mileage being achieved at that moment.

- *average fuel mileage.* The BCM is continuously receiving and updating its information regarding fuel used and distance traveled. This information is stored in memory. If, however, the reset button on the FDC panel is pressed, previous distance traveled and fuel used information will be erased from memory and new data will be accumulated. If the AVG button is pressed, the BCM will calculate fuel mileage for the distance traveled since the last reset.

- *fuel range.* When the RANGE button is pressed, the BCM checks remaining fuel from fuel level information, calculates fuel used during the last 25 miles, then calculates and displays the distance that can be traveled on the remaining fuel.

- *fuel used.* When the FUEL USED feature is requested, the BCM calculates and displays the fuel used since the last reset.

Temperature Display. Upon request the BCM displays outside temperature as indicated to it by the outside temperature thermistor. If outside temperature is not requested, the CCP will display inside temperature.

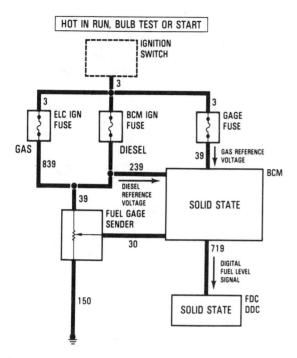

FIGURE 6-26   Fuel gauge circuit. *Courtesy of GM Product Service Training.*

Instrument Panel Display Dimming

The characters displayed by the FDC, the CCP, the radio, and the optional digital cluster are displayed by a group of small *vacuum fluorescent* (VF) glass tubes. *Anodes* (conductors on which positive potential voltage is placed) are placed on one side of the tube so that they form all of the vertical and horizontal bars of alpha and numerical characters. The anodes are coated with a fluorescent material. A series of thin, tungsten-coated wire strands are placed opposite the anodes on the other side of the tube. This serves as the *cathode* (a conductor through which current with negative voltage is passed). Between the anode and the cathode, a fine wire mesh called the *grid* is placed. The tube *(ampule)* is evacuated and filled with argon or neon gas.

As current is passed through the cathode, it becomes hot and causes the tungsten to give off a cloud of electrons. The BCM applies an amplified 16-volt positive potential to selected anodes. The electrons given off by the cathode are attracted to and bombard the energized anodes. As they strike an anode's fluorescent material coating, it glows and thus provides the digital display. The more electrons that strike the fluorescent material, the brighter it glows.

During daylight hours the VF displays are adjusted by the BCM to full brightness for maximum visibility. As sunlight fades the VF displays need less brightness. When the park lights are turned on (they come on with the headlights), that signal is fed to the BCM, Figure 6-27. Upon being alerted that the lights are

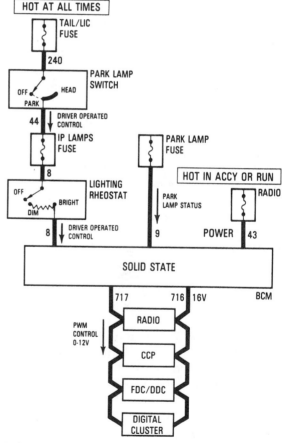

FIGURE 6-27   Vacuum fluorescent display dimming circuit. *Courtesy of GM Product Service Training.*

on, the BCM looks to the signal provided by the headlight switch rheostat, which the driver can turn to select VF display brightness.

The BCM controls display brightness by controlling a pulse width–modulated voltage to the grid in the ampule. When a modulated voltage is applied to the grid, it attracts or filters some of the electrons given off by the cathode and prevents them from striking the energized anodes.

Diagnostic Testing and Fail-Soft Actions

The BCM monitors many of its circuits, such as the blower motor and the cooling fan motors, for proper operation. It also has sensors specifically used to report problems such as the low-pressure switch in the A/C accumulator. If the BCM sees a fault (voltage in a given circuit or from a sensor is not within a range expected by the BCM) in any of its related circuits, it will record a code in its diagnostic memory. It may turn on the service air-conditioning light, and it may also initiate a fail-safe action, depending on the circuit the fault is in, Figure 6-28. For example, each time that the BCM turns the A/C compressor clutch on, it checks the A/C low-pressure switch first. If the switch is open (low pressure), it will not engage the clutch. If the switch stays open for thirty seconds, code F48 will be set and the following fail-safe actions are initiated:

- The service air-conditioning light will be turned on for thirty seconds. The light will come on again each time the driver selects AUTO on the CCP and the next time the ignition is turned on.
- The compressor clutch will be disabled until the ignition is turned off.
- The ECC system will be put into ECON instead of AUTO if the AUTO mode is selected.

It should be noted that fail-safe actions are intended to protect system components and

BCM DIAGNOSTIC CODES

CODE	CIRCUIT AFFECTED
▼ F10	OUTSIDE TEMP SENSOR CKT
▼ F11	A/C HIGH SIDE TEMP SENSOR CKT
▼ F12	A/C LOW SIDE TEMP SENSOR CKT
▼ F13	IN-CAR TEMP SENSOR CKT
▼ F14	DIESEL COOLANT SENSOR CKT
▼ F30	CCP TO BCM DATA CKT
▼ F31	FDC/DDC TO BCM DATA CKT
▼ F32	ECM-BCM DATA CKT'S
▼ F40	AIR MIX DOOR PROBLEM
▼ F41	COOLING FANS PROBLEM
☑ F46	LOW REFRIGERANT WARNING
☑ F47	LOW REFRIGERANT CONDITION
☑ F48	LOW REFRIGERANT PRESSURE
▼ F49	HIGH TEMP CLUTCH DISENGAGE
▼ F51	BCM PROM ERROR

☑ TURNS ON "SERVICE AIR COND" LIGHT
▼ DOES NOT TURN ON ANY LIGHT

COMMENTS:
F11 TURNS ON COOLING FANS WHEN
    A/C CLUTCH IS ENGAGED
F12 DISENGAGES A/C CLUTCH
F14 & F32 TURN ON COOLING FANS
F30 TURNS ON FT. DEFOG AT 75° F
F41 TURNS ON "COOLANT TEMP/FANS"
    LIGHT WHEN FANS SHOULD BE ON
F47 & F48 SWITCHES FROM "AUTO"
    TO "ECON"

FIGURE 6-28 BCM diagnostic codes. *Courtesy of Cadillac Motor Car Division.*

that the codes that initiate fail-safe actions cannot be set while the system is in diagnostics. Therefore, if a code is cleared while the system is in diagnostics and the system with a fault is operated while in diagnostics, the fail-safe actions will not occur and damage to a component might occur.

SYSTEM DIAGNOSIS AND SERVICE

The self-diagnostic function of the DFI system works basically the same way as other

General Motors' computer control systems, but it is more elaborate. When introduced in mid-1980 (the same system was used through 1981), its self-diagnosis process was like that of a CCC system. Since then, however, the system's capability in that area has greatly expanded. As the ECM and/or BCM control the systems for which they are responsible, they continuously monitor operating conditions for possible faults. They compare operating conditions against expected, preprogrammed operating standards. By doing so certain circuit and component faults can be identified. When a fault is identified, a two-digit trouble code number with either an E or an F prefix is stored in memory. Codes with an E prefix are stored by the ECM, Figure 6-29; codes with an F prefix are stored by the BCM, Figure 6-28. When the system is put into diagnostics, this code number is displayed on either the CCP or the FDC, depending on the year and model.

Types of Trouble Codes

A trouble code can be stored in either of two ways: as a current code (sometimes referred to as a hard code because it currently exists) or as a history code (representing a fault that existed at one time but does not exist now). When the system is put into diagnostics, all of the codes are shown once in numerical order from low to high; then they are shown again. This time, however, only the current codes are shown. From 1982 through 1984, DFI systems show all codes in the first two passes and show only current codes in a third pass.

Service Lights and Fail-safe

Faults that the ECM recognizes (E codes) and that require urgent attention bring on the service now light. Faults that the ECM recognizes and that need to be called to the driver's attention but are less urgent bring on the service soon light. Certain F codes pertaining to the ECC system bring on the service air-conditioning light.

ECM DIAGNOSTIC CODES

	CODE	MALFUNCTION
■■	E12	NO DISTRIBUTOR SIGNAL
☐	E13	OXYGEN SENSOR NOT READY (CANISTER PURGE)
☐	E14	SHORTED COOLANT SENSOR CIRCUIT
☐	E15	OPEN COOLANT SENSOR CIRCUIT
■■	E16	GENERATOR VOLTAGE OUT OF RANGE (ALL SOLENOIDS)
☐	E18	OPEN CRANK SIGNAL CIRCUIT
☐	E19	SHORTED FUEL PUMP CIRCUIT
■■	E20	OPEN FUEL PUMP CIRCUIT
☐	E21	SHORTED THROTTLE POSITION SENSOR CIRCUIT
☐	E22	OPEN THROTTLE POSITION SENSOR CIRCUIT
☐	E23	EST/BYPASS CIRCUIT PROBLEM (AIR)
☐	E24	SPEED SENSOR CIRCUIT PROBLEM (VCC)
☐	E26	SHORTED THROTTLE SWITCH CIRCUIT
☐	E27	OPEN THROTTLE SWITCH CIRCUIT
☐	E28	OPEN THIRD OR FOURTH GEAR CIRCUIT
☐	E30	ISC CIRCUIT PROBLEM
■■	E31	SHORTED MAP SENSOR CIRCUIT (AIR)
■■	E32	OPEN MAP SENSOR CIRCUIT (AIR)
■■	E34	MAP SENSOR SIGNAL TOO HIGH (AIR)
☐	E37	SHORTED MAT SENSOR CIRCUIT
☐	E38	OPEN MAT SENSOR CIRCUIT
☐	E39	VCC ENGAGEMENT PROBLEM
☐	E40	OPEN POWER STEERING PRESSURE CIRCUIT
■■	E44	LEAN EXHAUST SIGNAL (AIR & CL & CANISTER PURGE)
■■	E45	RICH EXHAUST SIGNAL (AIR & CL & CANISTER PURGE)
☐	E47	BCM - ECM DATA PROBLEM
■■	E51	ECM PROM ERROR
▼	E52	ECM MEMORY RESET INDICATOR
▼	E53	DISTRIBUTOR SIGNAL INTERRUPT
▼	E59	VCC TEMPERATURE SENSOR CIRCUIT
▼	E60	TRANSMISSION NOT IN DRIVE
▼	E63	CAR SPEED AND SET SPEED DIFFERENCE TOO HIGH
▼	E64	CAR ACCELERATION TOO HIGH
▼	E65	COOLANT TEMPERATURE TOO HIGH
▼	E66	ENGINE RPM TOO HIGH
▼	E67	CRUISE SWITCH SHORTED DURING ENABLE

ECM AND CRUISE CONTROL COMMENTS:

■■	TURNS ON "SERVICE NOW" LIGHT
☐	TURNS ON "SERVICE SOON" LIGHT
▼	DOES NOT TURN ON ANY TELLTALE LIGHT
( )	FUNCTIONS WITHIN BRACKETS ARE DISENGAGED WHILE SPECIFIED MALFUNCTION REMAINS CURRENT (HARD)

E16 & E24 DISABLE VCC FOR ENTIRE IGNITION CYCLE

E24 & E67 DISABLE CRUISE FOR ENTIRE IGNITION CYCLE

CRUISE IS DISENGAGED WITH CODE(S) E16, E51 OR E60 - E67

FIGURE 6-29   ECM diagnostic codes. *Courtesy of Cadillac Motor Car Division.*

If a fault that can cause damage to a circuit or component or that can result in unacceptable performance during continued operation occurs, the ECM or BCM will activate a fail-safe action. The fail-safe action is designed to either deactivate the faulty component or compensate for its malfunction.

Diagnostic Procedure

Before beginning a diagnostic procedure on a DFI-equipped vehicle, you should first be sure that you:

- have at least a general understanding of the operation of the ECM and/or the BCM and any of the subsystems in which the problem or problems seem to exist.
- are at least familiar with the self-diagnostic features of the system.
- have access to a service manual that contains the diagnostic charts.
- do a thorough visual inspection of the system in which the fault seems to exist and any related systems. Check for loose or broken vacuum lines, pinched wires, loose wire connections, and so forth.

Diagnostic Guide. Having complied with the suggestions above, continue your diagnostic procedure by answering the questions asked in the next four paragraphs.

1. *Are the on-car diagnostics working?* If the service air-conditioning, service now, or service soon lights are not working properly or if either of the data display panels fails to fully light up or respond properly, the self-diagnostic system check, Figure 6–30, will refer to the appropriate chart for correcting the problem.

2. *Are trouble codes displayed?* If any trouble codes are displayed, go to the trouble code chart of the same number. If the fault is intermittent (the code was not repeated on the second pass or third pass for earlier models), be sure to observe any notes concerning intermittent codes in the text that

accompanies each trouble code chart. The procedure for displaying the trouble codes is explained later in this section. If no trouble codes are stored, proceed to the next paragraph.

3. *Do all switch tests pass?* After the trouble codes have been displayed, perform the switch test as explained under **Switch Test** in this section. If any switch tests do not pass, go to the code chart identified by the same number as the diagnostic display number (under **Switch Tests** in Figure 6–31) for the switch that did not pass. If all of the switches do pass, go to the next paragraph.

4. *Is the fuel system controlling correctly?* To answer this use the field service mode, which is a part of the DFI system check, Figure 6–32. If the fuel system is not controlling properly, the DFI system check will refer to the appropriate chart. If the fuel system is controlling properly, the DFI system check will refer to the symptom charts.

Using these simple steps at the beginning of the diagnostic procedure saves time and often prevents unnecessary parts replacement that results from misdiagnosis.

Entering Diagnostic Mode. To enter diagnostics turn the ignition on, then simultaneously depress the off and warmer buttons on the CCP. Hold them until all display panel segments of both display panels light up, Figure 6–33. If any of the digit segments do not light up (they should all make 8's), the panel will have to be replaced. An 8 being displayed by a panel with two segments inoperative can appear as a 3.

Display of Trouble Codes. After the segment check ends, the FDC displays 8.8.8 followed by ..E. The ..E indicates that the first pass of the ECM codes are coming up next and that all ECM codes will be displayed, Figure 6–33. After the first pass, ..EE is displayed and indicates that the second pass of the ECM codes is coming where only the hard codes will be dis-

SELF-DIAGNOSTIC SYSTEM CHECK

- IF YOU HAVE NOT REVIEWED THE BASIC INFORMATION
  ON HOW TO USE THE COMPUTOR SELF-DIAGNOSTICS,
  GO TO THE INTRODUCTION OF THIS SECTION.

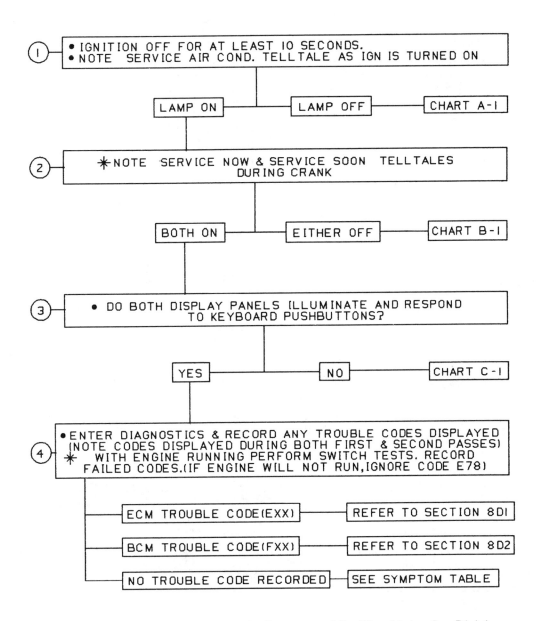

① • IGNITION OFF FOR AT LEAST 10 SECONDS.
   • NOTE SERVICE AIR COND. TELLTALE AS IGN IS TURNED ON

    LAMP ON ── LAMP OFF ── CHART A-1

② ✳ NOTE SERVICE NOW & SERVICE SOON TELLTALES
      DURING CRANK

    BOTH ON ── EITHER OFF ── CHART B-1

③ • DO BOTH DISPLAY PANELS ILLUMINATE AND RESPOND
      TO KEYBOARD PUSHBUTTONS?

    YES ── NO ── CHART C-1

④ • ENTER DIAGNOSTICS & RECORD ANY TROUBLE CODES DISPLAYED
   (NOTE CODES DISPLAYED DURING BOTH FIRST & SECOND PASSES)
   ✳ WITH ENGINE RUNNING PERFORM SWITCH TESTS. RECORD
     FAILED CODES.(IF ENGINE WILL NOT RUN,IGNORE CODE E78)

    ECM TROUBLE CODE(EXX) ── REFER TO SECTION 8D1

    BCM TROUBLE CODE(FXX) ── REFER TO SECTION 8D2

    NO TROUBLE CODE RECORDED ── SEE SYMPTOM TABLE

FIGURE 6-30   Self-diagnostic system check. *Courtesy of Cadillac Motor Car Division.*

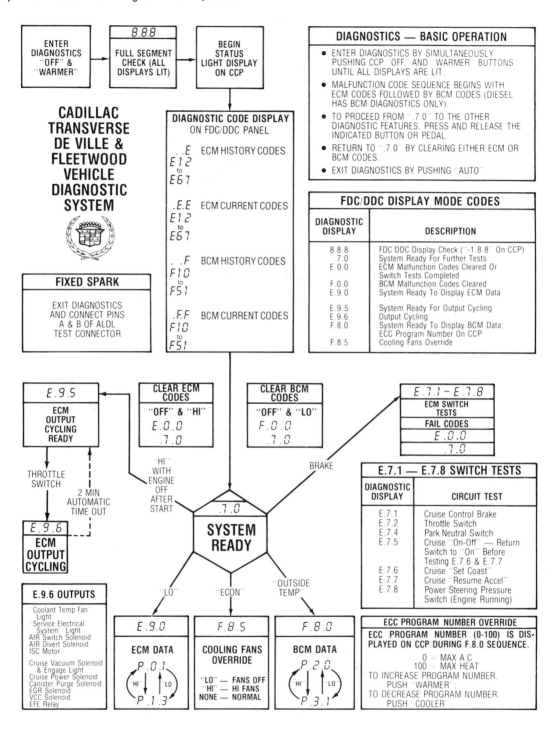

FIGURE 6-31   Diagnostic chart, part 1. *Courtesy of Cadillac Motor Car Division.*

DFI SYSTEM CHECK

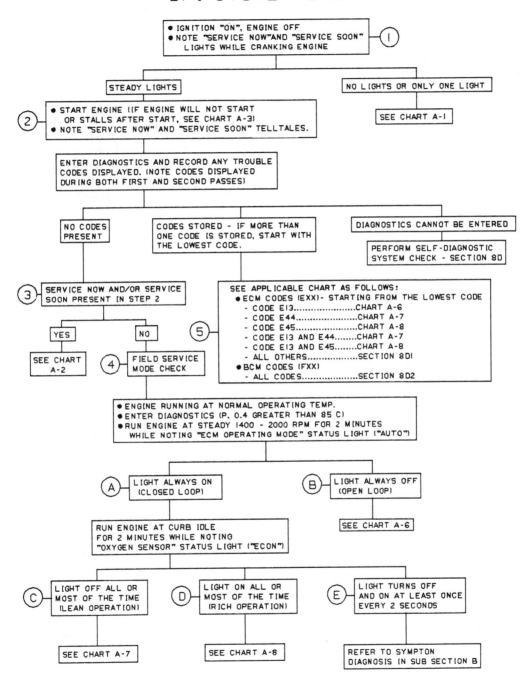

FIGURE 6-32   DFI system check. *Courtesy of Cadillac Motor Car Division.*

FIGURE 6-33 Display panel in segment check

played. If all stored ECM codes are intermittent, the ..EE will not be displayed and the second pass will not occur. If code E51 (a PROM fault) is displayed, it will continue to be displayed until the diagnostic mode is exited or until the fault is repaired and the code is cleared. While code E51 is displayed, the system will not advance to another diagnostic feature.

If no ECM codes are stored, the 8.8.8 will be followed by ..F, indicating that BCM codes are coming up. The same sequence occurs. The first pass displays all stored BCM codes followed by ..FF, and the second pass displays only hard codes. If no BCM hard codes are stored, the ..FF will not be displayed and the second pass will not occur.

After all stored codes are displayed or if there are no stored codes, .7.0 will be displayed. This indicates that the system is ready for the next self-diagnostic feature to be selected. There are several different diagnostic features that can be selected from this point:

- ECM switch tests
- ECM data display
- ECM output cycling
- BCM data display

- ECC program override
- cooling fan override
- exit diagnosis or clear codes and exit diagnosis

ECM Switch Tests. The engine must be running to do this test sequence. When the FDC displays .7.0, depress and release the brake pedal; the FDC display should advance to E.7.1. When it does, cycle the cruise control brake switch by tapping the brake pedal again within ten seconds. As the switch is cycled, the ECM monitors its operation to be sure that it is functioning properly. If the switch does not cycle within ten seconds, the ECM will record it as having failed to function. At the end of ten seconds or after the switch has cycled properly, the display advances to the next switch test code. The same sequence is repeated for each code and switch listed under **Switch Test** in Figure 6-31.

After the switch test sequence is completed, the ECM displays on the FDC the code of any switch that did not pass. The failed switch code remains on display until the switch circuit is repaired and retested. After the test sequence has been completed and all switches have passed, E.0.0 will be displayed, followed by .7.0 (the system is ready for another diagnostic feature selection).

ECM Data Display. This mode is often referred to as *engine parameters.* In this mode the ECM displays the values it is receiving from selected sensors or switches. It can be entered by pressing the lo button on the CCP while .7.0 is displayed on the FDC. When this mode is entered, E.9.0 appears on the FDC. The parameter codes, the parameters they represent, the value range, and the units in which the parameters are expressed are shown in Figure 6–34 under **Engine Data Display.** To advance to the next higher parameter code number, press the hi button on the CCP. Pushing the lo button returns to the next lower parameter code number.

These parameter values can be used as follows:

- They can be compared to those of a vehicle known to be operating properly to see if they are what they should be.
- They can be watched to see that they change as operating conditions change (for instance, the TPS parameter should change as the throttle is moved).
- Some of them are called for in the diagnostic charts, where their values are interpreted as good or at fault.

This diagnostic mode can be exited at any time and the display returned to the selection point, .7.0, by clearing the ECM or BCM codes (press off and hi for ECM codes or off and lo for BCM codes).

ECM Output Cycling. In this diagnostic mode, the ECM cycles all of its output actuators every three seconds with exception of the cruise control power valve, which cycles continuously. Cycling continues for two minutes; then the system automatically reverts to ECM Output Cycling Ready (E.9.5 displayed on the FDC). While the actuators are being cycled, you can observe that each is operating properly. To enter this output cycling mode, .7.0 must be displayed and the engine must be running.

- Turn the cruise on/off switch to the on position. This allows the cruise outputs to be cycled.
- Turn the engine off. Within two seconds turn the ignition on.
- Press hi on the CCP (E.9.5 appears on the FDC).
- Open and close the ISC throttle switch by depressing and releasing the throttle. The ECM output cycling mode is activated (E.9.6 is displayed).

This diagnostic mode can be exited at any time and the display returned to the selection point, .7.0, by clearing the ECM or BCM codes (press off and hi for ECM codes or off and lo for BCM codes).

BCM Data Display. As shown in Figure 6–34 under **BCM Data Display,** this mode displays the values sent to the BCM from most of its information sources as well as the value of some of its outputs. Figure 6–34 also shows the maximum range of each parameter and the units of measurement.

In order to go into the BCM data display mode, .7.0 must be displayed on the FDC. Depress and release the outside temp button on the CCP. The FDC displays F.8.0 briefly and then goes to parameter code P.2.0 or P.3.1 (the first and last parameters in the sequence). The parameter code is displayed for one second, then the parameter value is displayed for nine seconds. This sequence repeats with the same parameter until you select another one. To advance to another parameter, push the hi button on the CCP; to move back to a lower parameter code number, push lo on the CCP.

These parameter values can be used as follows:

- They can be compared to those of a vehicle known to be operating properly to see if they are what they should be.
- They can be watched to see that they change as operating conditions change (for instance, the commanded air mix

E.9.0 ENGINE DATA DISPLAY

PARAMETER NUMBER	PARAMETER	PARAMETER RANGE	DISPLAY UNITS
P.0.1	Throttle Position	-10 - 90	Degrees
P.0.2	MAP	14 - 109	kPa
P.0.3	Computed BARO	61 - 103	kPa
P.0.4	Coolant Temperature	-40 - 151	°C
P.0.5	MAT	-40 - 151	°C
P.0.6	Injector Pulse Width	0 - 99.9	ms
P.0.7	Oxygen Sensor Voltage	0 - 1.14	Volts
P.0.8	Spark Advance	0 - 52	Degrees
P.0.9	Ignition Cycle Counter	0 - 50	Key Cycles
P.1.0	Battery Voltage	0 - 25.5	Volts
P.1.1	Engine RPM	0 - 6370	RPM ÷ 10
P.1.2	Car Speed	0 - 255	MPH
P.1.3	ECM PROM I.D.	0 - 255	Code

F.8.0 BCM DATA DISPLAY

PARAMETER NUMBER	PARAMETER	PARAMETER RANGE	DISPLAY UNITS
P.2.0	Commanded Blower Voltage	-3.3 - 18.0	Volts
P.2.1	Coolant Temperature	-40 - 215	C
P.2.2	Commanded Air Mix Door Position	0 - 100	%
P.2.3	Actual Air Mix Door Position	0 - 100	%
P.2.4	Air Delivery Mode 0 = Max A C 4 = Off 1 = A C 5 = Normal Purge 2 = Intermediate 6 = Cold Purge 3 = Heater 7 = Front Defog	0 - 7	Code
P.2.5	In-Car Temperature	-40 - 102	°C
P.2.6	Actual Outside Temperature	-40 - 93	°C
P.2.7	High Side Temperature (Condenser Out)	-40 - 215	°C
P.2.8	Low Side Temperature (Evaporator In)	-40 - 93	°C
P.2.9	Actual Fuel Level	0 - 19.0	Gallons
P.3.0	Ignition Cycle Counter	0 - 99	Key Cycles
P.3.1	BCM PROM I.D.	0 - 255	Code

ECM PROM I.D.

ECM PROM I.D. is Parameter .1.3 of Engine Data and is displayed as a numerical code as follows:

X X X

FINAL DRIVE RATIO

2 = 3.33:1
(2.97:1 Effective Ratio)

EMISSIONS SYSTEM
1 = Federal
2 = California
3 = Export
4 = Altitude

ECM PROM CALIBRATION
Number varies with individual calibration.

BCM PROM I.D.

BCM PROM I.D. is Parameter .3.1 of BCM Data and is displayed as a numerical code as follows:

X X X

ENGINE SYSTEM

Blank = Gas
1 = Diesel

BCM PROM CALIBRATION
Numbers vary with individual calibration.

ECM STATUS LIGHT DISPLAY					
LIGHT ON	IN 4th GEAR	VCC ENABLED	CLOSED THROTTLE	RICH	CLOSED LOOP
LIGHT OFF	NOT IN 4TH GEAR	VCC DISABLED	OPEN THROTTLE	LEAN	OPEN LOOP
INDICATOR	〔⊞〕	〈⊞〉	Off	Econ	Auto
FUNCTION	4TH GEAR INPUT	VCC OUTPUT	THROTTLE SWITCH INPUT	OXYGEN SENSOR INPUT	ECM OPERATING MODE

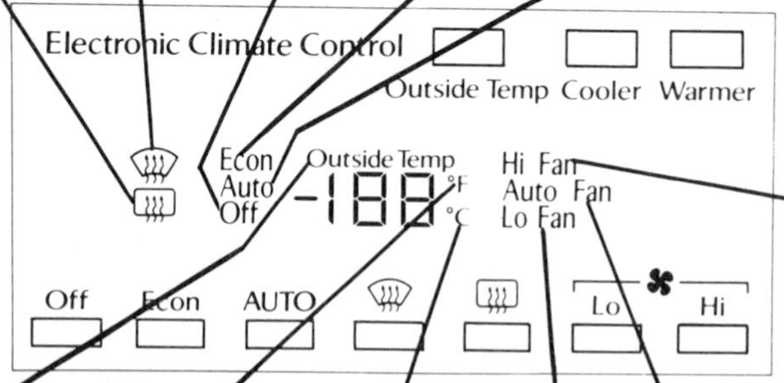

Electronic Climate Control

Outside Temp Cooler Warmer

Econ Outside Temp Hi Fan
Auto -188 °F Auto Fan
Off °C Lo Fan

Off Econ AUTO ⊞ ⊞ Lo ✿ Hi

BCM STATUS LIGHT DISPLAY	FUNCTION	A/C CLUTCH OUTPUT	COMPRESSOR LOW PRESSURE SWITCH INPUT	HEATER WATER VALVE OUTPUT	A/C-DEF MODE DOOR OUTPUT	COOLING FANS STATUS	UP/DOWN MODE DOOR OUTPUT
	INDICATOR	Outside Temp	°F	°C	Lo Fan	Auto Fan	Hi Fan
	LIGHT ON	ENERGIZED	OPEN (LOW PRESSURE)	CLOSED (NO WATER FLOW)	A/C	FANS RUNNING	UP
	LIGHT OFF	DE-ENERGIZED	CLOSED	OPEN	DEF	FANS OFF	DOWN

FIGURE 6–34   Diagnostic chart, part 2. *Courtesy of Cadillac Motor Car Division.*

door position should change as the selected temperature is changed).

- Some of them are called for in the diagnostic charts, where their values are interpreted as good or at fault.

This diagnostic mode can be exited at any time and the display returned to the selection point, .7.0, by clearing the ECM or BCM codes (press off and hi for ECM codes or off and lo for BCM codes).

ECC Program Override. During the BCM data display mode, the CCP displays a two-digit number that is the ECC program number. This number represents the level of heating or cooling effort by the ECC system as explained in the BCM section of this chapter. The program number automatically changes by the BCM as the inside temperature and/or the selected temperature changes.

The automatic calculation of the program number can, however, be overridden by pushing the warmer or cooler buttons (lower right of Figure 6–34). While in the BCM data display mode, holding the warmer button pushed in manually forces the program number up to a maximum of 100; 100 commands maximum heating. Pressing and holding the cooler button drives the program number down to a minimum of 0, which commands maximum cooling.

This override of the automatically calculated program number allows you to set it at any place between 0 and 100 and observe the reaction of the ECC system and the BCM data parameters. Once the program number has been overridden, it continues to be so until the BCM data display mode is exited.

Cooling Fan Override. This feature allows you to override the automatic control of the cooling fans' speed and manually command either high fans or fans off. To activate this feature the FDC must be displaying .7.0. Depress and release the econ button on the CCP. The FDC displays F.8.5. To command high fans, press and hold the hi button on the CCP until the fans achieve full speed. Releasing the hi button returns the fans to BCM automatic control. To command fans off, press and hold the lo button on the CCP until the fans stop. Releasing the lo button returns the fans to BCM automatic control. This feature can be exited at any time and the display returned to the selection point, .7.0, by clearing the ECM or BCM codes (press off and hi for ECM codes or off and lo for BCM codes).

Exit Diagnosis. Exit diagnosis by pushing auto on the CCP.

ECM/BCM Status Light Display. While in diagnostics the ECC mode indicators on the CCP are used to indicate the mode in which selected systems operated by either the ECM or BCM are operating (bottom portion of Figure 6–34). The operational mode is indicated by whether or not the designated light is on.

REVIEW QUESTIONS ⸻

1. How is the ECM's learning erased?

2. How is its learning restored?

3. The ECM monitors generator voltage to be sure that it is between _____ and _____ volts.

4. What is the function of the P/S switch?

5. Name at least one non-engine-related function that the ECM controls.

6. What is the acceptable fuel pressure for a DFI system?

7. What is the function of the black wire in the four-wire distributor connector?

8. List nine sensors or switches that are important in order for the ISC motor to operate in all driving conditions.

9. How does the ISC motor respond during deceleration?

10. List three criteria that must be met before the ECM will apply the TCC or VCC.

11. In addition to using a negative back-pressure EGR valve, how is the EGR flow rate controlled?

12. List three things that turn off the cruise control on a DFI system.

13. List six functions of the BCM.

14. What two inputs does the BCM use to control the coolant fans?

15. What does the term *fail-soft* mean?

16. How is the DFI system put into diagnostics?

17. After all of the trouble codes are displayed, .7.0 is displayed. What does this indicate?

18. What does the switch test accomplish?

19. What occurs during the ECM data display mode?

20. What is the most practical use of the ECM out cycling mode?

21. While in diagnostics what is the function of the status light display?

ASE-type Questions. (Actual ASE test questions will probably not be as product specific.)

22. A Cadillac with DFI is brought into the shop running very rough. Technician A says that the cause could be a blown injector fuse. Technician B says this cannot be the problem because if the injector fuse were blown, no fuel would be injected into the manifold and the engine would not run at all. Who is correct?
    a. A only
    b. B only
    c. both A and B
    d. neither A nor B

23. Technician A says that when a DFI system is running in module mode, ignition timing is being controlled by the pickup coil and HEI module. Technician B says that when the system is operating in module mode, the engine is at cranking speed or a system failure has occurred. Who is correct?
    a. A only
    b. B only
    c. both A and B
    d. neither A nor B

24. Technician A says that if voltage is applied to the divert valve solenoid, AIR system air will be directed to the air cleaner. Technician B says that under those conditions the air will be directed to the switching valve. Who is correct?
    a. A only
    b. B only
    c. both A and B
    d. neither A nor B

25. Technician A says that the viscous converter clutch works just like the torque converter clutch and is nothing more than a name change. Technician B says that the VCC works similarly to the TCC but that it is designed to allow a small amount of converter slippage even when applied. Who is correct?
    a. A only
    b. B only
    c. both A and B
    d. neither A nor B

26. Technician A says that the EFE system on rear-wheel drive DFI vehicles is not computer controlled and is designed to prevent throttle blade icing. Technician B says that the EFE system on front-wheel drive DFI vehicles is computer controlled and is designed to aid fuel evaporation during engine warm-up. Who is correct?
    a. A only
    b. B only
    c. both A and B
    d. neither A nor B

GLOSSARY

In addition to the following terms, refer to Figure 6–35 for a list of the abbreviations used in this chapter.

ACCUMULATOR   A reservoir located between the evaporator (under dash unit) and the compressor of an air-conditioning system and that allows freon gas and any remaining liquid to separate.

AMPULE   A strong glass tube that forms the body of an electrical component such as a vacuum tube.

ANODE   A conductor with positive voltage potential applied.

ASTROROOF   A term used by Cadillac to identify a power window in the roof of a vehicle.

CATHODE   A conductor with negative voltage potential applied.

FAIL-SOFT   A feature designed into a system and that provides limited continuous operation in the event of certain component failure.

MODULATED DISPLACEMENT   A system in which a series of ECM-controlled electromechanical devices can disable the rocker arms of selected cylinders and thus prevent the valves from opening. The engine operates on eight, six, or four cylinders, depending on power requirements.

PROPAGATION   The spread of the flame front across the combustion chamber of an engine as used in automotive terminology.

VACUUM FLUORESCENT   A glass tube containing anodes and a cathode. The anodes are

A/C	Air-conditioning
AIR	Air Injection Reaction
ALDL	Assembly Line Data Link
BARO	Barometric Pressure
CCP	Climate Control Panel
DFI	Digital Fuel Injection
ECC	Electronic Climate Control
ECM	Electronic Control Module (controller)
EECS	Evaporative Emissions Control System
EFE	Early Fuel Evaporation
EGR	Exhaust Gas Recirculation
EST	Electronic Spark Timing
FDP	Fuel Data Panel
HEI	High Energy Ignition
ISC	Idle Speed Control
MAP	Manifold Absolute Pressure
N.C.	Normally Closed
N.O.	Normally Open
O_2	Oxygen (sensor)
PCV	Positive Crankcase Ventilation
PFI	Port Fuel Injection
P/N	Park/Neutral
PROM	Programmable Read-only Memory (engine calibrator)
TBI	Throttle Body Injection (unit)
THERMAC	Thermostatic Air Cleaner
TPS	Throttle Position Sensor
TVS	Thermal Vacuum Switch
VIN	Vehicle Identification Number
VSS	Vehicle Speed Sensor
WOT	Wide-open Throttle

FIGURE 6-35   Explanation of abbreviations

coated with a fluorescent material and are positioned so that they form all of the segments of alpha and numerical characters. As current is passed through the cathode, it gives off electrons. If the anodes have positive voltage applied, the electrons given off by the cathode will be attracted to them and will strike the fluorescent material and cause it to glow.

FORD'S ELECTRONIC ENGINE CONTROL IV

Objectives

After studying this chapter, you will be able to:

- list the three operational strategies of the EEC IV system.
- identify at least five of the major sensors.
- identify at least five of the major actuators.
- identify the three types of fuel metering devices.
- identify the controlled components most directly concerned with emissions rather than engine performance or mileage.
- identify the major components of the diagnostic procedure.

The electronic engine control IV (EEC IV) system is the fifth generation of Ford's computerized engine control systems. It was first introduced in late 1982. By 1985 the EEC IV system was the only computerized engine control system Ford was using except for some microprocessor control unit (MCU) applications on 5.8-liter police and Canadian applications. By 1985 the EEC IV system was also used on most Bronco, Ranger, E-series, and F-series light trucks.

The EEC IV system's primary function is to control the air/fuel ratio. This can be achieved through the use of a carburetor, single-point injection (which Ford calls central fuel injection) or multipoint injection (which Ford calls electronic fuel injection [EFI]). The throttle body unit for CFI is shown in Figure 7-1, and the EFI unit is shown in Figure 7-2. In order to control these and in some cases other functions, the computer obtains information about engine and vehicle operating parameters from several sensors, Figure 7-3.

ELECTRONIC CONTROL ASSEMBLY (ECA)

The EEC IV system computer (ECA) has more capability than any previous Ford engine control computer, Figure 7-4. It has a keep-alive memory (an expansion of its random access memory) that enables it to store codes related to faults that it has observed but that are no longer present (this was not true for the first application on the 1.6-liter engine). The num-

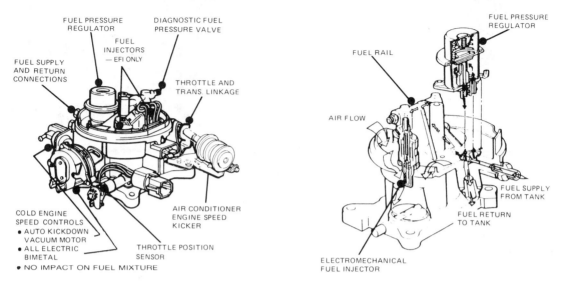

FIGURE 7-1   Central fuel injection unit. *Courtesy of Ford Motor Company.*

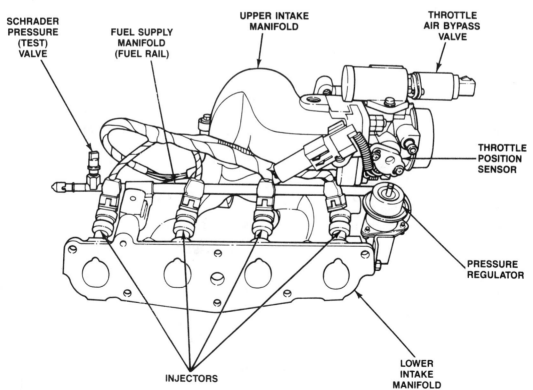

FIGURE 7-2   Multipoint injection manifold (1.6 L). *Courtesy of Ford Motor Company.*

Inputs

Engine coolant temperature
Manifold absolute pressure and/or
  barometric pressure
Throttle position
Engine speed and crankshaft position
Exhaust gas oxygen
EGR valve position

The following sensors are
unique to specific engines:

Air charge temperature
Vane air temperature
Vane airflow
Idle tracking switch
Transmission position
Inferred mileage sensor
Knock sensor
Ignition diagnostic monitor
Clutch engaged switch
Brake on/off switch
Power steering pressure switch
Air-conditioning clutch
Ignition switch

Outputs

Air/fuel mixture control device,
  carburetor solenoid or injector
Ignition timing control
Idle speed control
Thermactor airflow control
Canister purge control
EGR flow control
Torque converter clutch control
Turbocharger boost control
A/C and cooling fan controller module
Wide-open throttle A/C cut-off
Inlet air temperature control
Variable voltage choke
Temperature-compensated accelerator
  pump
Shift indicator light
Fuel pump relay

Electronic
Control
Assembly
(ECA)

Power
Relay

All components are not used on any one system.

FIGURE 7-3   EEC IV system overview

ber of codes it can recognize is greatly increased compared to the controllers of earlier Ford systems.

Engine Calibration Assembly

The ECA's calibration assembly contains the necessary programming to fine-tune the ECA's engine calibration commands to each vehicle's weight, axle ratio, and transmission application. Unlike many other systems' computer, the calibration assembly is an internal part of the ECA and is not serviced separately.

Operating Modes

The ECA operates the engine in one of several operational modes or strategies.

Base Engine Strategy. This is the mode that the ECA uses to control warm engine cali-

bration (around 88° C [190° F] or more) through the wide range of operating conditions that normally occur. This mode is broken down into four submodes:

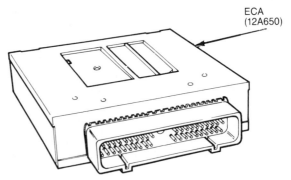

FIGURE 7-4   EEC IV electronic control assembly. *Courtesy of Ford Motor Company.*

- cranking
- closed throttle operation
- part throttle operation (closed loop)
- wide-open throttle (WOT)

Input information enables the ECA to recognize these operational modes. Having recognized the existing mode, the ECA issues calibration commands to the appropriate actuators, which produce the best results in terms of emissions, fuel mileage, and driveability.

If the engine threatens to stall while in the base engine strategy mode, the ECA will employ an underspeed response feature.

MPG Lean Cruise. When predetermined criteria are met during cruise conditions on some engine applications beginning in 1988, the ECA will take the system out of closed loop to a leaner air/fuel ratio. This is done in an attempt to achieve better fuel economy.

Modulator Strategy. Operating conditions that require significant compensation in order to maintain good driveability cause the ECA to modify the base engine strategy. Conditions that cause such modifications are:

- cold engine
- overheated engine
- high altitude

Limited Operational Strategy (LOS). When a component failure prevents the system from operating in a normal strategy, the ECA enters an alternate strategy designed to protect other system components, such as the catalytic converter, and still provide enough driveability to keep the vehicle operating until it can be repaired. If the ECA's central processing unit (CPU) fails, the ECA will operate in a fixed mode: no spark advance, no EGR, and air from the *Thermactor* (Ford's air injection system) will be dumped to the atmosphere.

Adaptive Strategy. This feature of the ECA was added in 1985 and refers to the ECA's learning ability that enables it to learn from past experiences. For instance, while the system is in closed loop, the processor observes the average cycling limits of the fuel metering device (carburetor or injectors), which is required to maintain a 14.7 to 1 air/fuel ratio. It then compares this to the open-loop cycling program. If there is a significant difference, a modified open-loop program will be stored in the keep-alive memory (KAM). This does not make the open-loop mode maintain a 14.7 to 1 air/fuel ratio. It does, however, make the open-loop mode more reflective of the driving conditions that that particular car experiences. The adaptive strategy also enables the ECA to adapt to changes in sensor inputs as a result of wear.

Because adaptive strategy learning is stored in the KAM, it is lost each time battery power is removed from the KAM portion of the ECA. Loss of adaptive strategy memory can result in a noticeable change in engine performance, such as surging, hesitation, high idle speed, or reduced fuel economy. Normal driving with all component systems working properly should restore learning within 5 miles, although there are reports of improvement continuing over much longer adjustment intervals. In 1986 the EEC IV system's ECA had an expanded adaptive strategy capability.

Neutral Idle Strategy. Beginning in 1987, neutral idle strategy has been programmed into the ECA for many EEC IV engine applications that is designed to reduce HC emissions during extended idle. When the engine comes to an idle speed, the ECA will check the neutral drive switch. When or if it sees that the transmission is in neutral, the ECA will start a timer function. After 60 seconds of idling, the ECA will start to gradually increase idle speed. It will increase idle speed by a maximum of 80 RPM over a time span of approximately 3 minutes. At the same time that it starts increasing idle speed, it will also start retarding spark timing. Timing will be retarded a maximum of 5° over a period of 60 seconds. These adjustments will be held until the engine is either taken off idle or shut off.

Power Relay

A power relay is used to supply operational power to the ECA, Figure 7–5. The control coil of the relay is powered by the ignition switch. When the ignition is turned on, the normally open contact points of the relay are closed and thus supply power to the ECA. The control coil of the relay is in series with a diode. If the battery's polarity is reversed, the diode will block current flow through the control coil and

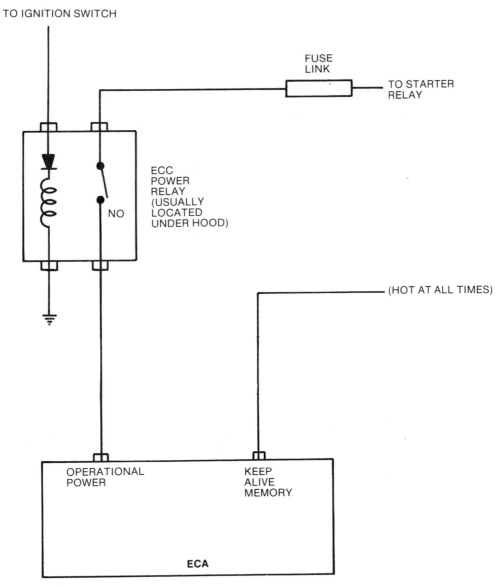

FIGURE 7-5   Power relay

will prevent the contact points from closing. This protects the power circuits of the ECA from damage resulting from reversed polarity.

Beginning in 1986 on some engine applications and becoming more widespread in 1987, the power relay has been integrated into a unit called the **Integrated Controller.** This unit typically contains the fuel pump relay, the power relay, the cooling fan relay (Electro-Drive Fan), the High Speed Electro-Drive Fan relay and the solid state A/C cut-out relay.

The power to the KAM does not come from the power relay. It comes directly from a battery source and is always hot. If this power supply is disconnected, any service codes stored and all learned adaptive strategy will be erased. Keeping this circuit powered requires very little current (a few milliamps) and does not discharge the battery even if the vehicle is not operated for several days. Because the KAM is not protected by the power relay, its circuitry is designed to be less susceptible to damage resulting from reversed polarity. Care should always be taken, however, to prevent such mistakes from occurring.

INPUTS

Engine Coolant Temperature (ECT) Sensor

The ECT is a thermistor type, Figure 7-6. The information it provides influences the ECA's calibration commands concerning:

- air/fuel ratio
- idle speed
- EGR
- Thermactor air
- canister purge
- choke voltage
- temperature-compensated accelerator pump
- upshift light

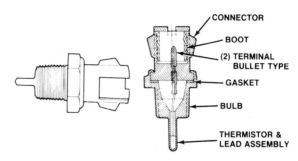

FIGURE 7-6 Engine coolant temperature sensor. *Courtesy of Ford Motor Company.*

Pressure Sensors

Most EEC IV engine applications use a manifold pressure sensor that also functions as a barometric pressure sensor. The barometric pressure reading occurs during key on/engine conditions such as the moment before the engine is cranked and during WOT operation. Engines equipped with vane airflow and vane air temperature sensors (such as the 1.6-liter and the 2.3-liter, turbocharged, multipoint injection–equipped engines) use a barometric pressure sensor and no manifold pressure sensor.

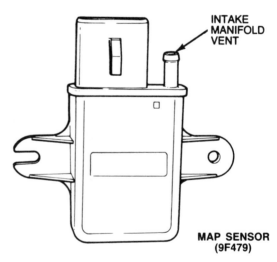

FIGURE 7-7 Manifold absolute pressure sensor. *Courtesy of Ford Motor Company.*

With the exception of the EGR pressure feedback sensor introduced in 1986, the pressure sensors used on EEC IV systems are frequency-generating devices, Figure 7–7. Reference voltage from the ECA is chopped up by the sensor to create a frequency signal that is sent back to the ECA. A solid-state switching device in the sensor converts the reference voltage to a frequency signal and changes the frequency in response to sensed pressure. If you were to apply a voltmeter directly to the sensor's output circuit, you would see a steady voltage signal (about 2.5 volts) that would change very little regardless of manifold pressure. Actually the signal does change (more or fewer pulses per second) in response to changes in manifold pressure, but the frequency is so rapid that a voltmeter cannot respond to the changes. The meter just gives an average voltage reading.

Manifold Absolute Pressure (MAP) Sensor. During lowest manifold pressure conditions (high vacuum), the sensor's signal frequency is 92 cycles per second (92 Hz). At WOT when manifold pressure is about equal to atmospheric pressure, the signal frequency is 162 Hz. The ECA interprets the sensor's signal as manifold pressure.

During ignition on/engine off conditions or during WOT operation, the ECA uses the MAP sensor's signal as a barometric pressure indication. The MAP sensor's input affects:

- air/fuel ratio
- EGR flow
- ignition timing

Barometric Pressure (BP) Sensor. The BP sensor is used only on turbocharged, multipoint injection applications where a MAP sensor is not used. It is nearly identical to the MAP sensor except that its pressure-sensing port is exposed to atmospheric pressure instead of being connected to manifold vacuum, Figure 7–8. Its signal frequency ranges from 122 Hz at lower atmospheric pressure to 162 Hz at higher atmospheric pressure. Its input affects:

- air/fuel ratio
- EGR flow

Throttle Position Sensor (TPS)

The TPS monitors throttle position and provides this information to the ECA. Two types of TPS have been used: a linear type and a rotary type.

Linear TPS. This is a potentiometer whose wiper moves in a straight line, Figure 7–9. The plunger, in contact with a cam on the throttle shaft, is pushed in as the throttle is opened, Figure 7–10. The linear-type TP sensor can be adjusted.

Rotary TPS. The rotary-type TPS is similar to the linear type except that its resistor forms part of a circle instead of being straight and the wiper pivots from the center of the circle. It is mounted on the side of the carburetor or throttle body so that the throttle shaft engages and drives the wiper, Figure 7–11. Beginning in 1985 all EEC IV applications use the rotary type. It is not adjustable; if its voltage output is

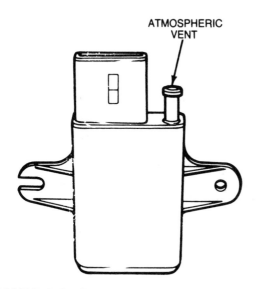

ATMOSPHERIC VENT

FIGURE 7–8 Barometric pressure sensor. *Courtesy of Ford Motor Company.*

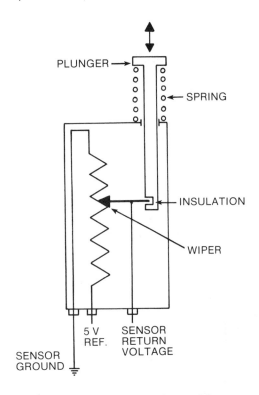

FIGURE 7-9 Linear throttle position sensor diagram

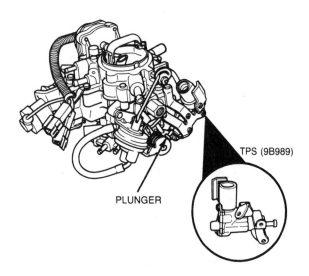

FIGURE 7-10 Linear throttle position sensor. *Courtesy of Ford Motor Company.*

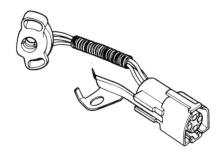

FIGURE 7-11 Rotary throttle position sensor. *Courtesy of Ford Motor Company.*

not within an acceptable range at a given throttle position, it must be replaced.

For either type, the ECA sends a 5-volt reference input to the TPS. At closed throttle the return voltage is about 12% of reference voltage; at WOT it is about 90% of reference voltage.

Profile Ignition Pickup (PIP) Sensor

The PIP signal is produced by a Hall-effect switch in the distributor, Figure 7-12. It provides information to the ECA concerning engine speed and crankshaft position. The circuit turned on and off by the Hall-effect switch is powered by the ECA (see **Hall-effect Switch** in Chapter 2).

PIP Signal

If the PIP signal were not generated by the Hall-effect switch, the engine would not run because there would be no ignition. If the PIP signal got to the TFI module but not to the ECA of a carbureted engine (see Figure 7-30), the engine would run because the TFI module could produce base timing from the PIP signal. If, however, the same condition occurred on an injected engine, the engine would not run because the ECA will not operate the injectors without the PIP signal.

Exhaust Gas Oxygen (EGO) Sensor

The EGO looks similar to the oxygen-sensing unit used on other systems and works exactly the same.

Air Charge Temperature (ACT) Sensor

The ACT sensor measures the temperature of the air that mixes with the fuel. It uses a thermistor as a temperature-sensing element just as the ECT does. On the ACT sensor, however, the sensing end of the sensor housing has openings that allow the air to come into direct contact with the thermistor coils, Figure 7-13.

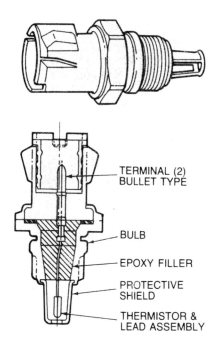

TERMINAL (2) BULLET TYPE

BULB

EPOXY FILLER

PROTECTIVE SHIELD

THERMISTOR & LEAD ASSEMBLY

FIGURE 7-13   Air charge temperature sensor. *Courtesy of Ford Motor Company.*

The ACT sensor is screwed into the intake manifold (into the air cleaner on 2.8-liter engines). The ECA uses ACT sensor information in the calculation of air/fuel mixture and spark timing. The ACT sensor is not used on all engine applications.

EGR Valve Position (EVP) Sensor

The EVP sensor is a linear potentiometer mounted on top of the EGR valve. It is connected to the EGR valve's diaphragm and pintle assembly and tells the ECA whether the EGR is open, closed, or somewhere in between, Figure 7-14. With the EGR valve fully closed, return voltage is about 16% of reference voltage; fully open the return voltage is about 86% of reference voltage. The ECA's processor can use preprogrammed values and other sensor data to convert EGR valve position to EGR flow rate for any given set of driving conditions. The EVP sensor's input is used to:

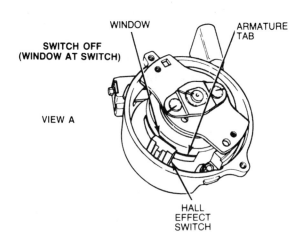

WINDOW

ARMATURE TAB

SWITCH OFF (WINDOW AT SWITCH)

VIEW A

HALL EFFECT SWITCH

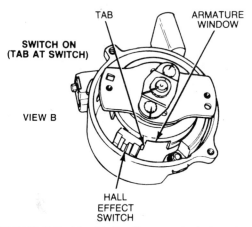

TAB

ARMATURE WINDOW

SWITCH ON (TAB AT SWITCH)

VIEW B

HALL EFFECT SWITCH

FIGURE 7-12   Hall-effect switch in distributor. *Courtesy of Ford Motor Company.*

FIGURE 7-14 EGR valve position sensor. *Courtesy of Ford Motor Company.*

- calculate air/fuel ratio
- calculate spark timing
- adjust EGR flow
- set a service code for an EGR valve that fails to open or close

Pressure Feedback EGR (PFE) Sensor. The PFE sensor (Figure 7-15) is a pressure-sensing

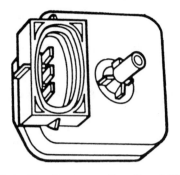

FIGURE 7-15 Pressure Feedback EGR (PFE) Sensor

voltage divider (functions as a potentiometer) similar to the ones used as MAP and BP sensors on the EEC III system. It senses exhaust pressure in a chamber just under the EGR valve pintle (Figure 7-16). This pressure causes the PFE to vary its output voltage signal to the

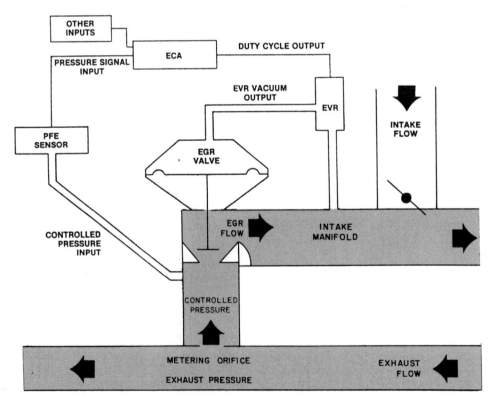

FIGURE 7-16

ECA. When the EGR valve is closed, the pressure in the sensing chamber is equal to exhaust pressure. When the EGR valve opens, pressure in this chamber will drop because of the restricting orifice that lets exhaust into the sensing chamber from the exhaust system. The more the valve opens, the more the pressure will drop. The PFE voltage signal tells the ECA how far the EGR valve is open. The ECA will use this information to fine tune its control of the electronic vacuum regulator (EVR), which controls vacuum to the EGR valve. This information also allows the ECA to more accurately control air-/fuel ratio and ignition timing.

Vane Meter

The vane meter is used on selected multipoint injection applications such as the 1.6-liter turbocharged and nonturbocharged and the 2.3-liter turbocharged engines. It fits between the air cleaner and the throttle body and measures the velocity of air flowing into the engine's induction system and the air's temperature. Both of these measurements are made by sensors within the vane meter.

Vane Airflow (VAF) Sensor. The air inlet opening of the vane meter is closed by a

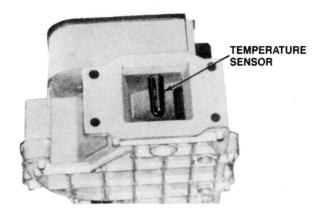

FIGURE 7–18   Vane air temperature sensor. *Courtesy of Ford Motor Company.*

spring-loaded door or vane, Figure 7–17. As the throttle valve is opened, air moving into the induction system forces the vane open; the more airflow, the wider the vane is opened. The vane is hinged on a pivot pin. As the vane is opened, it moves the wiper of a potentiometer, which acts as a sensor to the ECA. The greater the airflow, the higher the VAF sensor's signal voltage (less volt drop between reference voltage and sensor signal voltage).

As the vane is pushed open, its compensator flap is pushed into a specially designed cavity below the airflow passage. There is a slight lull in airflow between intake strokes, especially on four-cylinder engines; and the vane has a tendency to want to close slightly during those brief periods. The sealed space behind the compensator flap acts as a damper to reduce vane flutter.

Vane Air Temperature (VAT) Sensor. This sensor is essentially the same as the ACT sensor used on some other engine applications except that it is located in the vane meter's air inlet opening, Figure 7–18.

Speed Density Formula. The ECA compares inputs from the VAF, VAT, and BP sensors (or the MAP sensor, depending on engine application) along with those from TPS and EVP to a look-up chart. This way it can deter-

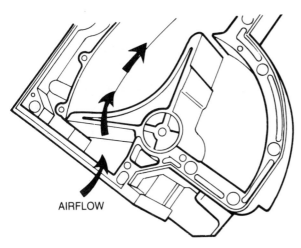

AIRFLOW

FIGURE 7–17   Vane airflow sensor. *Courtesy of Ford Motor Company.*

mine the engine's air intake volume and flow rate. This information, along with estimates of the engine's volumetric efficiency, is processed to determine exactly how much fuel should be injected. This is Ford's application of the speed density formula. (See **Measuring Air Mass** in Chapter 2 and **Speed Density Formula** in Chapter 5).

Mass Air Flow (MAF) Sensor. Beginning in 1988, selected engine applications will be equipped with a mass air flow (MAF) sensor instead of a MAP sensor. It is a Bosch hot-wire type similar to the one used by General Motors on selected applications (see the **Inputs** section of Chapter 5). The MAF output signal will vary from .5 volts to 4.75 volts with the voltage going up as air flow through it increases.

The MAF is housed in an aluminum body with an electronic module attached. It is located between the air cleaner and the throttle body. Proper alignment between the sensor's inlet and the air cleaner outlet is necessary in order for accurate measurement of mass air flow.

Idle Tracking Switch (ITS)

In the body of the idle speed control motor assembly (used on some carbureted and CFI applications) is a normally closed switch, Figure 7-19. When the throttle is closed, the throttle lever presses against a plunger that extends from the nose of the assembly. The pressure forces the plunger to move slightly back into the assembly and thus causes the switch to open. When the throttle is opened, the switch closes again. The ECA monitors the switch and

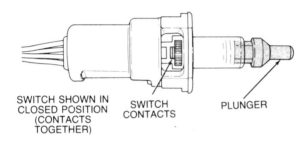

SWITCH SHOWN IN CLOSED POSITION (CONTACTS TOGETHER)   SWITCH CONTACTS   PLUNGER

FIGURE 7-19   Idle tracking switch. *Courtesy of Ford Motor Company.*

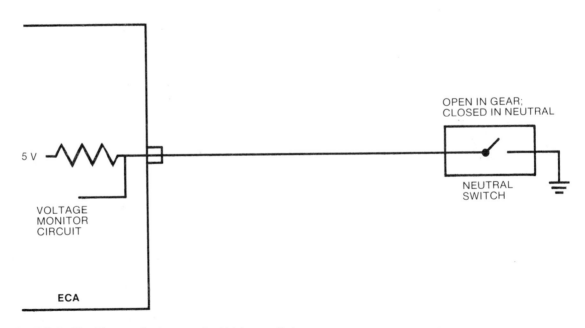

FIGURE 7-20   Transmission neutral/drive switch

can determine when the throttle is open or closed. It uses this information to control the ISC motor.

Transmission Neutral/Drive Switch

The ECA must know whether or not the transmission is in gear as this will affect some of its calibration commands. For instance, when an automatic transmission is in gear at idle, the ECA commands an increase in idle airflow to compensate for the additional load on the engine. If the throttle suddenly closes, the ECA will check to see if the transmission is in gear; if it is, the ECA will issue commands appropriate for deceleration conditions. A switch (neutral start switch for automatic transmission applications and neutral switch for manual transmission applications) is attached to the shift linkage to supply this information to the ECA, Figure 7-20.

Manual transmission applications can also have a clutch engaged switch in parallel with the transmission switch. It is closed when the clutch is disengaged (pedal depressed). If the transmission is in gear (neutral switch open) but the clutch is disengaged, the ECA will get a signal indicating that the engine is not connected to the drive wheels.

Inferred Mileage Sensor (IMS)

Beginning in 1985 some light truck EEC IV applications were equipped with an electronic module (IMS) that contained an E-cell, Figure 7-21. The IMS is powered by the power relay. After a specified amount of ignition on time, the E-cell opens the IMS's feedback circuit to the ECA. The ECA responds by changing programmed calibrations to allow thermactor air to be directed to the exhaust manifold for longer periods of time and changes in EGR flow rate to compensate for engine wear.

The IMS should not be confused with the emission maintenance warranty/extended useful life (EMW/EUL) module of which there are two types, each used on truck applications.

The early type contained an E-cell that was depleted after the ignition had been on for a total time that equated to about 60,000 miles. When the E-cell was depleted, it caused an indicator light on the instrument panel to be turned on. This alerted the driver that the EGR valve and EGO sensor were due to be replaced. On later models the E-cell in the module was replaced by a small microprocessor that could be reset. Replacing the EGR valve or EGO sensor at regular intervals is not required on other EEC IV system applications.

Knock Sensor (KS)

The EEC IV applications that use a KS use a piezoelectric type, Figure 7-22.

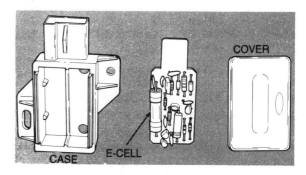

FIGURE 7-21 Inferred mileage sensor. *Courtesy of Ford Motor Company.*

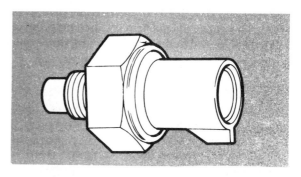

FIGURE 7-22 Knock sensor. *Courtesy of Ford Motor Company.*

Ignition Diagnostic Monitor (IDM)

A tach signal is fed from the negative side of the ignition coil to the ECA. A 22-kilohm resistor is placed in the signal wire to protect the ECA from high-voltage surges from the coil. The ECA compares the tach signal to the PIP signal as a way of monitoring proper ignition system operation. The IDM cannot be used on all engines.

Vehicle Speed Sensor (VSS)

The vehicle speed sensor, a magnetic pulse generator driven by the transmission speedometer gear, produces 16 AC signals per revolution (128,000 signals per mile) to the ECA. The ECA will modify this signal, convert it to a vehicle speed value, and store it in memory. In addition to vehicle speed control, this information may be used, depending on vehicle application, in the control of transmission converter lock-up, coolant fan control, and to identify deceleration conditions.

Brake On/Off (BOO) Switch

Vehicles equipped with the A4LD transmission, such as Rangers and Broncos, use the BOO switch (in the stop lamp switch assembly) to signal the ECA when the brakes are applied. The A4LD transmission features an ECA-controlled torque converter clutch. The ECA releases the clutch when the switch closes. If the brakes are applied during idle and the transmission is in gear, the ECA may raise idle speed. During a prolonged idle with the brakes applied, the ECA may disengage the A/C compressor clutch.

Power Steering Pressure Switch (PSPS)

The power steering switch is in the high-pressure side of the power steering system. At a pressure of 400 PSI to 600 PSI, the switch is closed. If the ECA sees this switch close during idle conditions, it will raise idle speed to compensate for the additional load on the engine.

Air-conditioning Clutch (ACC) Signal

On many engine applications, when the A/C is turned on, the same voltage signal that is sent to the A/C compressor clutch is also sent to the ECA. This signal alerts the ECA to raise idle speed slightly.

Ignition Switch

When the ignition is turned on, the power relay is activated. The power relay then supplies power to the ECA and to most of the circuits that the ECA controls. The relay's power to the ECA acts as an input and signals the ECA to start its programmed functions.

OUTPUTS

Air/Fuel Mixture Control

The EEC IV system employs three different types of fuel metering devices: carburetors referred to as feedback control carburetors, throttle body injectors referred to as central fuel injection, and port fuel injection referred to as multipoint injection.

Feedback Control (FBC) Carburetors. Three different carburetors have been used on EEC IV engine applications. They each use a duty-cycled solenoid to control the air/fuel mixture, although each uses a slightly different FBC solenoid. Each FBC solenoid is powered by the power relay, Figure 7–23; and the ECA controls the ground circuit.

The Motorcraft 2150A-2V, Figure 7–24, uses a solenoid that as it cycles alternately introduces and blocks air from the air cleaner into the vacuum passages, which provide bleed air to the idle and main metering systems. The air introduced by the solenoid is in addition to the air admitted by the fixed air bleeds typically used in most carburetors. If the solenoid were turned off completely (zero duty cycle), additional air would be blocked and the air/fuel mixture would be full rich. If the solenoid were turned on continuously (100% duty cycle), the

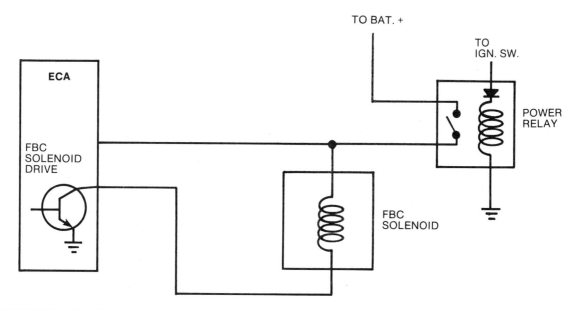

FIGURE 7–23   FBC solenoid circuit

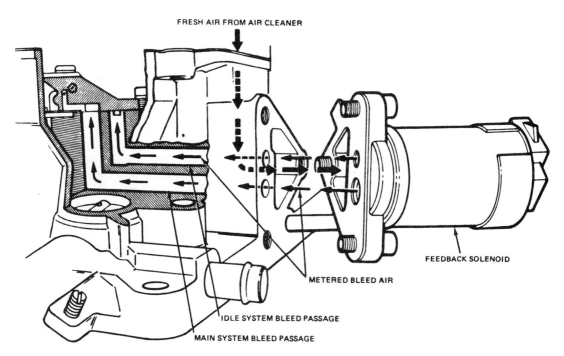

FIGURE 7–24   FBC solenoid (2150-2V carburetor). *Courtesy of Ford Motor Company.*

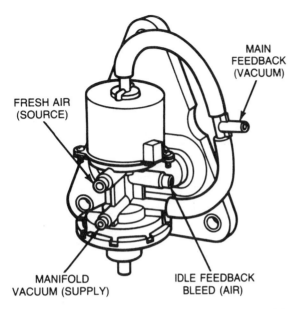

FIGURE 7–25   Remote duty-cycle solenoid.
Courtesy of Ford Motor Company.

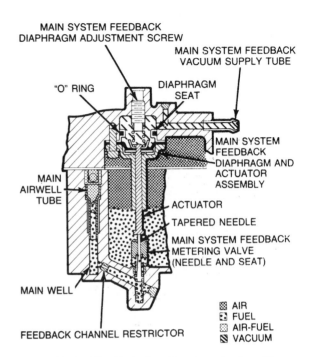

FIGURE 7–27   Main feedback metering. *Courtesy of Ford Motor Company.*

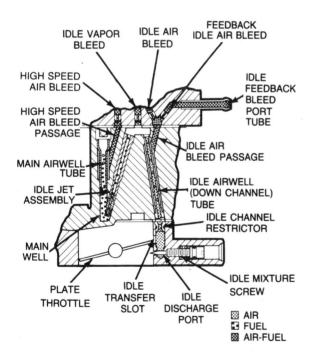

FIGURE 7–26   Idle feedback metering. *Courtesy of Ford Motor Company.*

air/fuel mixture would be full lean. The duty cycle is controlled by the ECA. In open loop the duty cycle is at a constant value, depending on coolant temperature, engine load, engine speed, and throttle position. In closed loop it varies constantly, depending on information being supplied by the EGO sensor.

The Carter YFA-1V carburetor is equipped with an FBC solenoid that looks different than the one used on the 2150A, but it works in the same way.

The Holley 6149-1V carburetor is controlled by a remote mounted FBC solenoid, Figure 7–25. Its control of the idle circuit air/fuel ratio is much like that of the other two solenoids. Its duty cycle controls the amount of air introduced to the idle speed circuit, Figure 7–26. The main metering mixture control, however, is somewhat different.

The solenoid controls vacuum to a diaphragm in the air horn section of the carburetor, Figure 7–27. The diaphragm is part of a fuel control valve assembly. It can easily be mistaken for a power valve assembly. A spring holds the diaphragm down. Extending down from the diaphragm into the fuel bowl is a rod with a foot at the bottom. The foot engages and pushes down on the upward extending tip of a valve (main system feedback metering valve) in the floor of the fuel bowl. With this valve pushed down (open), additional fuel flows into the main metering system. When the ECA wants to lean out the air/fuel mixture, it cycles the solenoid so that more vacuum is applied to the fuel control assembly diaphragm. This causes it to lift and allows the valve in the fuel bowl floor to close at least partially.

Central Fuel Injection (CFI). Two types of CFIs are used on EEC IV applications: a high-pressure system and a low-pressure system. Each uses a throttle body unit mounted on the manifold in place of a carburetor. The solenoid-operated injectors are opened by an electrical command (pulse) from the ECA. The injector's control circuit is similar to that of the FBC solenoid in Figure 7–28; power is supplied by the power relay and ground is controlled by the ECA. Fuel is sprayed from a single injector (low-pressure system, Figure 7–26) or from two injectors (high-pressure system, Figure 7–1) positioned directly over the throttle valves. During cranking, the high-pressure CFI system pulses both injectors simultaneously in response to each PIP signal. After the engine has started, they are pulsed alternately at a fixed frequency. Because the fuel pressure is constant and the injector always opens the same amount (about 0.01 inch), air/fuel mixture is controlled by the injector pulse width.

In 1986 a slight variation of the high-pressure CFI system was introduced on the 3.8-liter engine. Each of the two injectors has its own EGO sensor and logic circuit. A problem such as a leaking exhaust manifold on one side of the engine only affects the performance of half of the cylinders.

Multipoint Injection (MPI). On most MPI system applications, the injectors are pulsed in two groups. The two groups are pulsed alternately, with each group being pulsed once per engine cycle during normal operation. They are pulsed simultaneously during cranking. In 1986 the 5-liter engine featured sequential fuel injection. Each injector is pulsed individually in the firing order in coordination with its respective cylinder's intake stroke.

Fuel Supply System

Fuel Pump. The low-pressure CFI system uses a fuel pump mounted in the fuel tank. It is controlled by the fuel pump relay and delivers fuel to the fuel-charging assembly (throttle body unit) by way of the fuel pressure regulator.

The CFI high-pressure and the MPI systems use one of two fuel delivery systems. Most use a single high-pressure tank-mounted electric pump that feeds about 60 PSI to the pressure regulator. Vehicles with a greater distance between the fuel tank and the engine can use two electric fuel pumps to assure adequate fuel pressure to the injectors. A low-pressure pump is mounted in the tank. It feeds low-pressure fuel to an in-line high-pressure pump, which

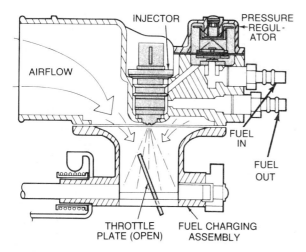

FIGURE 7–28 Low-pressure central fuel injection unit. *Courtesy of Ford Motor Company.*

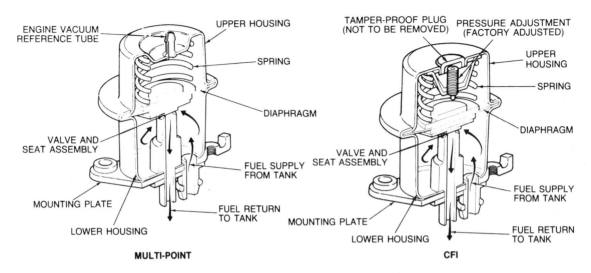

FIGURE 7-29 Fuel pressure regulator. *Courtesy of Ford Motor Company.*

boosts the pressure to about 60 PSI and sends it to the pressure regulator.

Either of the high-pressure pump systems generates a pressure of about 100 PSI if the fuel line is blocked. At this point the pump is turned off by a pressure relief switch.

The carbureted EEC IV system applications use a conventional mechanical fuel pump.

Fuel Pressure Regulator. Fuel pressure is controlled by a pressure regulator similar to those described in previous chapters, Figure 7-29. Manifold vacuum is routed to the MPI system pressure regulator so that it varies fuel pressure in response to throttle position. The pressure ranges from 30 PSI during high-vacuum (low manifold pressure) conditions to about 40 PSI at WOT (high manifold pressure). On turbocharged applications the higher manifold pressure during boost conditions causes the fuel pressure to go as high as 50 PSI.

The CFI systems do not expose manifold vacuum to the pressure regulator, and the pressure stays fairly constant. The high-pressure CFI system maintains from 39 PSI to 41 PSI. The low-pressure system (used on the 2.3-liter HSC engine) maintains about 14.5 PSI at the injector tip.

Fuel Pump Relay. The fuel pump relay controls the fuel pump and is controlled by the ECA, Figure 7-30. When the ignition is turned on, the ECA grounds the fuel pump relay coil. If it does not get a PIP signal indicating that the engine is being cranked, within one to two seconds it will turn off the relay. Once the engine is started, the ECA keeps the fuel pump relay activated until the ignition is turned off or until it stops receiving the PIP signal.

In series with the fuel pump relay is an inertia switch, Figure 7-31. Its purpose is to disable the fuel pump if the vehicle is involved in an accident. The inertia switch consists of a steel ball sitting in a metal bowl. A magnet holds the ball in the bottom of the bowl. In the event of a jolt severe enough to dislodge the ball, the ball strikes a lever above it. The lever triggers an overcentering device that causes the contacts of the inertia switch to open and break the circuit that powers the fuel pump. This also causes a reset button to pop up. Pushing the reset button back down closes the contacts again. This should be checked anytime a fuel-injected Ford vehicle that has been running fine suddenly fails to start. The inertia switch of a parked car can be triggered by hav-

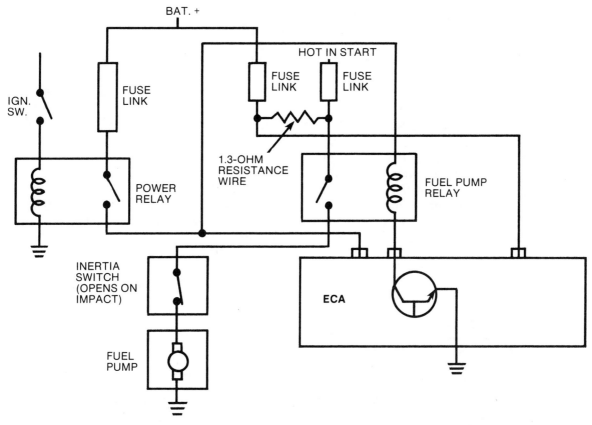

FIGURE 7-30   Fuel pump control circuit

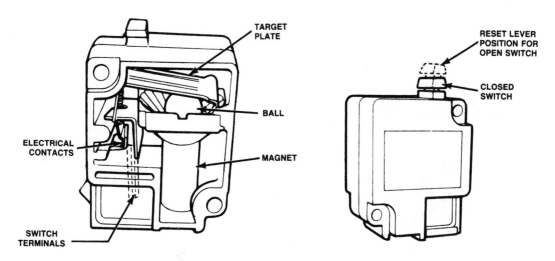

FIGURE 7-31   Inertia switch. *Courtesy of Ford Motor Company.*

ing another car back into it without any other damage occurring.

Thick Film Integrated (TFI-IV) Ignition System

The TFI-IV ignition system takes its name from the TFI module attached to the distributor housing. Thick film refers to the type of chip on which the module's circuit is located. Its function is to turn the primary ignition circuit off and on and to control primary circuit dwell. The spark output command from the ECA is the module's signal to open the primary circuit. If the spark output command fails to arrive, the TFI module will open the primary circuit in re-

sponse to the PIP signal; this results in base timing only.

Spark Output (SPOUT). The ECA receives the PIP signal from the Hall-effect switch in the distributor by way of the TFI module, Figure 7-32. From this signal it determines engine RPM and crankshaft position. The ECA then modifies this signal (changes its timing) to achieve the best ignition timing results considering the engine's speed, temperature, load, atmospheric pressure, EGR flow, and air temperature. The modified signal is sent to the TFI module as the spark output (SPOUT) signal or command. Upon receiving this signal, the TFI module opens the primary ignition circuit and

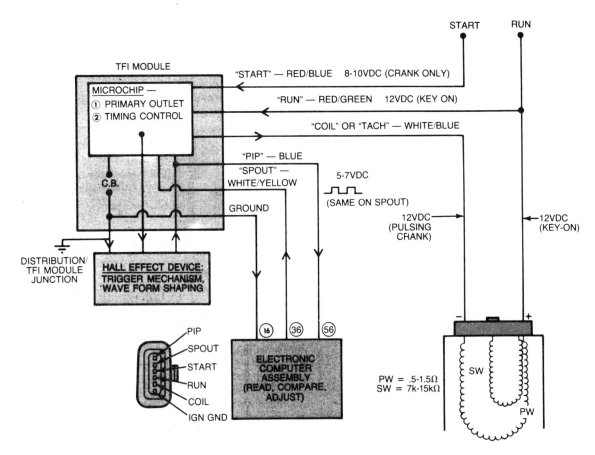

FIGURE 7-32   TFI ignition circuit. *Courtesy of Ford Motor Company.*

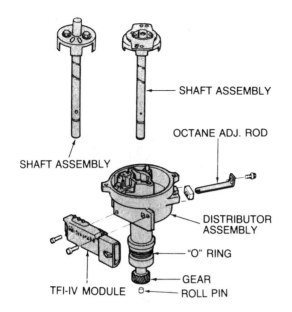

FIGURE 7-33 Octane adjustment rod. *Courtesy of Ford Motor Company.*

thus causes the coil to generate the high secondary voltage needed to fire the spark plug. If the SPOUT signal fails to arrive at the TFI module, the module will operate the ignition from the PIP signal that equates to base timing.

E Core Coil. The TFI-IV ignition system uses an E core coil that is different from the conventional coil used on prior Ford Motor Company applications. The coil core resembles two Es turned face to face. The coil windings, which are molded in epoxy, are wound around the Es' center horizontal bars. Because the center bars are shorter than the ones at the top and bottom, they do not quite touch each other and therefore form a small air gap between them. The smaller air gap (compared to the distance between the ends of a conventional coil core) helps the coil develop a much stronger magnetic field. With no primary ballast resistor and the small core air gap, the E core coil draws high primary circuit amperage (11 amps with the engine running and 18 amps

while cranking) and can put out as much as 50,000 volts.

Octane Adjustment Rod. The TFI-IV distributor has an octane adjustment feature. It has a rod inserted from the outside and that changes its spark advance capability, Figure 7-33. The standard rod has zero adjustment. In some cases where spark knock becomes a problem, Ford Motor Company has issued a technical bulletin that authorizes the replacement of the standard rod with a 3° or a 6° retard rod.

Knock Sensor Response. On systems using a knock sensor, the strength of the knock sensor's signal is proportional to the severity of the spark knock. When a knock signal is received by the ECA, it retards timing 1/2° per engine revolution until the knock disappears.

Idle Speed Control

Three different systems are used for idle speed control among various EEC IV applications.

Throttle Kicker. Throttle kicker is a vacuum-actuated device containing a diaphragm, a spring, and a plunger. It is mounted on the carburetor or CFI unit. On applications where the throttle kicker is used, normal idle speed is controlled by a conventional idle stop screw. However, when vacuum is applied to the diaphragm, the plunger is extended and opens the throttle blade slightly to increase idle speed. It is sometimes referred to as a vacuum-operated throttle modulator (VOTM).

Vacuum to the throttle kicker is controlled by a solenoid controlled by the ECA. Part of the throttle kicker solenoid (TK solenoid) assembly is a normally closed vacuum valve, Figure 7-34. The term *normally closed* means that a device such as a switch or a valve is designed to remain closed unless it is opened by some force, in this case a solenoid. *Normally open* means just the opposite. When the solenoid is energized, the vacuum valve opens and allows vacuum to the throttle kicker. The throttle

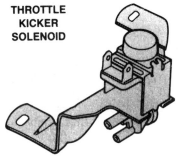

THROTTLE KICKER SOLENOID

FIGURE 7-34   Throttle kicker solenoid. *Courtesy of Ford Motor Company.*

kicker is activated during the following operating conditions:

- when engine temperature is below a specified temperature
- when engine temperature is above a specified temperature
- when the A/C compressor clutch is engaged
- when the vehicle is above a specific altitude

DC Motor Idle Speed Control (ISC). The ISC motor is a small, reversible, electric motor. It is part of an assembly that includes the motor, a gear drive, and a plunger, Figure 7-35. When the motor turns in one direction, the gear drive extends the plunger; when the motor turns in the opposite direction, the gear drive retracts the plunger. The ISC motor is mounted so that the plunger can contact the throttle lever. The ECA controls the ISC motor and can change the polarity applied to the motor's armature in order to control the direction it turns,

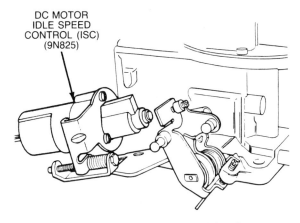

FIGURE 7-35   ISC motor assembly. *Courtesy of Ford Motor Company.*

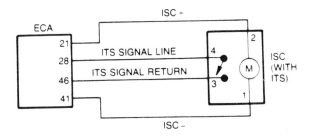

FIGURE 7-36   ISC motor circuit. *Courtesy of Ford Motor Company.*

Figure 7-36. When the idle tracking switch is open (throttle closed), the ECA commands the ISC motor to control idle speed. The ISC provides the correct throttle opening for cold or warm engine idle.

When the ITS is closed (throttle open), the ECA commands the ISC motor to fully extend the plunger to act as a dashpot. In other words, when the throttle tries to close again (deceleration), the plunger stops the throttle just before it finishes closing. The pressure on the plunger opens the ITS and the motor in turn retracts the plunger to normal idle position. This delay in throttle closing allows additional air into the manifold to compensate for the fuel that starts to evaporate from the manifold walls and floor as a result of the high vacuum.

When the ignition is turned off, the ECA commands the ISC motor to fully retract the plunger. This allows the throttle to fully close to prevent dieseling. When the engine stops, the ECA commands the plunger to a fully extended position in preparation for the next start-up.

Throttle Air Bypass Valve Solenoid. This idle speed control device is used on multipoint injection applications. It consists of an ECA-controlled solenoid that operates a pintle-type valve, Figure 7–37. This bypass valve assembly is attached so that when it is open air is allowed to pass from in front of the throttle blade through the pintle valve to the manifold side of the throttle blade. The ECA is programmed to control this valve by duty cycling the solenoid. As the solenoid is pulsed, the valve opens. The ECA cycles the solenoid to achieve the idle speed it wants.

During a cold start, the ECA holds the solenoid at a 100% duty cycle (valve held wide open). During cranking, hot or cold, the bypass valve provides sufficient air so that it is not necessary for the operator to touch the throttle. It also acts as a dashpot during deceleration.

Thermactor Air Management

The Thermactor system delivers air to the exhaust manifold or to the catalytic converter to aid in the reduction of HC, CO, and NO_x. During operating conditions that require a rich air/fuel mixture, Thermactor air is dumped into the atmosphere to avoid overheating the catalytic converter. One such condition for EEC IV and other Ford electronic engine control systems is during idle, as Ford tends to program their systems for a rich idle mixture in order to maintain a good idle quality.

There are two types of Thermactor systems used on EEC IV applications. One is ECA-controlled and will be discussed here. The other is a pulse injection system that uses the negative pressure existing between exhaust pulses within the exhaust manifold (most effective on four-cylinder engines) to draw air from the air cleaner past a check valve into the exhaust manifold. This system is sometimes referred to as Thermactor II and is not ECA controlled.

Thermactor Air Bypass (TAB) and Thermactor Air Divert (TAD) Valves. The pump supplies air to the bypass valve, which can either direct it to the divert valve or dump it into the atmo-

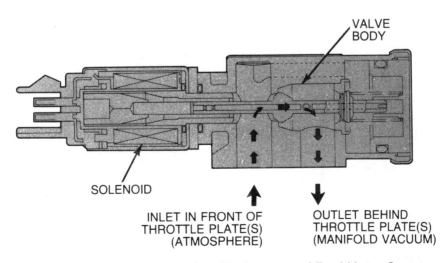

FIGURE 7–37   Throttle air bypass valve solenoid. *Courtesy of Ford Motor Company.*

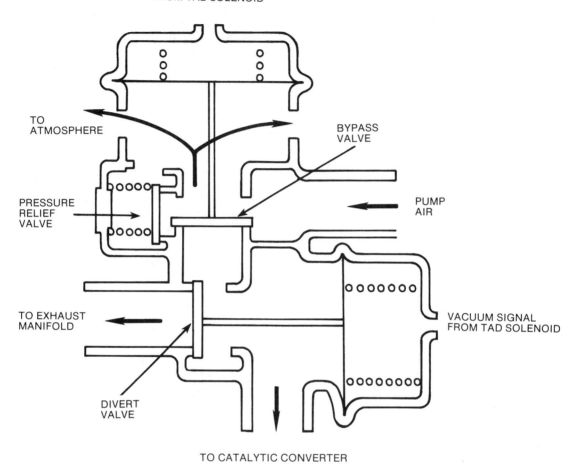

VACUUM SIGNAL
FROM TAB SOLENOID

TO
ATMOSPHERE

BYPASS
VALVE

PRESSURE
RELIEF
VALVE

PUMP
AIR

TO EXHAUST
MANIFOLD

VACUUM SIGNAL
FROM TAD SOLENOID

DIVERT
VALVE

TO CATALYTIC CONVERTER

FIGURE 7-38   Combination TAB/TAD valve

sphere, Figure 7-38. The divert valve can either direct the air upstream to the exhaust manifold or downstream to the catalytic converter. The valves are controlled by vacuum diaphragms that are controlled by ECA-controlled solenoids.

TAB and TAD Solenoids. The TAB and TAD solenoids control whether or not vacuum is applied to the TAB and TAD valves. If the engine is started with its coolant temperature below 10° C (50° F), the ECA will leave the TAB

solenoid circuit ungrounded. Vacuum to the bypass diaphragm is blocked, and the diaphragm spring holds the valve in the down position and thus bypasses pump air to the atmosphere. When engine temperature reaches 10° C, the ECA grounds the TAB solenoid circuit, Figure 7-39. Vacuum is routed to the diaphragm, which in turn lifts the bypass valve; and air is directed to the divert valve. At the same time, the ECA grounds the TAD solenoid; and it directs vacuum to the divert valve diaphragm. The divert valve is pulled back to close off the passage to the catalytic converter passage and direct the air to the exhaust manifold. When the coolant reaches approximately 88° C (190° F), the system goes into closed loop, the ECA deactivates the TAD solenoid, and air is directed to the catalytic converter.

Some operating conditions put the system back into bypass mode to protect the catalytic converter from overheating. They are:

- idle
- failure of the EGO sensor signal to switch between rich and lean frequently enough (indicating that air/fuel mixture is not being controlled or that the EGO sensor is cooling)
- acceleration or WOT operation

Canister Purge

Three types of canister systems are used on EEC IV-controlled engines. Some fuel-injected engines (both CFI and multipoint) use a constant purge system with no ECA control. Some engines, especially those equipped with carburetors, use an in-line, ECA-controlled, solenoid-operated vacuum valve. The third system uses an ECA-controlled solenoid that controls vacuum to a ported vacuum switch (temperature-operated vacuum control valve), which in turn controls a canister purge valve and an exhaust heat control valve.

In-line Canister Purge (CANP) Solenoid. This is a simple, normally closed, solenoid-controlled vacuum valve that is placed in the purge line connecting the intake manifold to

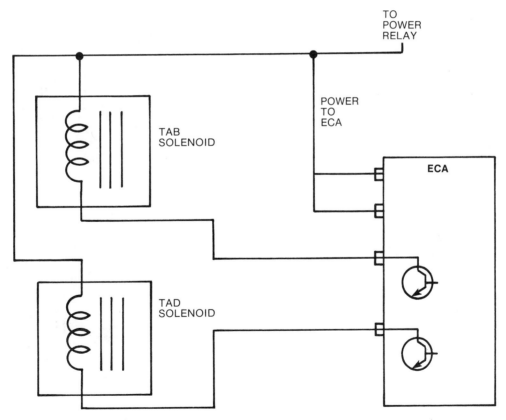

FIGURE 7-39 TAB/TAD solenoid control circuit

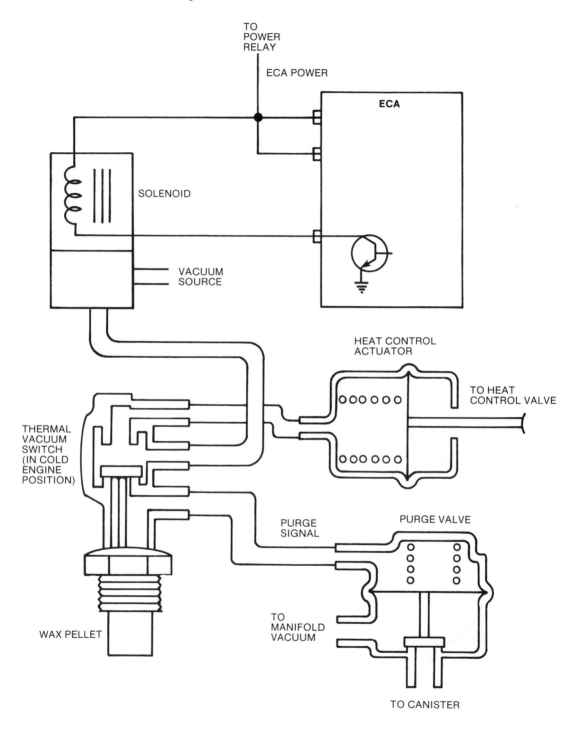

FIGURE 7-40   Canister purge/heat control system

the charcoal canister. When the engine has reached sufficient operating temperature, the ECA duty cycles the solenoid's circuit; the duty cycle depends on the vehicle's operating conditions. This allows the canister to be purged at a controlled rate.

Canister Purge/Heat Control System. This system (limited application) gets dual use out of one ECA function. The ECA controls a normally closed, solenoid-operated vacuum valve. When the solenoid is activated, vacuum is routed to the center port of a thermal vacuum switch, Figure 7–40. When the engine is cold, the thermal vacuum switch directs the vacuum to the top port, which is connected to the heat control valve actuator. The actuator closes the heat control valve and thus forces exhaust gases to pass through the heat crossover and warm the intake manifold plenum.

When the engine warms, the wax in the lower portion of the thermal vacuum switch expands and forces the valve up. It closes the top port and opens the lower port to the vacuum signal. The heat control valve opens; then the purge control valve opens and allows the charcoal canister to be purged.

EGR Control

The ECA controls four slightly different EGR systems.

EGR Control (EGRC) and EGR Vent (EGRV) Solenoids. This system is a carryover from EEC III and is designed so that the amount of EGR flow can be controlled (valve open, closed, or anywhere between). A position sensor (EVP) on top of the EGR valve tells the ECA what position the EGR valve is in. Two solenoids are used to control vacuum to the EGR valve, Figure 7–41. The normally closed control solenoid allows manifold vacuum to the EGR valve when it is energized. The normally open vent solenoid allows atmospheric pressure into the vac-

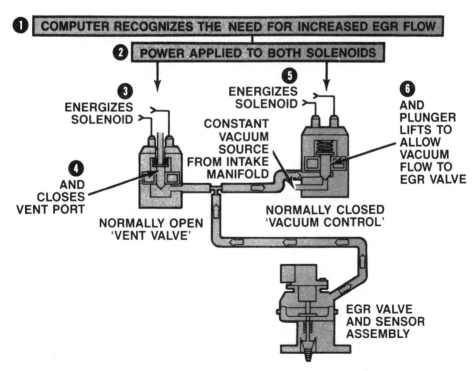

FIGURE 7–41   EGR control/EGR vent solenoids. *Courtesy of Ford Motor Company.*

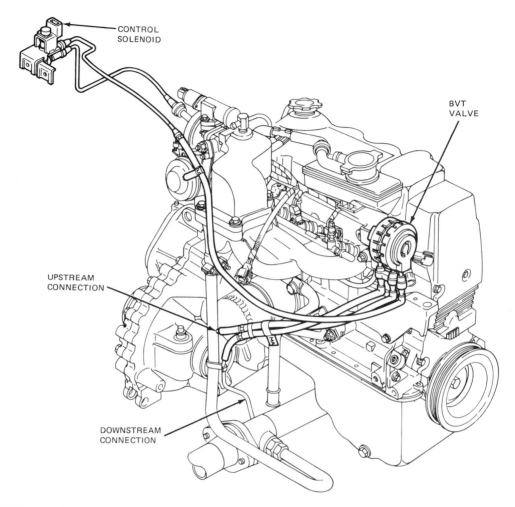

FIGURE 7-42   Back-pressure variable transducer. *Courtesy of Ford Motor Company.*

uum line when it is not energized. To open the EGR valve, the ECA grounds the circuits for both solenoids; the control solenoid routes vacuum to the EGR valve, and the vent solenoid blocks the vent passage. The EVP sensor reports EGR valve position to the ECA, which in turn manipulates the two solenoids to achieve whatever EGR valve position it wants for existing operating conditions. Both solenoids are constantly being cycled to achieve and maintain desired EGR valve position.

EGR Shutoff Solenoid. This system uses one vacuum control solenoid that operates like

the vacuum control solenoid just described except that a ported vacuum source is used. When the solenoid is energized, vacuum is applied to the EGR valve; when it is not energized, vacuum is blocked. With this system the EGR valve may or may not have an EVP sensor. Those with the sensor work much like the EGRV and EGRC solenoids combined into one, while those without the sensor are not modulated; they are either open or closed.

EGR Shutoff Solenoid with Back-pressure Transducer. This system uses one normally closed vacuum control solenoid like the system just described. The strength of the ported vacuum signal to the EGR valve is controlled, however, by an exhaust back-pressure variable transducer. The transducer is tied into the EGR valve's vacuum supply line and controls an air bleed into the line, Figure 7–42. Pressure signals are piped to the transducer from the exhaust system. The back-pressure signal comes from a tap in the tube that routes exhaust gas to the EGR valve. The control pressure signal comes from a point just downstream from the orifice that meters exhaust gas into the EGR exhaust supply pipe. These two pressures act on a valve in the transducer to position it to control the air bleed. When exhaust pressure is low and vacuum is high, the air bleed opens and causes the vacuum signal to the EGR valve to be weak. As exhaust pressure increases and manifold vacuum decreases, the air bleed becomes more closed and causes the vacuum signal to the EGR valve to be stronger. In this way the transducer controls how far the EGR valve opens and the resulting EGR flow. This system does not use an EVP sensor.

EGR Vacuum Regulator (EVR) Solenoid. This system, introduced in 1985, operates like the EGR cut-off solenoid system except that the solenoid has a bleed orifice and can bleed off EGR vacuum. It restricts EGR flow during idle, cold engine operation, and WOT operation. It uses an EVP sensor.

Converter Clutch Overdrive (CCO) Solenoid

Vehicles equipped with EEC IV control and an A4LD transmission feature an ECA-controlled torque converter clutch. The ECA operates a solenoid located in the transmission, Figure 7–43. When the ECA is satisfied with operating conditions, it grounds the solenoid circuit. The CCO solenoid then moves a valve in the transmission valve body and thus causes transmission fluid to be routed into the torque converter so as to apply the converter clutch. Its operation is similar to that described in the **Outputs** section of Chapter 3.

In order for the ECA to apply the CCO, the following conditions must exist:

- engine warmed up to normal operating temperature
- engine not under heavy load
- part throttle cruise (engine within a pre-calibrated RPM range)

The solenoid is deactivated when the brakes are applied (BOO switch) or when the throttle position sensor provides a WOT signal.

Turbocharger Boost Control Solenoid

Turbocharged 2.3-liter engines with EEC IV control feature an ECA-controlled, boost control solenoid that allows boost pressure to rise from the normal 10 PSI to as much as 15 PSI. Boost pressure is controlled by a wastegate that opens at the criterion pressure and allows some exhaust gas to bypass the exhaust turbine wheel. The wastegate is controlled by a spring and diaphragm assembly (actuator). Intake manifold pressure is routed to the actuator diaphragm. At the criterion pressure, the dia-

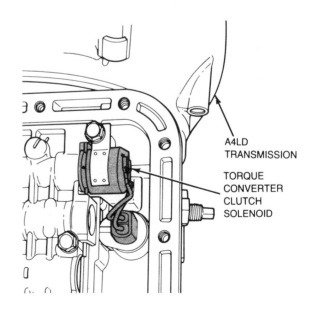

A4LD TRANSMISSION

TORQUE CONVERTER CLUTCH SOLENOID

FIGURE 7–43   Torque converter clutch solenoid. *Courtesy of Ford Motor Company.*

phram compresses the spring and moves a rod that opens the wastegate.

If the ECA is satisfied with existing operating conditions, it will activate the normally closed solenoid and cycle it at 40 Hz. When the solenoid is activated, it bleeds off pressure from the wastegate actuator and allows a higher intake manifold pressure before the wastegate is opened.

A/C and Cooling Fan Control

Some engine applications use a separate electronic control module to control the A/C compressor clutch and the engine cooling fan(s). Others, typically rear-wheel drive vehicles with traditionally mounted engines and belt-driven cooling fans, use a relay to disable the A/C clutch during WOT operation.

A/C and Cooling Fan Electronic Module. The module, Figure 7–44, is usually located in the passenger compartment and gets inputs from:

- the ECA (WOT signal)
- a coolant temperature switch (closes at 105°C [221°F])

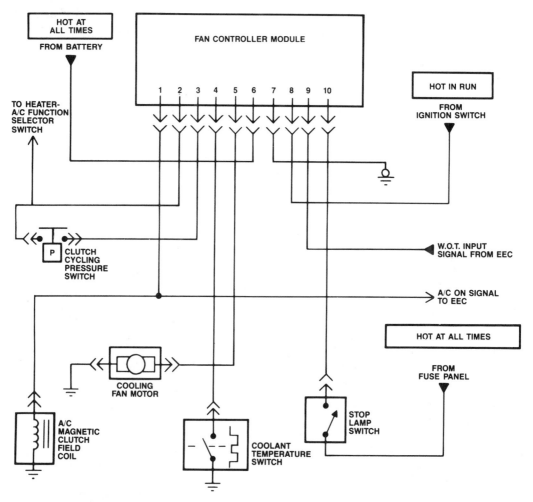

FIGURE 7–44   A/C and fan control module. *Courtesy of Ford Motor Company.*

- the brake switch (brakes applied)
- the A/C switch (A/C on)

The module has a ground terminal and gets power from the battery and the ignition switch.

When the A/C switch is turned on (provided the ignition is on), the module provides power to the compressor clutch. During WOT operation the ECA signals the module, which in turn disables the A/C clutch. If the WOT condition continues, the clutch will only be disabled for thirty seconds before normal A/C operation is resumed. If coolant temperature is below 105°C, the module will also disable the engine cooling fan in response to the WOT signal. If the coolant switch has closed, the module will override the WOT signal from the ECA by refusing to disable the cooling fan.

If the vehicle is equipped with automatic transmission and power brakes, depressing the brake pedal will cause the module to disable both the A/C clutch and the cooling fan for five seconds.

Wide-open Throttle A/C Cut-off Relay. The normally closed WOT cut-out relay is in series with the A/C switch and the clutch coil. During WOT operation the ECA grounds the WOT relay coil. The relay opens and thus disables the A/C clutch.

Inlet Air Solenoid (IAS)

A solenoid is used on some engine applications to control vacuum to the snorkel-mounted actuator that controls the air cleaner snorkel door. This function has traditionally been controlled by a bimetal vacuum valve mounted in the air cleaner. When the engine is cold, the ECA energizes the solenoid and vacuum is routed to the actuator. The actuator moves the door to close the snorkel opening to ambient air. Heated air is drawn, instead, from a shroud around the exhaust manifold, through a duct, to an opening in the bottom of the snorkel. The heated air helps evaporate the fuel as it is introduced to intake manifold.

As the engine and the air around it warm up, the heated air mode is not needed. The ECA deactivates the solenoid, vacuum to the actuator is blocked, and the door moves to open the snorkel's ambient air inlet and close the heated air inlet.

Variable Voltage Choke

On some engine applications that use a feedback carburetor, an ECA-controlled choke relay is used to more precisely control choke on time, Figure 7–45. The choke cover contains an electric heat element that applies heat to the choke spring. When the engine is started at low temperature, the ECA duty cycles the relay every 2.5 seconds. When the relay contacts are closed, battery voltage is applied to the choke heater element. When the engine reaches 27°C (80°F), the ECA holds the relay on continuously (100% duty cycle). This allows the choke to stay on while the engine is at low temperature but turns it off quickly as the engine begins to warm up and can function without it.

Temperature-compensated Accelerator Pump (TCP) Solenoid

The 2.8-liter engine equipped with a 2150A carburetor uses a TCP solenoid to control accelerator pump discharge. When the engine is below 35°C (95°F), the ECA does not ground the solenoid and the accelerator pump functions normally. When the engine coolant temperature reaches the criterion temperature and the engine needs less fuel as the throttle is opened, the TCP solenoid is energized and vacuum is routed to the accelerator pump. With vacuum applied the pump cannot deliver as much fuel to the accelerator pump discharge nozzle.

In 1987 the 4.9-liter engine was introduced with multi-point injection instead of a carburetor. Because the 4.9-liter engine is not a cross-flow engine (the exhaust valve is not across the combustion chamber from the intake valve) the injector is located just above the exhaust mani-

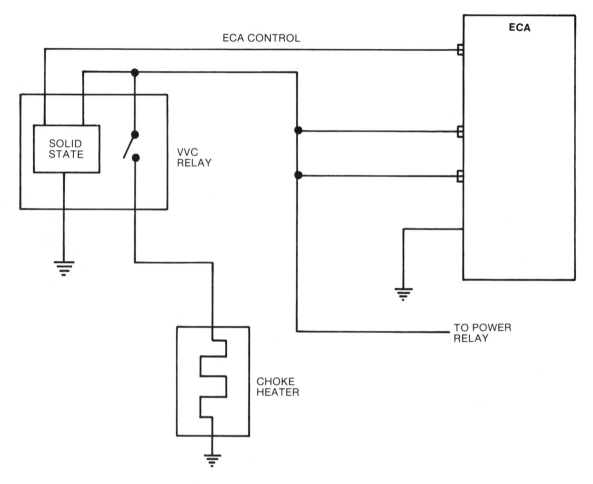

FIGURE 7-45   Variable voltage choke circuit

fold. This exposes the injector and its connecting fuel line to a great deal of heat, especially during the heat soak that follows a hot engine shut off. Exposing the injector to this much heat could result in fuel vaporation in the injector and adjacent fuel line or injector seizing. Either condition could result in hard starting, rough running following start-up or no start at all. To offset this possibility, an injector cooling system was put on the 4.9 liter engine.

A tube that runs almost the length of the engine has been installed along the side of the engine, just above the injectors (Figure 7–46). Jets from the tube point to each of the first five injectors (the number six injector suffers less from exhaust manifold heat). A small blower motor, when turned on, blows air into the tube. The far end of the tube is closed off so the air is directed at the injectors. The fan is powered by an injector blower relay that is controlled by a module (Figure 7–47). The module is connected to a temperature-sensitive switch that is attached to the fuel rail. When the engine is turned off and fuel rail temperature reaches

170° F (77° C), the switch triggers the module, which then turns the fan on. The module will allow the fan to run for a maximum of 15 minutes. It will turn the fan off if the temperature switch opens or if the ignition is turned on before the 15 minutes has expired. This system is not part of, nor is it controlled by, the EEC system.

Shift Light Indicator

Some EEC IV-controlled vehicles with manual transmissions feature a shift indicator light on the instrument panel to indicate to the operator when to shift into the next higher

gear in order to obtain maximum fuel economy. The ECA uses information concerning engine speed, manifold vacuum, and engine temperature to control the light. When the ECA sees an engine speed greater than 1,800 RPM with light engine load, it grounds the indicator light circuit. The ECA will not turn on the light while coolant temperature is low. When the transmission goes into top gear, a switch in series with the ECA and the indicator light opens to prevent the light from coming on again.

VEHICLE SPEED CONTROL (VSC)

On limited applications in 1986, the cruise control system was integrated into the EEC IV system. In 1988, this integration became more universal for EEC IV systems. If the vehicle speed control system is activated, the ECA will compare vehicle speed to commanded speed and will control two solenoids that control vacuum to the speed control servo. The servo controls throttle position. These solenoids—speed control vacuum control (SCVAC) and speed control vacuum vent (SCVNT)—control vacuum to the servo just like the EGRC and EGRV solenoids (discussed earlier in this chapter) control vacuum to the EGR valve.

Top Speed Limiting

On some applications of the 5.0 liter engine beginning in 1988, the ECA is programmed to control top vehicle speed. This is done to limit top speed to a value that is within the rated limits of the vehicle's original equipment tires. It is achieved by reducing injector pulse width.

Programmed Ride Control (PRC). Beginning in 1987, selected Ford vehicles were equipped with a PRC system. A switch on the instrument panel allows the operator to select either a plush or firm ride. The shock absorbers

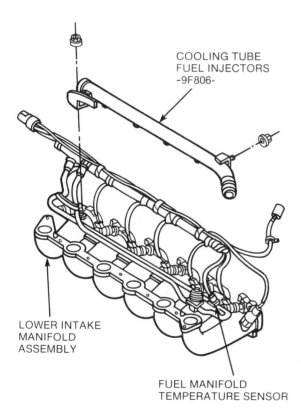

COOLING TUBE FUEL INJECTORS -9F806-

LOWER INTAKE MANIFOLD ASSEMBLY

FUEL MANIFOLD TEMPERATURE SENSOR

FIGURE 7-46   4.9L EFI engine fuel injector cooling tube. *Courtesy of Ford Motor Company.*

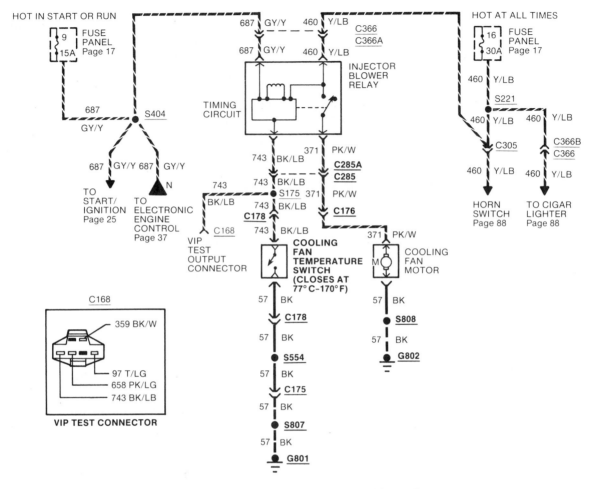

FIGURE 7-47 Cooling fan control circuit. *Courtesy of Ford Motor Company.*

have adjustable damping valves, which are adjusted by rotating a small shaft that extends up to the top of each shock absorber. A small electric motor mounted on the top of each shock absorber turns the shaft. The motor is controlled by a PRC module. If, while operating in the plush mode, certain operating conditions occur, the module will automatically adjust the shock absorbers to the firm position. Some of the information that identifies those conditions is supplied to the PRC module by the EEC IV, ECA.

Information supplied directly to the PRC module from sensors is:

- hard braking—supplied by a pressure switch in the brake system
- hard cornering—supplied by a sensor on the steering shaft

Information supplied by the EEC IV, ECA is:

- more than 90% of WOT
- speeds above 83 miles per hour

SYSTEM DIAGNOSIS AND SERVICE

Malfunction Indicator Light (MIL)

Beginning in 1987, a limited number of Ford Motor Company vehicles were produced with a Malfunction Indicator Light. The MIL is in the instrument panel and works much the same as the Check Engine Light found on General Motors vehicles. When the ECA detects a fault in one of the circuits it monitors, it will turn the light on. In Self-Test the MIL will also blink out the service codes.

The first step for diagnosing an EEC IV system is to verify or identify the driveability complaint, then go to the Diagnostic Routine index in the Diagnostic Routine section of the service manual and find the listing that most nearly identifies the driveability complaint. From that point the procedure provides a step-by-step approach to finding the cause of the problem.

Pay particular attention to the pretest inspection and test preparation steps. Malfunctions in any of the engine's non–EEC IV systems can prevent the EEC IV system from

functioning properly and can cause inappropriate service codes to be set.

Equipment Hookup. All EEC IV–equipped vehicles have a *Self-Test* connector somewhere in the engine compartment, Figure 7–48. This connector can be connected to an analog voltmeter; the *Star Tester,* which can be purchased through a Ford dealer, Figure 7–49; or to one of several other test instruments manufactured by test equipment manufacturers. The ECA displays service codes on the test instrument as a digital readout (or as needle sweeps on the analog voltmeter). A timing light should be hooked up at this time also. Consult the service manual or the test equipment manufacturer's instructions for specific directions on how to conduct the Self-Test.

Service Codes

The ECA monitors most of its input and output circuits. For any given set of driving conditions, it expects to see voltage values within a specified range on each monitored circuit. If the voltage is too high or too low, it will be interpreted as a fault (open or short) and a two-digit service code number representing that fault will be recorded in the ECA's memo-

---

EEC IV Diagnostic Features

One of the outstanding features of the EEC IV system is its self-diagnostic capability. Like other systems, however, it is not totally self-diagnosing. The diagnostic features built into the ECA, combined with the diagnostic procedures in the service manual, will be very effective if the procedures are performed by a knowledgeable technician. The procedures seem confusing at first because of their extent and complexity, but studying and practicing the procedures make them a powerful and effective tool in diagnosing EEC IV system complaints.

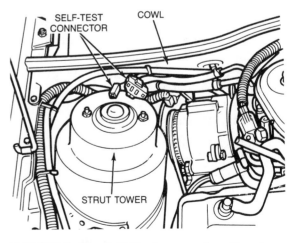

FIGURE 7-48   Self-Test connector. *Courtesy of Ford Motor Company.*

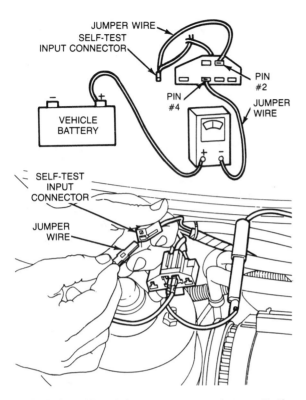

FIGURE 7-49 Voltmeter connection to Self-Test connector. *Courtesy of Ford Motor Company.*

ry. If the fault does not recur within the next twenty ignition cycles, the ECA will erase the recorded service code. The ECA reads the service code to the technician through a special test instrument or an analog voltmeter when it is put into the Self-Test mode.

Types of Codes. Six different types of service codes are stored by the ECA.

- *Fast codes:* These are codes that are useful only to test equipment at the vehicle assembly plant and are used to test each vehicle on the assembly line. They contain the same information as the other service codes but occur much too fast for most test equipment to be able to read them. They can be seen as a slight needle quiver on an analog voltmeter or

a flicker of the LED in the harness of the Star Tester.
- *On-demand codes* (sometimes referred to as Hard Codes): These codes refer to faults that exist at the time the Self-Test is being conducted.
- *Separator code:* This code (represented by the number 10) only occurs during the Key On/Engine Off segment of the Self-Test and indicates to the technician that the On-Demand codes are completed and that memory codes are coming next.
- *Memory codes* (sometimes referred to as intermittent codes or continuous codes): These codes refer to a fault or faults that the ECA detected during normal vehicle operation but that are not present during the Self-Test.
- *Dynamic response code:* This code, which only occurs during the engine running segment of the Self-Test and is also represented by the number 10 (the meaning of code 10 depends on the Self-Test segment in which it appears), tells the technician that he has fifteen seconds to quickly press the accelerator pedal to the floor and let it up. This allows the ECA to check the TPS, MAP, and the VAF for proper response. It is often referred to as the "Goose Test."
- *Engine identification codes:* These codes have no useful purpose for the technician. Like the fast codes, they are intended for testing in the assembly plant. They indicate to the factory test equipment how many cylinders the engine has. The number 20 means four cylinders, 30 means six cylinders, and 40 means eight cylinders.

Reading Service Codes with a Voltmeter.
Reading service codes displayed on an analog voltmeter requires close attention because it is easy to miscount when higher numerical codes are presented. It involves counting needle sweeps and estimating the time lapse between

sweeps. A code 21 (hard code) followed by a code 10 (separator code) and then an 11 (no memory code stored) appears as follows: the needle sweeps upscale and back, then immediately swings up and back again. It swings near full scale on a 12-volt scale. A two-second pause indicates that the first digit is completed and the two sweeps indicate that the digit is 2. After the two-second pause, one sweep occurs and is followed by a four-second pause. The single sweep indicates that the second digit is 1. The four-second pause indicates that the code is completed. After the four seconds, the 21 is repeated. A six- to nine-second pause precedes the separator code, which is shown as a single needle sweep. No needle sweeps occur for the second digit of a code if the second digit is zero. The separator code is not repeated. A second six- to nine-second pause precedes the memory codes. A single sweep followed by a two-second pause and another single sweep indicates a code 11. A four-second pause occurs and the 11 is repeated. The Engine Running segment of the Self-Test is presented in the same format except that a six- to twenty-second pause separates the engine identification code from the dynamic response code and a four- to fifteen-second pause separates the dynamic response code from the hard codes, Figure 7–50.

Quick-Test/Self-Test

The *Quick-Test* is a comprehensive set of procedures designed to help the technician identify problems within the EEC IV system. As presented in the Ford Service Manual, it includes preparation for testing (visual inspection and warming up the engine), Diagnostic Routines for symptoms without codes, the Self-Test, and Pinpoint Tests to locate the exact cause of faults that have been identified. The Self-Test is the portion of the Quick-Test in which the ECA tests itself and its related circuits and presents service codes for any faults it finds.

Key On/Engine Off. This is the first segment of the Self-Test. It is activated by connecting the single-lead portion of the Self-Test connector to pin 2 of the multiple-lead part of the Self-Test connector and turning the igni-

SELF-TEST OUTPUT CODE FORMAT
ENGINE-RUNNING

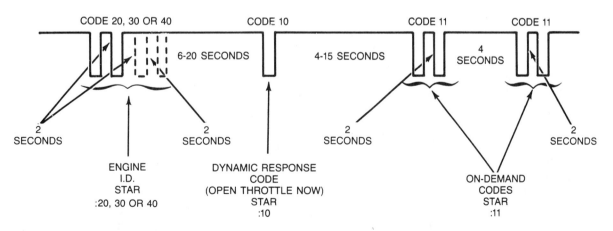

FIGURE 7–50 Typical service code display during Engine Running segment of Self-Test. *Courtesy of Ford Motor Company.*

tion on. If a voltmeter is being used to read codes, this will be done with a jumper lead; if a digital readout instrument such as the Star Tester is being used, this connection will be made by manipulating a button or switch as per test instrument instructions. In this segment the ECA tests first the input circuits. Any faults detected are reported as On-Demand Codes. A code 11 (system pass code) indicates that no input circuit faults have been found. Any Hard Codes that are identified should be repaired using the appropriate service manual before proceeding to the Computed Ignition Timing Check. After the repair is made, repeat this test segment to verify the repair.

Continuous Self-Test (still in Key On/Engine Off segment). The On-Demand Code or the code 11 are followed by the separator code, (10), which means that the On-Demand Codes are completed and memory codes (intermittent faults) are coming next. If no memory codes are stored, a code 11 will be displayed. If memory codes are displayed, they should be recorded by the technician; but no attempt to repair them should be made at this point.

Output Cycling Test (still in Key On/Engine Off segment). At this point, if the throttle is moved to WOT and released, the ECA will go into the Output Cycling Test in which it will turn most of the actuators on. If the throttle is moved to WOT and released again, it will turn the actuators off. This can be repeated as many times as desired. This step, however, is not normally used unless called for by one of the Pinpoint Routines, a later part of the Quick-Test. If the Output Cycling Test is called for in one of the Pinpoint Routines and to determine if one of the actuators is working properly, the Key On/Engine Off Test can be re-entered. It is only in this segment that the output cycling test can be initiated.

Computed Timing Check. This is the second segment of the Self-Test and is designed to test the ECA's ability to control ignition timing. It is initiated by having the ignition off for at least ten seconds, activating the Self-Test, and

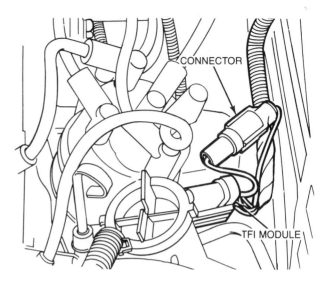

FIGURE 7–51   In-line base timing connector. *Courtesy of Ford Motor Company.*

starting the engine. Any codes displayed at this time should be ignored. In this segment the ECA holds the timing at 20° above base timing for two minutes. If, for example, the base timing is 10° BTDC, the timing will read 30° BTDC. Check the timing with the timing light that was connected while the test equipment was being hooked up.

Be sure that the in-line base timing connector, Figure 7–51, is reconnected (from having checked base timing) before the Computed Timing Check is started.

Engine Running Test. The third segment of the Self-Test tests actuators (output devices) and some sensor circuits for hard faults under dynamic conditions. Before going into this segment, the Self-Test should be deactivated and the engine started and run at 2,000 RPM for two minutes to warm the EGO sensor. Turn the engine off and wait ten seconds. To activate the Engine Running Segment, activate the Self-Test and restart the engine. The next display to appear is the Engine Identification Code followed by the Dynamic Response Code (10).

The technician now has fifteen seconds to move the throttle to WOT and release it. The ECA next issues fast codes followed by the On-Demand Codes. If no hard faults are observed during the Engine Running segment, a code 11 will be issued.

Continuous Monitor Test. This test, which is the last of the Self-Test segments and is often called the *Wiggle Test,* is used to locate intermittent faults. It can be used in either of two modes: Engine Off or Engine Running. To initiate the Engine Off Mode, turn the ignition on with the test instrument connected but the Self-Test deactivated. Begin wiggling (pulling, twisting, etc.) the wires of the sensor circuit indicated by the intermittent fault code obtained during the Key On/Engine Off segment. Bump the sensor; manipulate any of its moving parts; apply heat from a heat gun to any thermistors involved, and so on. If this manipulation reproduces the intermittent open or short to which the ECA was responding when it set the code, the LED in the Star Tester harness will either flicker off or stay off. If a voltmeter is being used, the needle will pulse or swing upscale and stay there.

The Engine Running Mode of the Wiggle Test is automatically entered two minutes after the last code was displayed in the Engine Running segment of the Self-Test if the engine is not turned off.

Cylinder Balance Test. This is a feature of the Self-Test on the sequential injection 5.0 liter engine that tests cylinder balance by momentarily stopping injection on each cylinder, one at a time, while the ECA records the RPM loss of each cylinder. The number of any cylinder that does not cause a significant RPM loss will be displayed on the test instrument connected to the test lead. In 1988 this feature was improved so that a less severe cylinder miss could be identified. Rather than responding only to a larger RPM drop as the earlier test strategy did, it now has the capability of making the same test in three passes with each successive pass responding to a smaller RPM loss. After each pass is completed, the operator must depress the accelerator to a part throttle position to initiate the next pass.

Pinpoint Procedures. This is the last part of the Quick-Test. It provides directed steps to locate the exact cause of faults that have resulted in codes being set. If the Quick-Test steps are being followed in the applicable service manual, a *Pinpoint Procedure* will be referred to for each service code requiring corrective action. The Pinpoint Procedures also provide directions that aid in finding the cause of driveability problems that do not have accompanying service codes. The symptoms of such problems are listed with suggested things to check.

Repairs should always be made in the order in which their codes are presented. After the repair the Self-Test should be repeated. The applicable service manual should be used for each vehicle application to be sure that all of the Quick-Test steps match the engine, vehicle, and model year. For instance, the 1983 1.6-liter EFI engine does not store memory codes.

Failure Mode Effects Management (FMEM). FMEM is a strategy programmed into the ECA, beginning in mid-1986, that is intended to maintain a good engine performance in the event of a failure of one of the most critical sensors. Prior to 1987, such a sensor failure would either stop the engine or put the system into Limited Operational Strategy in which the engine would run with no-spark advance and a fixed air/fuel mixture. With FMEM Strategy, the ECA will in most cases replace the failed sensor's input with a substitute value from another sensor. It will also set a code 98 in its Continuous Self-Test memory. Code 98, for 1987, indicates that a fault exists and that FMEM is operating.

The operator may not know that a fault exists, except for the following:

- idle speed may be fixed
- the cooling fan may be on all the time

Failed Sensor	Action Taken
MAP	Substitute TPS value. If TPS is faulty, use a fixed constant value. Disable Adaptive Learning. Disable EGR. Fixed idle speed.
TPS	Substitute fixed values during cranking, MAP value after.engine starts.
ECT	Substitute ACT value during warm-up, fixed value after warm-up. Disable EGR. Disable Adaptive Learning. Fixed idle speed.
ACT	Substitute ECT value during warm-up, fixed value after warm-up. Disable EGR. Disable Adaptive Learning. Fixed idle speed.
EVP	No substitute. Disable EGR.

FIGURE 7-52   FMEM strategy

- California vehicles are equipped with a Malfunction Indicator Light on the instrument panel that will come on when the fault occurs

All 1987 and later EEC IV vehicles feature FMEM Strategy. Figure 7–52 shows the sensors included in FMEM Strategy and the action taken in response to each one's failure.

Self-Test Improvements. Several improvements have been made in the Self-Test over the years. Some new codes have been added and some codes have been made more reliable. The Self-Test and the continuous monitor have the ability to check more functions. As a result, code numbers may not always have the same meaning from year to year.

Service Tips

The following suggestions should be adhered to when servicing an EEC IV system:

- Pinpoint Procedures shall not be used unless you are directed to do so by the

Quick-Test. Each Pinpoint Procedure assumes that a fault has been detected in the circuit being considered, and to use them without being directed to by the Quick-Test can result in unnecessary repair procedures and replacement of nondefective parts. Do not replace parts unless specific test procedures indicate that they are defective.

- Do not measure voltage or resistance or connect a test light at the ECA harness connector unless directed to do so by a specific test procedure.
- Isolate both ends of a circuit and turn the ignition off before testing for shorts or continuity unless directed otherwise by a specific test procedure. Treat solenoids and switches the same way.
- An open is defined as any resistance that reads greater than 5 ohms, and a short is defined as any path to ground that reads less than 10,000 ohms, unless otherwise specified.

MAP/BP Relocation. Beginning in 1987, MAP and/or BP sensors have to be repositioned so that they sit in a horizontal position with the vacuum line sloping downward to its manifold connection. This is to prevent fuel migration into the sensor. Fuel getting into these sensors has been responsible for many sensor failures.

REVIEW QUESTIONS _____

1. Name three types of fuel-metering devices used on the EEC IV system.

2. In what part of the ECA are the memory codes stored?

3. What type of signal does the MAP sensor produce?

4. Name at least three sensors used on EEC IV systems that are potentiometers.

5. Name at least three sensors used on EEC IV systems that are thermistors.

6. Name at least three sensors used on EEC IV systems that are switches.

7. Name at least one actuator on an EEC IV system that is duty cycled and one operated by pulse width.

8. What is the purpose of the IMS (inferred mileage sensor)?

9. What is the function of the PIP signal?

10. What is the function of the SPOUT signal?

11. Name three types of idle speed devices used on EEC IV systems.

12. What is the function of the TAD solenoid?

13. Although the EGR valve, Thermactor air, and the canister purge valve influence engine performance, they have a greater impact on _____ .

14. What is the function of the VVC (variable voltage choke)?

15. What is the function of the IAS (inlet air solenoid)?

16. What is the function of the TCP (temperature-compensated accelerator pump)?

17. List the order in which readable codes are presented during the Engine Running segment of the Self-Test.

18. Name two points during the Self-Test when the Wiggle Test can be initiated.

19. Name two different criteria for using the Pinpoint Test.

ASE-type Questions. (Actual ASE test questions will probably not be as product specific.)

20. Technician A is working on a vehicle equipped with an EEC IV system. He accidently reverses the polarity while hooking up a set of jumper cables. He fears that he has damaged the ECM. Technician B says that a diode in the power relay probably prevented damage from occurring to the ECA. Who is most likely correct?
   a. A only
   b. B only
   c. both A and B
   d. neither A nor B

21. Technician A says that if the PIP signal is lost on an injected EEC IV system, the engine will not run. Technician B says that if the SPOUT signal is lost, the engine will operate on base timing. Who is correct?
   a. A only
   b. B only
   c. both A and B
   d. neither A nor B

22. Of the four steps listed here (Diagnostic Routine, verify complaint, Pinpoint Procedure, Self-Test), technician A says they should be performed in the order in which they are listed. Technician B says they should be performed in the following order: verify complaint, Self-Test, Diagnostic Routine, and Pinpoint Procedure. Who is correct?
   a. A only
   b. B only
   c. both A and B
   d. neither A nor B

23. During the Key On/Engine Off portion of the Self-Test, technician A says that On-Demand Codes are presented followed by a separator code (10), which is followed by memory codes. Technician B says the On-Demand Codes represent faults currently present and that the memory codes represent intermittent faults. Who is correct?
   a. A only
   b. B only
   c. both A and B
   d. neither A nor B

24. Technician A says that of all the steps of the Self-Test, there are three that are intended to be used at the discretion of the technician or if directed to by one of the

Pinpoint Procedures. Technician B says that she can only recall two: the Output Cycling Test and the Wiggle Test. Who is correct?

a. A only
b. B only
c. both A and B
d. neither A nor B

GLOSSARY

A4LD TRANSMISSION   A light-duty, four-speed, overdrive transmission with a computer-controlled torque converter clutch for rear-wheel drive vehicles.

DIAGNOSTIC ROUTINE   A service manual section that lists possible causes of an identified driveability complaint. It is the beginning point of the Diagnostic Procedure once the complaint has been verified.

PINPOINT PROCEDURE   A service manual section that gives specific directions to find the solution of identified faults. This section should not be used until either the Diagnostic Routine or the Self-Test has identified it as the next step.

QUICK-TEST   The whole set of procedures that make up the diagnostic procedure.

SELF-TEST   The self-diagnostic portion of the quick-test in which the computer tests itself and its circuits for faults.

STAR TESTER   A tester designed to test MCU and EEC IV systems that activates and deactivates the Self-Test procedure and displays a digital readout of the service codes.

THERMACTOR   The name Ford Motor Company uses to identify its air injection pump for exhaust emissions reduction.

THICK FILM INTEGRATED (TFI)   A term that identifies the type of chip used in the ignition module that is attached to the side of the distributor for EEC IV system applications.

FORD'S EEC I, EEC II, AND EEC III

Objectives

After studying this chapter, you will be able to:

- list the engine applications of the EEC I, II, and III systems.
- identify the controlled functions of the EEC I system.
- identify the controlled functions of the EEC II system.
- identify the controlled functions of the EEC III system.
- describe the self-diagnostic capability of the EEC I and II systems.

EEC I was introduced in 1978 and used on the 1978 and 1979 5.0-liter Lincoln Versailles. EEC II was introduced in 1979 but was limited to full-size Fords sold in California and Mercurys sold in all fifty states with 5.8-liter engines. EEC III was introduced in 1980 with expanded application but was still limited to V8 engines. In 1981 central fuel injection replaced the feedback carburetor as part of the EEC III system on some engine applications. In 1982 EEC III was introduced on some light truck applications.

The component parts of each of these earlier EEC systems are listed in this chapter. Many of them, however, are incorporated in the EEC IV system and were discussed in Chapter 6. Only the components or functions that are different from those found in the EEC IV system are discussed here.

ELECTRONIC ENGINE CONTROL (EEC) I

When considering computerized engine control systems, the most significant factor about the EEC I system is that it does not control air/fuel mixture. The engine is fitted with a variable venture (VV) carburetor, model 2700, which has no capability for electronic mixture control. No EGO sensor is used, and there is no closed-loop mode.

Electronic Control Assembly (ECA)

The ECA is similar to that used in the EEC IV system except that it:

- was designed to perform fewer functions.
- has no self-diagnostic capability.

- has an externally mounted calibration assembly that calibrates the unit to specific vehicle values such as curb weight, axle ratio, high altitude use, and so forth, Figure 8–1.
- has a fuel octane adjustment switch that can be set to retard ignition timing either 3° or 6° if spark knock occurs (on EEC IV this feature was located on the distributor).
- provides a reference voltage of about 9 volts to its sensors.

Default Mode. If a failure occurs in the ECA, it stops sending commands to the three actuators it controls. The ignition timing stays at base timing, the thermactor system dumps its air into the atmosphere, and the EGR valve is left in the closed position.

As with the EEC IV system, the ECA is powered by a power relay that also protects it

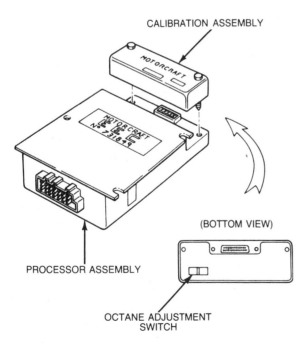

CALIBRATION ASSEMBLY

(BOTTOM VIEW)

PROCESSOR ASSEMBLY

OCTANE ADJUSTMENT SWITCH

FIGURE 8–1   EEC I ECA and calibration assembly. *Courtesy of Ford Motor Company.*

from reversed polarity. Both the ECA and the power relay are located on the passenger compartment side of the instrument panel, near the brake pedal.

Inputs

The ECA receives inputs from the following sensors:

- engine coolant temperature (ECT) sensor
- manifold absolute pressure (MAP) sensor
- barometric pressure (BP) sensor
- throttle angle position (TAP) sensor
- crankshaft position (CP) sensor
- inlet air temperature (IAT) sensor
- EGR valve position (EVP) sensor

The ECT, TAP, and EVP are essentially the same units as those used on the EEC IV system, although we find the word *angle* added to the name of the throttle position sensor. The IAT sensor is located in the air cleaner and is basically the same sensor that the EEC IV uses as either an ACT or a VAT.

MAP Sensor. The MAP sensor consists of an *aneroid* that moves the wiper of a potentiometer. The aneroid is an accordionlike capsule that contains a gas (so it can be compressed and expanded) at atmospheric pressure. As the pressure outside the aneroid increases, it compresses; as the outside pressure decreases, the aneroid expands. The aneroid is placed in a housing connected to manifold vacuum, Figure 8–2. As manifold vacuum changes, the expansion or contraction of the aneroid moves the wiper of the potentiometer to provide a signal to the ECA that is proportional to manifold pressure.

BP Sensor. The BP sensor works like the MAP sensor except that it is open to atmospheric pressure instead of being connected to manifold vacuum.

CP Sensor. The CP sensor consists of a four-lobe pulse ring on the back of the crankshaft and a probe with a permanent magnet and a coil of wire at its tip. Essentially the pick-

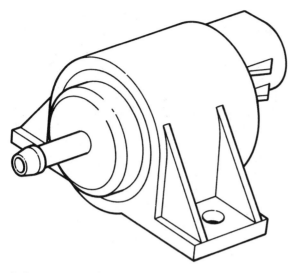

FIGURE 8-2   Aneroid manifold absolute pressure sensor. *Courtesy of Ford Motor Company.*

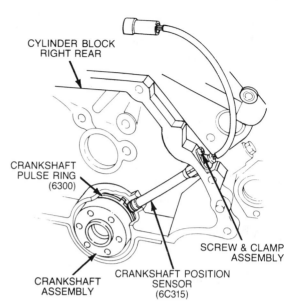

CYLINDER BLOCK
RIGHT REAR

CRANKSHAFT
PULSE RING
(6300)

SCREW & CLAMP
ASSEMBLY

CRANKSHAFT POSITION
SENSOR
(6C315)

CRANKSHAFT
ASSEMBLY

FIGURE 8-3   Crankshaft position sensor. *Courtesy of Ford Motor Company.*

up coil has been removed from the distributor and placed at the rear of the crankshaft, Figure 8-3. The probe is inserted into the rear of the block so that a small air gap exists between its tip and the pulse ring. As the crankshaft turns and as each lobe passes under the magnet in the probe tip, the air gap is suddenly made even smaller. This causes a sudden increase in the strength of the magnetic field being emitted from the pole of the magnet. The sudden change in the magnetic flux causes a voltage signal to be produced in the coil wound around the magnet. This steady stream of signals is transmitted to the ECA, and each one means that a piston is at 10° before top dead center.

Outputs

The EEC I system controls three functions:

- ignition timing
- Thermactor air control
- EGR flow control

Ignition Timing. To calculate ignition timing, the ECA evaluates coolant temperature, inlet air temperature, engine speed, throttle position, and engine load (intake manifold pressure). If temperatures are low, more spark advance will be provided; if they are high, less will be provided. For example, when inlet air temperature exceeds 90° F, spark advance decreases to avoid spark knock. If, however, the engine starts to overheat during idle, spark advance will be increased to raise engine speed and provide more coolant flow. As throttle opening increases, especially during rapid increases, spark advance is decreased. Spark advance is also decreased in response to an increase in manifold pressure even though throttle position does not change as might occur if the vehicle started up a hill and thus caused the engine to lose speed. The ECA's spark timing command is sent to the standard Duraspark II ignition module, which, upon receiving the signal, interrupts primary coil current flow to fire the appropriate spark plug. The ECA can provide as much as 30° of spark advance as opposed to the approximately 20° provided by pre-EEC systems.

A newly designed distributor that directs the spark from the coil to the appropriate cylinder is also used. The new distributor housing features a notch in its mounting flange that can index on a lug on the engine block so that once installed the distributor's base timing is set and cannot be changed, Figure 8-4. An enclosure for mounting a future miniaturized ignition module is also added. (This feature did not carry over to EEC III and was not used until the EEC IV system, when it was redesigned.)

To prevent crossfiring inside the distributor cap, the cap and rotor were given a unique design. (The increased spark advance, without the rotor being advanced as would occur if a centrifugal advance was used, increased the possibility of crossfire. This design was not carried over to the EEC IV system.) Alternate spark plug terminals in the cap extend down to a lower level. The rotor, Figure 8-5, is equipped with two electrode blades, a high one at one end and a low one at the other. The other end of each electrode blade forms a pickup arm that provides the same function as the spring contact found on most rotors. The center terminal, which receives input voltage from the coil, transmits the voltage to a center electrode plate instead of directly to the rotor, Figure 8-4.

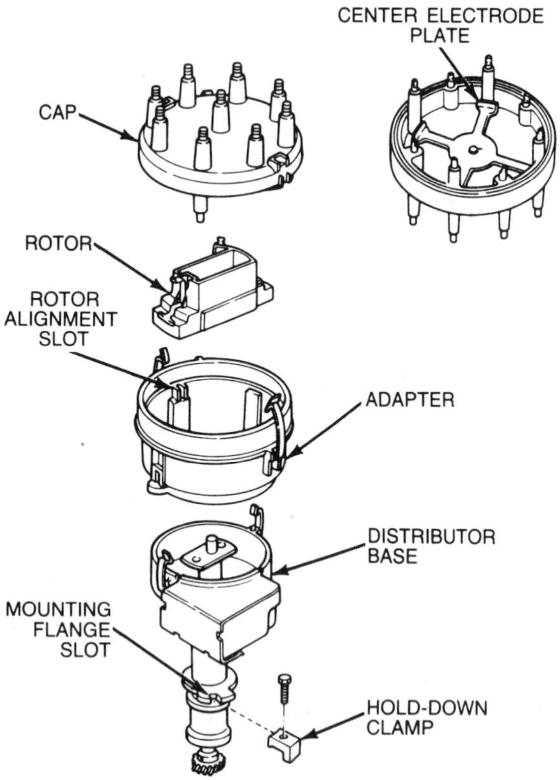

FIGURE 8-4   EEC I distributor assembly. *Courtesy of Ford Motor Company.*

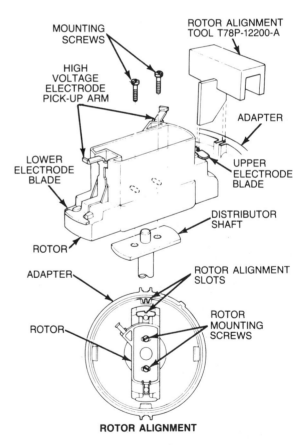

FIGURE 8-5   EEC I distributor rotor. *Courtesy of Ford Motor Company.*

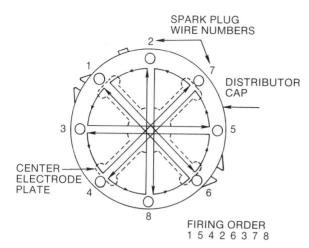

SPARK PLUG
WIRE NUMBERS

DISTRIBUTOR
CAP

CENTER
ELECTRODE
PLATE

FIRING ORDER
1 5 4 2 6 3 7 8

THE SOLID ARROWS INDICATE
A WIRE TERMINAL BEING
FIRED. THE HOLLOW ARROWS
INDICATE THE DIRECTION OF
ROTOR ROTATION.

FIGURE 8–6   Cylinder firing pattern of EEC I
distributor

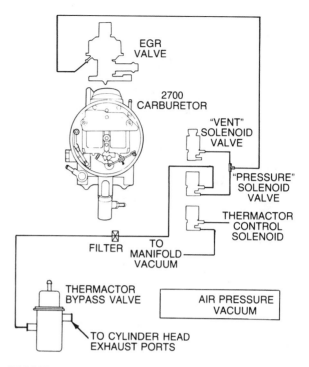

FIGURE 8–7   EGR and Thermactor system
control. *Courtesy of Ford Motor Company.*

From there the voltage is transmitted through a small air gap to the pickup arm, which is aligned with the center electrode plate.

By using this design, the next spark plug to be fired is almost directly across the cap (135°) from the one just fired instead of next to it. For example, let's assume that number one cylinder has just been fired from the lower electrode blade, Figure 8–6. The firing order is 1-5-4-2-6-3-7-8, and the distributor turns counterclockwise. The rotor turns 45° counterclockwise, and its lower electrode blade is pointed to the number three terminal. Its pickup arm, however, is between the spokes of the center electrode plate and does not receive the voltage surge from the coil. However, because the pickup arm for the upper electrode blade at the opposite end of the rotor is offset 45° from the rotor, it is aligned with the center electrode plate and picks up the coil voltage and transmits it to terminal number five through a second air gap. (The addition of a second rotor gap should add 2 killivolts to 3 killivolts to the required firing voltage.) Spark plug voltage continues to be delivered alternately from opposite ends of the rotor each 45° of rotor rotation.

Thermactor Air Control. The EEC I system uses a single-bed catalytic converter. The Thermactor system is therefore not designed to supply air to the converter during warm engine operation. Air is supplied to the exhaust manifolds during low-temperature operation and is dumped into the atmosphere after normal operating temperature is reached. The ECA controls these two operational modes with a solenoid-operated vacuum control valve (Thermactor control solenoid). This is a normally closed valve that controls manifold vacuum to the Thermactor bypass valve, Figure 8–7. The Thermactor pumps air to the bypass valve.

When underhood temperature is low (determined by air temperature in the air cleaner),

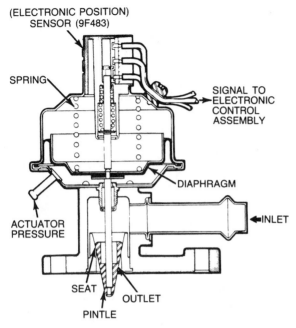

FIGURE 8-8   Pressure-operated EGR valve. *Courtesy of Ford Motor Company.*

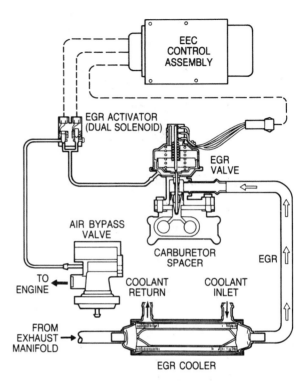

FIGURE 8-9   EGR cooler assembly. *Courtesy of Ford Motor Company.*

the ECA grounds the Thermactor control solenoid circuit. The solenoid opens the vacuum valve, and vacuum is routed to the bypass valve. The bypass valve then routes Thermactor air into the exhaust manifolds. When the engine warms up, the ECA de-energizes the solenoid. Vacuum is blocked from the bypass valve, which then dumps Thermactor air to get rid of it. The ECA also dumps during WOT operation.

EGR Flow. The EEC I EGR valve is opened by air pressure instead of vacuum, Figure 8-8. The pressure is supplied by the Thermactor from a port on the bypass valve, Figure 8-9. Exhaust gases are supplied to the EGR valve by an external tube that connects to an exhaust manifold. To improve driveability characteristics and to extend EGR valve life, the temperature of the EGR gases is reduced by routing the supply tube through a heat exchanger (EGR cooler). Coolant from a heater hose is circulated through the EGR cooler.

EGR flow is controlled by the ECA through two solenoid-operated valves, the normally closed pressure valve and the normally open vent valve, Figure 8-10. The ECA monitors sensor input information and determines what EGR flow rate is needed. For example, assume that an increased EGR flow is needed. The ECA energizes both solenoids. The normally closed pressure valve opens to allow pressure from the Thermactor system to be applied to the lower side of the EGR valve. The normally open vent valve closes to prevent the pressure from escaping to the atmosphere. The EGR valve opens wider until the EVP sensor indicates that the EGR valve has reached the desired position. When this occurs the ECA de-energizes the pressure solenoid. It closes and causes the existing pressure under the EGR valve diaphragm to be trapped and the EGR valve to be held in a fixed position. This condition is maintained until operating conditions change (throttle position changes, intake manifold pressure changes, and so forth) or until some of the trapped pressure leaks out and the EVP sensor indicates to the ECA that the EGR valve is no longer in the desired position. If the ECA wants to decrease EGR flow, it will de-energize both solenoids and thus close the pressure source and vent existing pressure. As barometric pressure decreases, EGR flow is reduced. EGR flow is cut off when coolant temperature is below 21° C (70° F) or above 110° C (230° F). Normally each of the two control solenoids is cycled at a rapid rate to maintain desired EGR valve position.

System Diagnosis and Service

The ECA used in the EEC I system has no self-diagnostic capability. Special test equipment and procedures have been developed and are listed in the service manual. To conduct tests special tester Rotunda No. T78L-50-EEC-I (or its equivalent) is placed in series in the ECA harness (a disconnect in the harness is provided for this). All signals going into or out of the ECA must now pass through the tester. A special DVOM (Rotunda No. T78-50) connects to the first tester and provides readings for each test in the test sequence.

The first of two test sequences is performed after the engine has been warmed up and with the engine ignition off. Eight test readings are taken to check the condition of the sensors, the power relay, the Thermactor control solenoid, and part of the ECA's circuitry. In the second test sequence, the technician checks the spark advance, EGR valve position, and Thermactor air mode with the engine running at 1,600 RPM in park with vacuum applied (from an external source) to the MAP sensor. This sequence checks the remaining ECA circuitry.

The results of each reading and check are recorded and compared to test result limits provided with the tester. If all results are within limits, the system is functioning properly; if not, diagnostic procedures are provided in the service manual for finding the fault within the identified circuit.

Rotor Alignment. Proper alignment of the distributor rotor is critical to proper engine performance. Having the rotor out of alignment with the distributor cap terminal does not affect timing but does shorten spark duration and

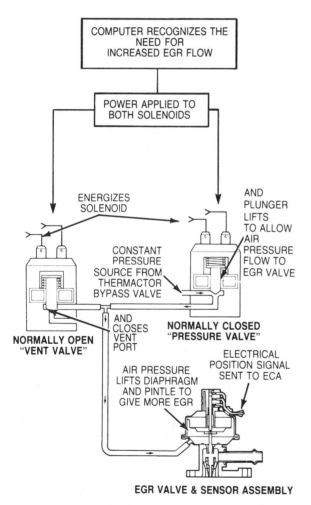

FIGURE 8-10   EGR control system. *Courtesy of Ford Motor Company.*

produce poor driveability. Notches in the upper electrode blade and the distributor body are provided so that a special tool, Figure 8-5 can be used to align the rotor. This is true for EEC I, EEC II, and EEC III systems even though the EEC III distributor uses a slightly different rotor design.

Rotor Alignment

Technicians who regularly service Ford products have learned that correct rotor alignment is critical to good engine performance. It is one of the first things they check in response to complaints of poor engine performance on vehicles with EEC I, II, or III systems.

EEC II

The EEC II system is similar to EEC I; however, it has some minor differences in its sensors and significantly more responsibility. The ECA and power relay are the same except for more internal circuitry in the ECA for its additional functions. It also provides a 9-volt reference voltage to its sensors. With the addition of an electronically controlled air/fuel mixture control device and an exhaust oxygen sensor, the system can operate in a closed-loop mode. The default mode is now called *limited operational strategy* (LOS), as we see in EEC IV. In this mode (caused by an electrical failure) all actuators are left in their deactivated mode.

Inputs

The sensors for EEC II include:

- engine coolant temperature (ECT) sensor
- manifold absolute pressure (MAP) sensor
- barometric pressure (BP) sensor
- throttle position (TP) sensor
- crankshaft position (CP) sensor

- exhaust gas oxygen (EGO) sensor
- EGR valve position (EVP) sensor

ECT. All temperature-related information is taken from the ECT, which has not changed; and the IAT is not used on EEC II.

MAP and BP Sensors. The MAP and BP sensors work like those on the EEC I system except that they are both located in one housing called a *B/MAP sensor,* Figures 8-11 and 8-12.

TP Sensor. The TP sensor is the same except that the word *angle* has been dropped from its name.

CP Sensor. The CP sensor is the same as that on the EEC I except that it has been moved to the front of the crankshaft and is shielded to protect it from electrical interference, Figure 8-13.

EGO Sensor. The EGO sensor first appeared on the EEC II systems and has been carried over to EEC III and EEC IV.

EVP Sensor. The EVP sensor has not changed.

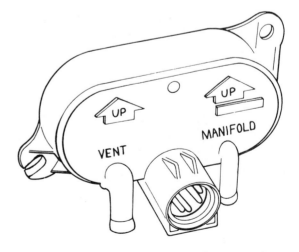

FIGURE 8-11   BARO and MAP sensor. *Courtesy of Ford Motor Company.*

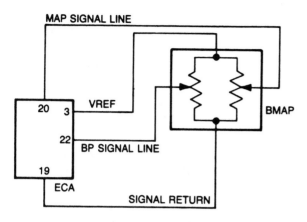

FIGURE 8-12   BARO and MAP sensor circuit. *Courtesy of Ford Motor Company.*

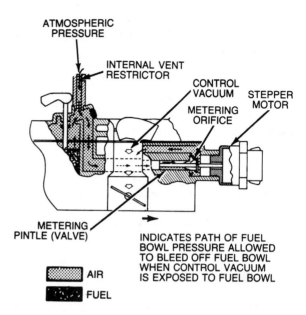

FIGURE 8-14   7200 feedback carburetor, back-suction system. *Courtesy of Ford Motor Company.*

Outputs

The functions under the control of the ECA in the EEC II system are:

- air/fuel ratio
- ignition timing
- idle speed control
- EGR flow
- Thermactor air control
- canister purge

All of these functions are either new to the EEC system or are at least slightly different from their application on the EEC I system.

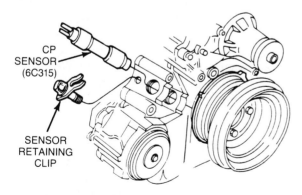

FIGURE 8-13   EEC II crankshaft position sensor. *Courtesy of Ford Motor Company.*

Air/Fuel Ratio. The fuel bowl of a carburetor must be vented through either a balanced vent or an external vent to atmospheric pressure. The atmospheric pressure forces the fuel from the bowl through either the main metering or idle speed system and into the airstream moving through the carburetor throat. The Model 7200 VV carburetor used on EEC II applications is designed to control the pressure in the fuel bowl and thus to control the amount of fuel forced into the airstream. This is done by putting a restrictor in the bowl vent and then providing a vacuum passage that allows carburetor vacuum (called control vacuum in this case) to draw air (pressure) from the bowl area, Figure 8-14.

The amount of pressure drawn from the fuel bowl area is controlled by a pintle valve controlled by a stepper motor. As the stepper motor extends the pintle valve, the vacuum passage is restricted. This permits more pressure to develop in the fuel bowl, and the air/fuel mixture becomes richer. When the pintle valve is retracted, fuel bowl pressure is reduced and the air/fuel mixture becomes leaner.

The stepper motor contains four separate coils that can be selectively energized by the ECA to make it run forward or backward. The coils, however, do not get a steady voltage applied to them. Instead they get a series of short pulses. Each pulse rotates the armature a specific number of degrees. As the armature turns, the pintle valve is either extended or retracted depending on which way the armature turns. The ECA can then issue pulses to the stepper motor to achieve whatever air/fuel ratio it wants. The pintle has an extension range of 0.4 inch, and 120 pulses are requried to move it through its range.

Ignition Timing. The EEC II ignition system features the same coil and distributor as does EEC I and a new *Duraspark* III ignition module. The Duraspark III module was designed specifically to work with the ECA.

Idle Speed Control. The idle kicker solenoid and idle kicker actuator first appeared on EEC II and were carried over to EEC III and some applications on EEC IV. They are discussed in Chapter 6.

EGR Flow. The EGR control solenoids, EGR vent (EGRV), and EGR control (EGRC) function just like those in EEC I; however, the EGRP (P for pressure) solenoid used in EEC I was renamed. The word *pressure* was replaced by *control.* This name change occurred because the EGRC solenoid on the EEC II system is connected to manifold vacuum and the EGR valve is opened by vacuum rather than Thermactor system pressure. This system was carried over to EEC III and some EEC IV applications.

Thermactor Air Control. The EEC II system uses a dual-bed catalytic converter and therefore requires an additional Thermactor system mode: the ability to direct air to the converter in addition to being able to direct it to the exhaust manifold or to the atmosphere. To do this, two control solenoids and a combina-

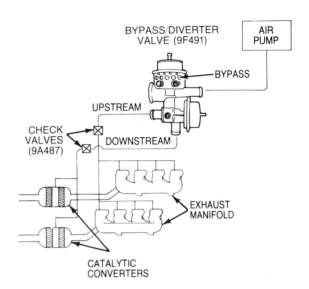

FIGURE 8–15   EEC II Thermactor control system. *Courtesy of Ford Motor Company.*

tion air management valve are used: a Thermactor air bypass (TAB) solenoid, a Thermactor air divert (TAD) solenoid, and a combination TAB/TAD air control valve, Figure 8–15. This system was carried over to EEC III and EEC IV and is discussed in Chapter 6.

Canister Purge Solenoid. This is a new function for EEC II. It consists of a normally closed solenoid-controlled vacuum valve. The valve is placed in the purge hose that connects the canister to the intake manifold. When the ECA grounds the solenoid's circuit, it opens and allows the canister to be purged. This canister purge system was carried over to EEC III and some applications of EEC IV and is discussed in Chapter 6.

System Diagnosis and Service

Like EEC I the EEC II system has no self-diagnostic capability, and its diagnostic procedure is similar to that of EEC I. It requires the use of special test equipment, listed in the service manual, or its equivalent.

EEC III

On carbureted EEC III applications, we see the same components as we do on the EEC II. The air/fuel mixture control device on the carburetor and the ECA have, however, been modified. The ECA is becoming more sophisticated, and we see the addition of electronic fuel injection to some EEC III applications. The fuel injection system was the throttle body type that Ford most often called central fuel injection. In literature pertaining to EEC III systems, the terms *electronic fuel injection* and *central fuel injection* are used interchangeably. In literature pertaining to EEC IV systems, electronic fuel injection usually refers to multipoint fuel injection.

ECA

For EEC III the ECA is designed to operate in one of three operating strategies: the base engine strategy, the modulator strategy, and the limited operational strategy.

The base engine strategy covers normal operating conditions, which are divided into four modes:

- cranking
- closed throttle operation
- part throttle operation
- wide-open throttle operation

The ECA determines the operating conditions that are occurring from sensor input and selects the appropriate actuator commands according to its specific vehicle calibration programming.

The modulator strategy is used to compensate for conditions that require more extreme calibration commands. These conditions are:

- cold engine
- overheated engine
- high altitude

The limited operational strategy is like that of the EEC II system on carbureted vehicles. For CFI vehicles the LOS keeps the injectors operating but at a fixed pulse rate that produces a full, rich condition.

Inputs

The sensors for EEC III receive a 9-volt reference from the ECA. The only difference between the sensors for EEC III and those for EEC II is that the EEC III system equipped with CFI has an air charge temperature sensor screwed into the number seven intake manifold runner. The ACT sensor was carried over to some EEC IV applications and is described in Chapter 6.

Outputs

With the following exceptions, all of the actuators for the EEC III system are exactly like those used for EEC II.

EEC III 7200 VV Air/Fuel Mixture Control. The method of fuel mixture control is a little different on EEC III than on EEC II. A stepper motor and pintle valve are used in the same way (the stepper motor looks a little different), but the pintle valve does not control fuel bowl pressure. Instead the pintle valve controls the amount of air that is allowed to bleed into the main metering system, a variable main metering air bleed, Figure 8–16. Air is allowed to

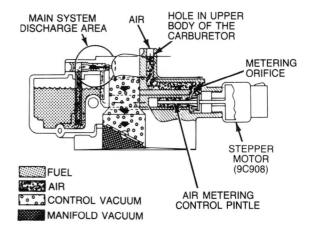

FUEL
AIR
CONTROL VACUUM
MANIFOLD VACUUM

FIGURE 8–16   7200 feedback carburetor, air control system. *Courtesy of Ford Motor Company.*

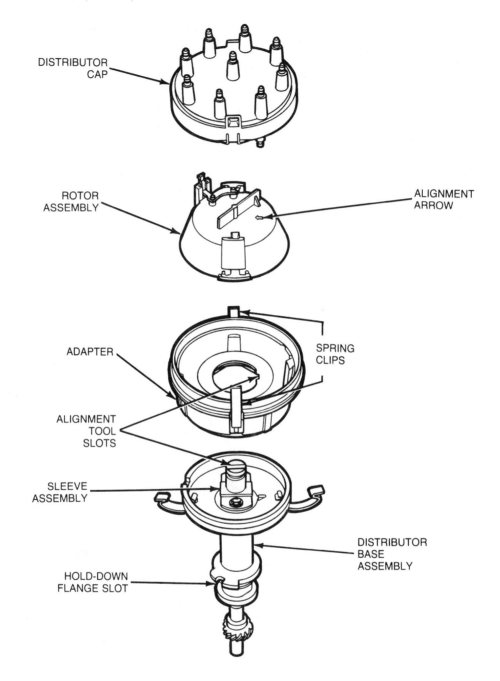

DISTRIBUTOR ASSEMBLY — SECOND GENERATION
(12127)

FIGURE 8–17   EEC III second-generation distributor. *Courtesy of Ford Motor Company.*

enter a passage at the top of the carburetor from inside the air cleaner. The passage leads to an orifice that the stepper motor and pintle valve control. When the pintle is extended, less air is allowed to pass through the orifice; when it is retracted, more air is allowed to pass. After passing through the orifice, the air is channeled past the throats to the other side of the carburetor. Here it is bled into the main metering system. The more air that is bled into the main metering system, the less fuel it introduces into the carburetor throat.

CFI Injectors. The fuel delivery system and the operation of the CFI injector solenoids for EEC III were carried over to some EEC IV applications and are described in Chapter 6.

Second-generation Distributor. Later EEC III systems have a second-generation distributor, Figure 8–17. It has the upper- and lower-level electrode blades, but the rotor is cone shaped instead of rectangular. The rotor's pickup arms and the cap's center terminal are redesigned.

CFI Canister Purge. Four different canister purge systems have been used on EEC III CFI systems. A service manual for the specific vehicle should be consulted for information on how a specific purge system operates.

System Diagnosis and Service

The ECA for EEC III has self-diagnostic capability. It does not, however, have a memory that enables it to record and report faults that occur during normal driving. It can only report faults, in the form of two-digit code numbers, that are present when the self-test is being run. A quick-test procedure (similar to that for EEC IV) containing the self-test is outlined in the service manual. All necessary test equipment is listed in the service manual.

REVIEW QUESTIONS _____

1. What engine and vehicle use the EEC I system?

2. List the controlled functions of the EEC I system.

3. Where is the CP (crankshaft position) sensor located on the EEC I system?

4. What is the signal-producing device for the MAP and BP (manifold absolute pressure and barometric pressure) sensors? What operates the signal-producing device?

5. What are the controlled functions of the EEC II system?

6. What are the controlled functions of the EEC III system?

7. What is the reference voltage used for EEC I, II, and III systems?

8. What is unique about the EGR valve operation on the EEC I system?

9. What fuel-metering device is used on EEC I?

10. What method does the stepper motor and pintle valve of the EEC II system use to control the air/fuel ratio?

11. What method does the stepper motor and pintle valve of the EEC III system use to control the air/fuel ratio?

12. What type of service codes are presented by EEC I, EEC II, and EEC III?

ASE-type Questions. (Actual ASE test questions will probably not be as product specific.)

13. Technician A says that EEC III systems use the 7200 VV carburetor with a stepper motor for air/fuel mixture control. Technician B says that EEC III systems use CFI for air/fuel mixture control. Who is correct?
    a. A only
    b. B only
    c. both A and B
    d. neither A nor B

14. An EEC III–equipped car is brought into the shop with complaints of lost power and rough running. Technician A says that rotor misalignment can be the problem. Technician B says that rotor misalignment can-

not be the problem because she has checked ignition timing and it is correct. Who is correct?
a. A only
b. B only
c. both A and B
d. neither A nor B

15. Technician A says that the EEC III system has self-diagnostic capability. Technician B says that the EEC III system has self-diagnostic capability but it does not have the ability to store codes for intermittent faults. Who is correct?

a. A only
b. B only
c. both A and B
d. neither A nor B

GLOSSARY

ANEROID   An accordionlike capsule that expands or contracts in response to external pressure changes.

DURASPARK   An electronic ignition system developed by Ford. Three generations of the system have evolved: Duraspark I, II, and III.

FORD'S MICROPROCESSOR CONTROL UNIT

Objectives

After studying this chapter, you will be able to:

- identify the applications of the MCU system.
- identify the operating modes of the MCU system.
- identify the operating modes in which open loop and closed loop are employed.
- describe the most commonly used sensing device in the MCU system.
- describe the self-diagnostic capability of the MCU system.

When it was designed, the microprocessor control unit (MCU) system's major purpose was to control air/fuel mixture and Thermactor air. That has not changed even though some additional functions have been added on some engine applications. The system uses a dual-bed catalytic converter first put into production in 1980 for California cars with 2.3-liter engines. In 1981 it was used on:

- passenger cars equipped with a 2.3-liter engine and Holley 6500, 2V carburetor (until 1983).
- California trucks with a 4.9-liter engine (until 1984).
- passenger cars equipped with a 4.2-liter engine (engine was discontinued after 1982).
- passenger cars equipped with a 5.0-liter engine (until 1984).
- passenger cars and trucks equipped with 5.8-liter engine (until 1984).

In 1982 and 1983, the MCU system was used on California passenger cars equipped with the 3.8-liter engine. In 1983 and 1984, it was used on 2.0- and 2.3-liter Ranger trucks. Its most recent applications have been on 1984, 1985, and 1986 police packages equipped with 5.8-liter engines.

There are two distinguishing characteristics of the MCU system: First, it does not control ignition timing; however, on some engine applications it has the ability to retard timing to prevent spark knock. Second, most of its sensors and actuators rely on engine vacuum. Figure 9–1 provides an overview of the MCU system.

MCU MODULE

The module for the MCU system is located under the hood and has no separate calibration unit, Figure 9–2. It is powered by the ignition

MCU System									
Inputs					**Outputs**				
Engine						**Engine**			
2.3	4.9	3.8	8 Cyl.			2.3	4.9	3.8	8 Cyl.	
X	X	X	X	Exhaust Oxygen Sensor		X	X	X	X	Fuel Control Device
X	X	X	X	Tach Signal Engine Speed		X	X	X	X	Thermactor Bypass Sol.
X	X	X	X	Low Coolant Temp. Sw.		X	X	X	X	Thermactor Divert Sol.
		X	X	Warm Coolant Temp. Sw.			X	X	X	Canister Purge Sol.
		X	X	High Coolant Temp. Sw.			X	X	X	Spark Retard
X	X	X	X	Closed Throttle Sw.				X	X	Throttle Kicker
	X	X	X	Crowd/Mid. Throttle Sw.						
X	X	X	X	WOT Sw.						
		X	X	Knock Sensor						

FIGURE 9-1   MCU system overview

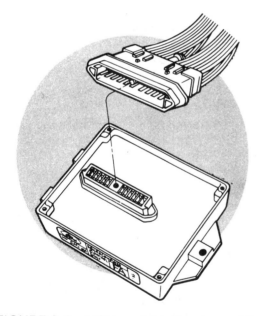

FIGURE 9-2   MCU module. *Courtesy of Ford Motor Company.*

switch (no power relay) when the switch is in the run position. No constant power supply is needed because the module has no keep-alive memory. The module operates in three fuel control modes.

Initialization Mode

When the ignition is turned to the run position, the module is activated and the initialization mode begins. The initialization mode provides a rich air/fuel ratio (between 12.8 and 13.2 to 1) for starting the engine. During warm- or moderate-temperature starts, this mode is maintained for just a few seconds after the engine is started and thus allows the induction system to stabilize. During colder starts (coolant temperature below about 10° C [50° F]) the initialization mode is maintained until the engine warms up to approximately 10° C coolant temperature.

Open-loop Mode

If after the initialization mode the engine and the EGO sensor are not warm enough to go into closed loop, the module will put the system into open loop. In open loop the air/fuel mixture is richer than 14.7 to 1 but is leaner than that provided in the initialization mode. In addition to the warm-up period, the system is in open loop during:

- idle.
- deceleration.

- heavy throttle operation (this can be defined as a throttle opening that produces a vacuum value of less than 10 inches Hg and includes high-speed cruise, acceleration, and WOT).

During open-loop operation, the MCU module sends a fixed command to the mixture control device. The command varies, however, from one open-loop operating condition to another (the command for an idle condition is different from that for a WOT or deceleration condition). During engine warm-up, the command changes as the engine coolant temperature goes up. In each of the open-loop operating conditions, the module commands an air/fuel mixture just rich enough to provide good driveability.

Closed-loop Mode

As the engine approaches normal operating temperature (measured by either a timer circuit in the module or a temperature switch, depending on module calibration), the module puts the system into closed loop. In order to go into closed loop, the coolant temperature must be at 53° C (125° F) or more, the exhaust gas oxygen sensor must be 346° C (650° F) or more, and the engine must be operating in a part throttle, light load condition. If brought to idle speed while in closed loop, the system will drop into open loop; however, some system will stay in closed loop for approximately three minutes before dropping into open loop.

INPUTS

Most of the MCU system's sensors are simple electric switches that when triggered by vacuum, temperature, or mechanical linkage provide an electrical signal to the module. For example, consider the low-temperature vacuum switch used on the 4.9-liter engine, Figure 9-3. Vacuum to this switch is controlled by a temperature-controlled, ported vacuum switch. When coolant temperature is below 35° C

(95° F), the ported vacuum switch allows vacuum to the low-temperature switch and its contacts are held open, Figure 9-4. When coolant temperature exceeds 35° C, vacuum to the low-temperature switch is blocked and the contacts close.

The module applies 12 volts through an internal resistor to the insulated contact of the low coolant temperature switch. The other contact is grounded. When vacuum is applied to the diaphragm, the circuit is not complete to ground and there is no voltage drop across the resistor. The voltage signal to the sensing circuit stands at 12 volts. When vacuum is cut off, the points close and the completed circuit to ground causes a voltage drop across the resistor. This drop causes the voltage signal to the sensing circuit to drop to near zero. This tells the module that coolant temperature has reached 35° C.

Exhaust Gas Oxygen (EGO) Sensor

The EGO sensor is the same as that used on other Ford systems as well as by some other manufacturers.

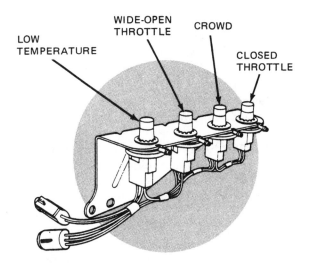

FIGURE 9-3 Vacuum switch assembly (4.9 L California engine). *Courtesy of Ford Motor Company.*

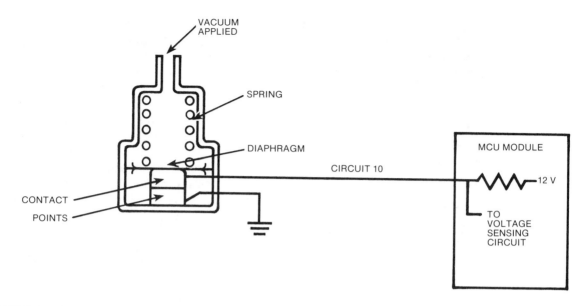

FIGURE 9-4   Vacuum-operated switch used as a low coolant temperature sensor

Tach Signal

The tach signal is obtained by running a wire from the distributor side of the ignition coil to the MCU module.

Low Coolant Temperature Switch

The low coolant temperature switch for the 4.9-liter engine was described in the example above. The low coolant temperature switch for another engine group, including the 2.3 liter, is similar except that it is mounted on top of the ported vacuum switch, Figure 9-5.

Mid- and Dual-temperature Switches

The 3.8-liter and eight-cylinder engine group use two switches that provide information about three different levels of coolant temperature, Figure 9-6. The dual-temperature switch (not vacuum operated) has one sliding and one stationary contact that are pushed together when the coolant reaches 13° C (55° F). The contacts are moved by a wax pellet in the nose of the switch. The pellet expands as the coolant temperature goes up. As the tempera-

ture continues to increase, the sliding contact continues to move. At 113° C (235° F) the contacts separate again.

The mid-temperature switch looks and functions like the low-temperature switch in

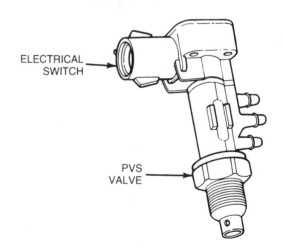

FIGURE 9-5   Low coolant temperature switch. *Courtesy of Ford Motor Company.*

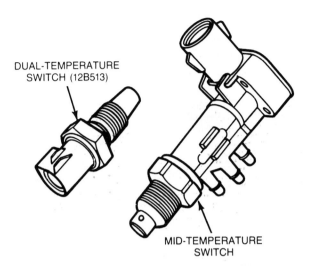

FIGURE 9-6 Mid- and dual-temperature switches. *Courtesy of Ford Motor Company.*

Figure 9-5 except that it closes at 53° C (128° F). When both switches are open, the module knows that coolant temperature is below 13° C. When the dual-temperature switch closes but the mid-temperature switch remains open, the module knows that the coolant temperature is between 13° C and 53° C. When the mid-temperature switch closes and the dual-temperature remains closed, the module knows that coolant temperature is between 53° C and 113° C. When the dual-temperature switch is open but the mid-temperature switch is closed, the module knows that coolant temperature is above 113° C.

Coolant temperature information affects all of the module's output decisions.

Throttle Position Switches

The 2.3- and 2.0-liter engine group uses a normally closed idle tracking switch that is mechanically opened by the throttle linkage when the throttle closes, Figure 9-7. The switch's opening and the resulting voltage rise in its circuit signal the module that the engine is at idle. This engine group also uses a vacuum-operated, normally closed WOT switch. The WOT switch gets manifold vacuum, but its vacuum signal is routed through a thermal vacuum switch in the air cleaner, Figure 9-8. If the engine is warm, vacuum gets to the WOT switch and holds it open. When the throttle is opened far enough for vacuum to drop to the switch's set point, it closes and the module knows to adjust the air/fuel mixture for WOT operating conditions.

The WOT switch also functions as a low-temperature indicator. When the engine is started cold and air temperature inside the air cleaner is low, the thermal vacuum switch in the air cleaner, which controls vacuum to the WOT switch, is closed and blocks vacuum to the WOT switch. Remember that the low coolant temperature switch used on the 2.3-liter engine group is closed at 35° C or more. As a result of the effectiveness of the heated air inlet system, air cleaner temperature reaches the thermal vacuum switch's set point before coolant temperature reaches 35° C. Therefore, if the WOT switch is closed and the low coolant temperature switch is still open, the module

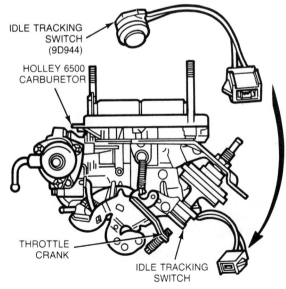

FIGURE 9-7 Idle tracking switch. *Courtesy of Ford Motor Company.*

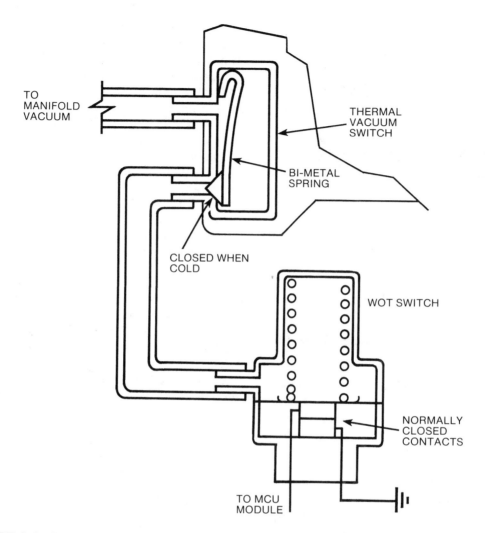

TO
MANIFOLD
VACUUM

THERMAL
VACUUM
SWITCH

BI-METAL
SPRING

CLOSED WHEN
COLD

WOT SWITCH

NORMALLY
CLOSED
CONTACTS

TO MCU
MODULE

FIGURE 9-8   Vacuum control to 2.3 L WOT switch

will know that engine temperature is below a value of about 10° C (50° F) as measured by coolant temperature. It can then command an air/fuel ratio that is appropriate for low-temperature starting and for the brief period following a cold start before the heated air inlet system raises inlet air temperature to the thermal vacuum switch's set point.

The throttle position switches for the 4.9-liter engine consist of three switches, Figure 9-3. They are the closed throttle switch, the *crowd* switch, and the WOT switch. The word *crowd* refers to a steady-state, heavy-throttle condition, but less than WOT, that produces 10 inches Hg or less of vacuum. These are part of a four-switch assembly; the fourth is the low coolant temperature switch discussed earlier. They are all normally closed.

The closed throttle switch is connected to ported vacuum. When the throttle is open, vac-

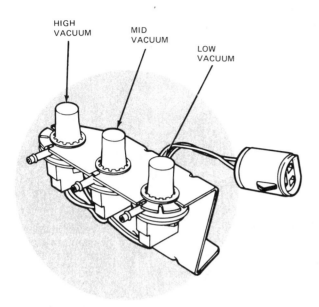

FIGURE 9-9   Zoned vacuum switch assembly (eight cylinder). *Courtesy of Ford Motor Company.*

KNOCK SENSOR (SOME 8-CYLINDER CALIBRATIONS)
(12A699)

FIGURE 9-10   Knock sensor. *Courtesy of Ford Motor Company.*

uum holds the switch open. When the throttle closes, vacuum is cut off and the switch closes.

The crowd switch and the WOT switch are each connected to manifold vacuum. As the throttle is opened, vacuum drops. When vacuum drops to about 10 inches Hg, the crowd switch closes. When vacuum drops to around 2 to 3 inches Hg, the WOT switch closes.

The throttle position switches for the eight-cylinder engine group (called *zoned vacuum switches*) consist of a three-switch assembly, Figure 9-9. They provide basically the same information for the eight-cylinder engine module that the other three switches provide for the 4.9-liter engine module.

Knock Sensor

Only those engines in the eight-cylinder engine group are equipped with knock sensors, Figure 9-10. The sensor contains a piezo-electric crystal that produces a voltage signal in response to the vibration caused by spark knock.

OUTPUTS

Fuel Control Devices

Holley 6500 (Early 2.3-liter Engine). The Holley 6500 carburetor was modified slightly for use on MCU applications. Its main metering jet is reduced in size so that it alone cannot provide a rich enough air/fuel mixture. It relies on what resembles a power valve circuit to provide the additional fuel needed to produce a 14.7 to 1 mixture for closed-loop operation or a richer mixture for open-loop operation, Figure 9-11. This is, however, not a power valve; it is a feedback system that is part of the main metering circuit. When its metering rod is pushed down by the foot above it, additional fuel is allowed to flow past it into the main metering circuit. The farther it is pushed down, the more fuel that is admitted.

The foot that positions the metering rod is controlled by a diaphragm and spring. Vacuum applied above the diaphragm raises the foot.

289

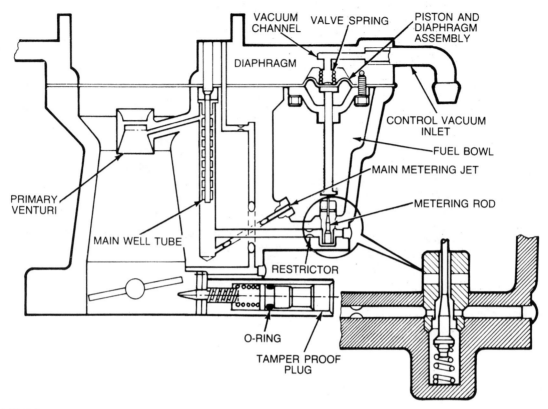

FIGURE 9-11   Feedback carburetor (6500 Holley). *Courtesy of Ford Motor Company.*

The metering rod is pushed up by its spring, and fuel flow is restricted. Vacuum above the diaphragm is controlled by a vacuum regulator solenoid, Figure 9-12. The solenoid is mounted away from the carburetor and in the right rear corner of the engine compartment with vacuum hoses going to its vacuum source and to the carburetor. One of the three hoses connected to the solenoid is vented to atmosphere. As the vacuum regulator solenoid is cycled, it alternately connects the vacuum chamber above the diaphragm to vacuum and to atmospheric pressure. The solenoid is cycled ten times per second by the MCU module, but the proportion of time its circuit is grounded or ungrounded within each cycle can be varied. In this way the module can control the amount of vacuum experienced above the diaphragm and the air/

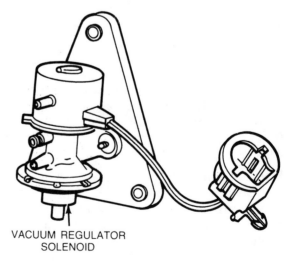

FIGURE 9-12   Vacuum regulator solenoid. *Courtesy of Ford Motor Company.*

fuel mixture fed to the engine. A vacuum of 5 inches Hg produces a full lean mixture, and 0 Hg produces full rich.

Carter YFA IV (4.9-liter and Late 2.3-liter Engine). The fuel control solenoid on this carburetor, Figure 9–13, is cycled in the same way as the one just described. As it is stroked, however, it controls the amount of air that is allowed to pass through the air bleed passage into the main metering system. This control solenoid and carburetor was carried over to some applications of the EEC IV system.

Motorcraft 7200 VV (3.8-liter and Eight-cylinder Engine Group). This device was called a *feedback carburetor actuator* and used a stepper motor to control the air bleed system. It was carried over to EEC III and is discussed in Chapter 7.

Thermactor Control

Thermactor air control is achieved with the TAB and TAD solenoids and a Thermactor air control valve. It is essentially like the system used on the EEC II and EEC III systems discussed in Chapter 7. The module's programming for control of Thermactor air, however, is somewhat different. Thermactor air is delivered as indicated during the following operating conditions:

- prolonged initialization mode—atmosphere (bypass mode)
- open-loop warm-up—exhaust manifold
- WOT or crowd (open loop)—exhaust manifold
- extended idle or deceleration (open loop)—bypass
- closed loop—catalytic converter

Canister Purge

The canister purge function is controlled by a normally closed solenoid-controlled valve. It is located in the purge hose that runs between the charcoal canister and the manifold vacuum source, just as in the EEC II and carbureted EEC III systems. The module activates the solenoid and allows purging to occur:

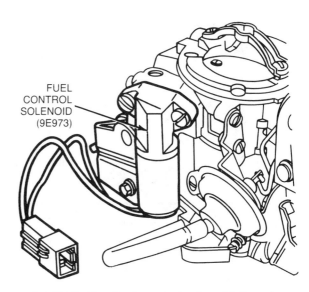

FUEL CONTROL SOLENOID (9E973)

FIGURE 9–13   Fuel control solenoid. *Courtesy of Ford Motor Company.*

- after a predetermined time has elapsed from when the engine was started.
- when the engine is within a predetermined RPM window.
- when the engine is within normal operating temperature limits.

Spark Retard

Two different spark retard systems are used on MCU systems.

4.9-liter Engine. This system does not have a knock sensor. The module is programmed to recognize conditions that are prone to produce spark knock. When it does, it eliminates vacuum advance by activating a solenoid-controlled valve, which then vents the vacuum advance hose to the distributor.

3.8-liter and Eight-cylinder Engine Group. Most of the engines in this group are equipped with a knock sensor and the universal ignition module. This module is one of three different ones used with various applications of the Duraspark II ignition system. It has the ability to retard ignition timing in response to a command from the MCU module.

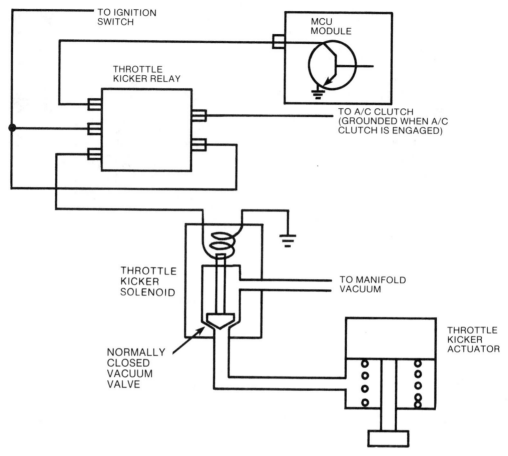

FIGURE 9-14   Throttle kicker control circuit

Throttle Kicker (TK)

The TK is similar to those used on other EEC systems. The TK solenoid controls vacuum to the actuator, Figure 9–14. When 12 volts is applied to the solenoid, vacuum is routed to the TK actuator. Vacuum causes the TK actuator to extend its plunger and slightly increase throttle opening when the engine is idling. Grounding the coil of the TK relay enables the relay to supply voltage to the solenoid. The relay can be grounded by either the A/C clutch being activated or by the MCU module. The TK actuator is actuated during idle conditions when:

- the engine is cold and warming up.
- the engine is overheated.
- the A/C compressor clutch is engaged.

SYSTEM DIAGNOSIS AND SERVICE

The MCU module has a self-diagnostic capability that is limited to faults observed by the module while it is in the diagnostic mode. It has no memory and therefore does not monitor and record service codes related to its sensor or actuator circuits during normal driving.

Functional Test

The Functional Test (occasionally referred to as Self-Test) is divided into three sections: Visual Inspection, Key On/Engine Off, and Engine Running.

Visual Inspection. The Visual Inspection is included as a part of the Functional Test to encourage the technician to check for obvious conditions such as loose wires, disconnected or broken vacuum hoses, faulty vacuum hose routing, low coolant temperature, faulty spark plugs or wires, and so on. Any of a number of problems such as these can cause the driveability problem for which the vehicle was brought in for service or can even cause the module to misread the results of the problem as a faulty sensor or actuator circuit. The mistaken service code produced by the module can likely result in needless testing by the technician and needless replacement of MCU system parts.

Key On/Engine Off. In this portion of the Functional Test, the module tests some of its sensor circuits electrically for proper operating condition. It is triggered by grounding the trigger terminal in the Self-Test connector, Figure 9–15, and then turning the ignition on. Service codes are read by connecting an analog voltmeter or a Star Tester to the output terminal and to ground.

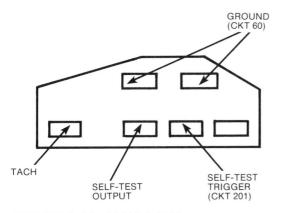

GROUND
(CKT 60)

TACH

SELF-TEST
OUTPUT

SELF-TEST
TRIGGER
(CKT 201)

FIGURE 9–15   MCU Self-Test connector

Engine Running. In this portion of the Functional Test, the module tests its input and output circuits for proper operation under dynamic conditions. The procedure for entering this portion of the test varies slightly depending on engine application and model year. Be sure to follow test procedure steps and to use the appropriate service manual.

Service Codes. Service codes are presented by sweeps of the voltmeter needle or by a digital readout on the Star Tester just as they are for the EEC III and EEC IV systems. No special tools are required for testing the MCU system with the possible exception of the Star Tester as an alternative to the voltmeter.

Service Procedures

Service procedures for circuits that have been identified by a service code are covered by **Subroutines** found in the service manual. Driveability problems with no associated service code are covered by the **Diagnostic Routine** section of the service manual.

REVIEW QUESTIONS

1. What are the three operating modes used by the MCU system?

2. What is the most common sensing device used by the MCU system?

3. What is the reference voltage used by the MCU system?

4. Indicate where Thermactor air should be going during the following operating conditions:
   warm-up
   WOT or crowd
   extended idle
   closed loop

5. What is the source of the tach signal?

6. What method is used to anticipate spark knock on the 4.9-liter engine?

7. What method is used to retard timing on the 4.9-liter engine?

8. Which MCU engines use a knock sensor?

9. What are the two segments of MCU's self-diagnostic mode?

10. What type of memory does the MCU module have?

ASE-type Questions. (Actual ASE test questions will probably not be as product specific.)

11. Technician A says that service codes obtained from an MCU system always refer to a problem that the computer perceives as currently existing. Technician B says that service codes obtained from an MCU system can refer to a problem the computer recognized while the car was being driven. Who is correct?
   a. A only
   b. B only
   c. both A and B
   d. neither A nor B

12. The Functional Test of the MCU system consists of three segments. In what order do these segments occur?
   a. Visual Inspection, Engine Running Test, and Key On/Engine Off
   b. Engine Running Test, Key On/Engine Off, and Visual Inspection
   c. Visual Inspection, Key On/Engine Off, and Engine Running Test
   d. Key On/Engine Off, Visual Inspection, and Engine Running Test

13. Technician A says that the Key On/Engine Off segment of the MCU system's Functional Test tests the sensor circuits electrically. Technician B says that the Engine Running Test segment of the MCU system's Functional Test tests the actuator circuits electrically. Who is correct?
   a. A only
   b. B only
   c. both A and B
   d. neither A nor B

GLOSSARY

CROWD   A steady-state, heavy-throttle condition that is less than full throttle.

10

CHRYSLER CORPORATION'S SINGLE-POINT AND MULTIPOINT FUEL INJECTION

Objectives

After studying this chapter, you will be able to:

- identify the operating modes of Chrysler's single-point and multipoint fuel injection systems.
- describe the function of the power module and how it interacts with the logic module.
- describe the two methods that each of these systems uses to control idle speed.
- describe the method of spark knock control used on the multipoint (turbo-charged) system.
- describe the three methods of turbocharger boost control.
- identify malfunctions within the system that cause the engine to stop.
- identify the test modes used to diagnose driveability problems.

In the early 1980s, Chrysler introduced an electronic fuel injection system on the Chrysler Imperial. Few of those systems were produced; therefore, they are not discussed in this text.

The first current fuel-injected systems by Chrysler Motors were introduced in 1984 on 2.2-liter engines. They were the single-point system—referred to by Chrysler as Electronic Fuel Injection (EFI)—and the multipoint system—referred to by Chrysler as Multi-Point Injection (MPI). The 2.2 liter MPI system is often called the "Turbo" system because it is equipped with a turbocharger. Since then, other engines have been equipped with either the EFI or the MPI system. The two systems are quite similar and, for the sake of time and space, will be treated in the same chapter. Variations between the systems will be pointed out.

Prior to 1986 the single-point system was a high-pressure system operating at a fuel pressure of about 36 PSI. In 1986 the single-point, high-pressure–style injector was replaced by a low-pressure–style injector, and the fuel pressure was reduced to 14.5 PSI. The pintle-style valve was replaced by a ball-shaped valve similar to those used by other low-pressure, single-point systems.

LOGIC MODULE/ POWER MODULE

The system is controlled by two separate modules. The logic module, which contains the central processor, is inside the passenger compartment, Figure 10–1. It receives most of its power from and works with the power module, which is located in the engine compartment. The logic module sends a 5-volt reference signal to its sensors and controls all system calibrations.

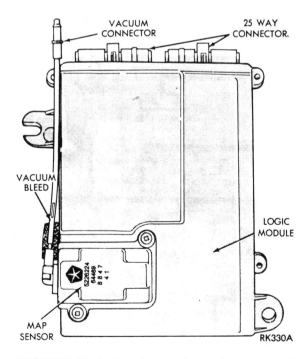

FIGURE 10–1   Logic module. *Courtesy of Chrysler Motors.*

The power module drives by ground side switching the ignition coil, the injector/injectors, and the automatic shutdown relay, Figure 10–2. Beginning in 1985 it controlled the charging system by providing ground for the alternator's field coil. These are the circuits that have relatively heavy current flows and have been separated from the logic module to protect it from electrical interference and heat buildup. The power module is located near the battery in such a way as to receive a bath of air to help cool it, Figure 10–3.

The power module is instructed by the logic module when to open and close the primary ignition circuit and when to turn on and off the injector/injectors. The power module also provides a constant 8 volts to the logic module and the distributor pickup unit.

When the ignition switch is in the run position, it powers the power module by way of a fused circuit. The circuit that supplies this power is commonly referred to by Chrysler literature as *J2*. The printed circuit board of the power module has a rail that acts as a fuse and protects the module against reversed polarity. If a reversed polarity condition occurs, the fuse will blow. A reserve circuit with a diode continues to supply J2 power to the power module but with a slight voltage drop resulting from the resistance of the diode.

Single-Module Engine Control (SMEC). In mid-1987 on some 3.0-liter engine applications, the Logic Module and the Power Module were placed as a logic board and power board into a single unit called a SMEC module. The SMEC is located behind the battery in the duct that supplies air to the air cleaner where the Power Module had been. In 1988, the SMEC was used on most 1988 electronic fuel injection applications.

The eight-volt regulator on the power board has been changed to supply 9.2 to 9.4 volts to the SMEC. This is due to the use of a different type of transistor (Metal Oxide Semiconductor Field Effect Transistor) on the power board that requires the higher voltage. The

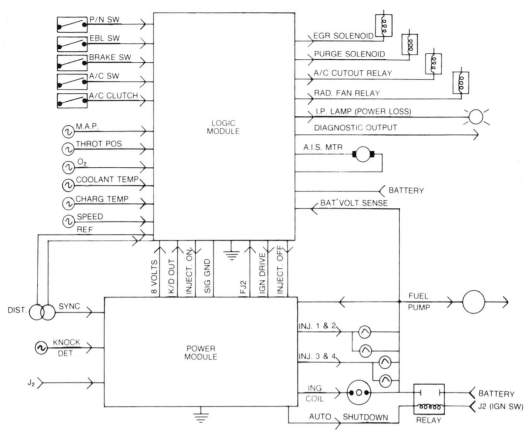

FIGURE 10–2   2.2 L turbo EFI system. *Courtesy of Chrysler Motors.*

battery temperature sensor that was formerly in the Power Module has been moved to the logic board.

Features

Automatic Shutdown (ASD) Relay. The ASD relay is powered by the ignition switch when the ignition is turned to the run position. When the power module grounds the ASD coil, the ASD provides power for the electric fuel pump, the ignition coil, and the injector(s). The fuel pump has a permanent ground and comes on automatically. The ignition coil and injector(s) wait for their circuits to be grounded by the power module. On later models the ASD is located inside the power module.

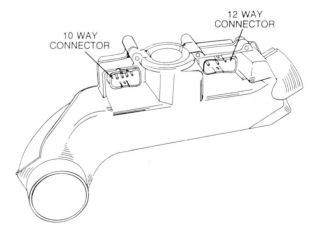

FIGURE 10-3   Power module. *Courtesy of Chrysler Motors.*

On-board Diagnostics or Self-diagnosis.
The logic module monitors its most vital input and output circuits for faults and records fault codes in its memory when it perceives an inappropriate voltage value.

Adaptive Memory. The logic module has the ability to modify some of its programmed calibrations concerning fuel metering to compensate for production tolerances and changes in barometric pressure.

Operating Modes

Starting. While the engine is being cranked, the injector pulse frequency is doubled. The single-point injector is normally pulsed twice per revolution, and the multipoint system normally pulses its injectors in pairs, with each pair being pulsed once each revolution. The double pulsing only occurs for a preprogrammed time interval (to avoid flooding). The pulse width is influenced only by coolant temperature. After the engine has started and if it is below a criterion temperature, the logic module will provide a fuel enrichment program based on coolant temperature and manifold pressure. The start-up enrichment program decays to base enrichment over a programmed time span.

Primer Function. When the 3.0-liter engine's ignition is turned on, all six injectors will be pulsed to spray fuel into the intake ports. This is intended to improve starting. It will occur regardless of engine coolant temperature, but if the engine is warm, less fuel will be injected; if the engine is cold, more fuel will be introduced.

Open Loop. Open loop occurs during the following conditions:

- coolant temperature is below normal operating temperature
- oxygen (O₂) sensor is below minimum operating temperature (315° C [600° F])
- O₂ sensor is not switching across a preset switch point of about 0.45 volt
- wide-open throttle (WOT) or acceleration

- when the multipoint system reaches 1 PSI or more of boost
- deceleration
- when the single-point system is at idle

Closed Loop. The following conditions must be met before the system will go into closed loop:

- coolant temperature must reach a preset value
- O₂ sensor must be producing a valid signal
- predetermined amount of time must have elapsed since engine was started

Limp-in Mode. If the logic module through its on-board diagnostics circuit receives a voltage signal (or no signal) from one of its most vital sensors and the signal does not fit within an acceptable range or does not change under specific conditions, it will consider the sensor inoperative and will put the system into limp-in mode. In this mode the logic module will substitute one or more other sensor values for the failed sensor and will turn on the power loss lamp on the instrument panel. Although the engine may not operate efficiently, it will continue to operate. Those sensors for which this will occur include the following (the indicated codes will be stored in memory):

- *manifold absolute pressure*—Throttle position and engine speed are used to create a simulated MAP value. (13 and/or 14)
- *throttle position*—The MAP is used to produce rough estimates of throttle position. (24)
- *coolant temperature*—Charge temperature (manifold air/fuel mixture temperature) is used as a substitute. (22)
- *charge temperature*—The engine usually functions effectively without this input. The power loss lamp is not turned on. (23)

Loss of the distributor reference causes the engine to stop.

INPUTS

Manifold Absolute Pressure (MAP) Sensor

This is a piezoresistive-type pressure sensor. The logic module uses it as a barometric pressure sensor during periods of ignition on/engine off or during certain other conditions on some models as described in the **Outputs** section of this chapter, Figure 10–4. The MAP is usually located on the logic module.

Some of the earlier applications had a small bleed hole in the vacuum line to the MAP to prevent condensation from collecting in the line. The MAP sensor was calibrated to allow for the vacuum leak.

Throttle Position Sensor (TPS)

The TPS is a rotary potentiometer attached to the end of the throttle shaft on the side of the throttle body, Figure 10–5.

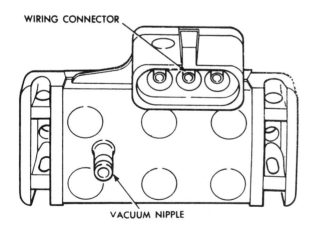

FIGURE 10–4   MAP sensor. *Courtesy of Chrysler Motors.*

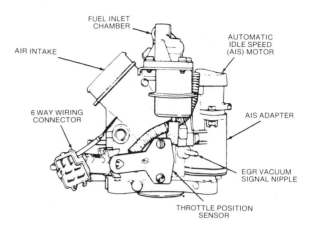

FIGURE 10–5   Throttle position sensor. *Courtesy of Chrysler Motors.*

Oxygen (O₂) Sensor

This is a single-wire O_2 sensor. Some applications use a three-wire sensor, and the additional wires power an internal heat element.

Coolant Temperature Sensor (CTS)

The CTS is a single-element thermistor that screws into the thermostat housing, Figure 10–6. It has a resistance from 11,000 ohms at –20° C (–4° F) to 800 ohms at 90.5° C (195° F); 195° F is considered full operating temperature for these systems.

Charge Temperature Sensor (Charge Temp)

The charge temp sensor is a single-element thermistor screwed into a runner of the intake manifold. It has about the same resistance range as and looks similar to the CTS. On earlier models its input is used during cold enrichment calculations; otherwise it serves as a backup to the CTS. On later model multipoint systems, its input is used in the control of air/fuel mixture (especially cold) and boost control.

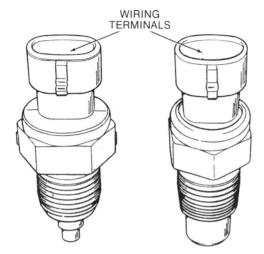

FIGURE 10-6   Charge sensor and coolant sensor. *Courtesy of Chrysler Motors.*

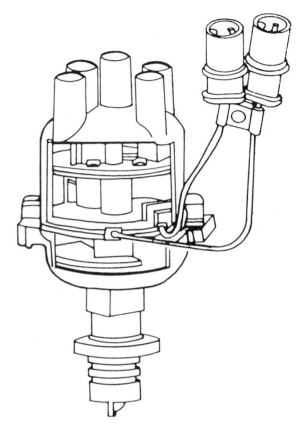

FIGURE 10-8   Turbo distributor with reference pickup and sync pickup Hall-effect switches. *Courtesy of Chrysler Motors.*

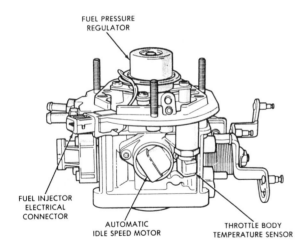

FIGURE 10-7   Low-pressure, single-point throttle body with throttle body temperature sensor. *Courtesy of Chrysler Motors.*

Throttle Body Temperature Sensor

The low-pressure, single-point system no longer uses a charge temperature sensor; rather, it uses a throttle body temperature sensor that measures fuel temperature in the throttle body, (Figure 10-7). This information is used to calculate air/fuel mixture during hot restarts.

Distributor Reference Pickup (REF Pickup)

A Hall-effect switch in the distributor generates the REF pickup signal from which the logic module determines engine speed and crankshaft position. On the single-point system, this same signal is supplied to the power module as well. Without this signal it does not keep the ASD relay grounded, which in turn removes power from the ignition coil, the injector, and the fuel pump. On the multipoint system, this signal is sent only to the logic module; but in response to this signal, the logic module relays to the power module exactly when to turn on the injectors.

Distributor Sync Pickup (SYNC Pickup)

Most multipoint system has a second Hall-effect switch assembly in its distributor and beneath the REF pickup, Figure 10–8. This Hall-effect switch has a shutter wheel with only one vane instead of four. Its one vane, however, occupies 180°, or one-half of the shutter wheel. The leading and trailing edges of the vane each provide a signal to both the logic module and the power module. The power module uses this input to determine which pair of injectors (1 and 2 or 3 and 4) to turn on when the on signal from the logic module arrives. The logic module uses the SYNC pickup signal along with other sensor information to help determine pulse width.

Optical Distributor. A 3.0-liter V-6 engine was introduced in 1987 with multipoint fuel injection. This was Chrysler's first naturally aspirated MPI-equipped engine. The computer uses a speed-density formula (does not use a vane airflow meter or a mass airflow meter) to calculate fuel quantity and is similar to the system used on the 2.2-liter multipoint injection system. Rather than using Hall-effect switches for the reference and sync pick-up signals, this engine uses a pair of optical sensors (Figure

10–9). A disc rotates with the distributor shaft. The disc has two sets of slits. The first set, closest to the center of the disc, consists of six slits. These are called "low data rate" slits. The second set, "high data rate," are near the outside circumference. These slits are spaced every two crankshaft degrees with the exception of one small, blank spot where there are no slits. This one blank section alerts the SMEC where number one cylinder is at in the engine cycle. The high data rate of this sensor allows the SMEC to monitor crankshaft acceleration and deceleration.

Two light-emitting diodes (LED) are placed on one side of the disc so that their light beams can pass through the slits, one for each set of slits. When one of the slits aligns with its respective LED, the light beam strikes a photo diode, of which there is one for each set of slits. When the light from one of the LEDs strikes one of the photo diodes, the photo diode generates a small voltage that is applied to the base of a transistor (Figure 10–10). Let's assume that we are talking about the photo diode that triggers the reference circuit. The transistor acts as a ground switch in the reference circuit that extends from the SMEC. The SMEC applies a constant 5 volts to the reference circuit. When the reference circuit transistor is turned on, current flows across resistor A in the SMEC through the transistor to ground. Voltage drop across resistor A causes the voltage at point B to be near zero. When the disc blocks the light beam from the LED, the switching transistor in the reference circuit is turned off. The reference has no path to ground and, there is no current flow across resistor A, so the voltage at point B is 5 volts. Each time the voltage at point B drops to near zero, the SMEC knows that the next piston in the firing order has reached a specified position in its cycle. The sync circuit works the same way with each signal meaning that 2 degrees of crankshaft travel have occurred.

At engine speeds below 1200 RPM, the SMEC will use the sync signal (high data rate)

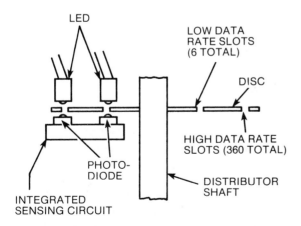

FIGURE 10–9   Optical-sensing device in distributor

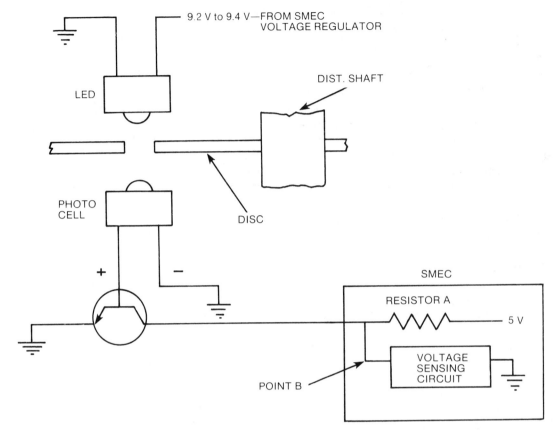

FIGURE 10–10   Reference optical sensor

for calculating ignition timing and injector pulsing. At speeds above 1200 RPM, it will use the reference signal for these calculations. This is because at lower engine speeds there is a significant, momentary RPM gain each time a spark plug is fired. However, due to variations in compression, spark plug condition, and so on, the RPM change may not be the same for each cylinder. The high data rate signal allows the SMEC to monitor these RPM changes and adjust ignition timing and injector pulse width accordingly. At higher engine speeds, there is much less variation in RPM between power impulses and such adjustments are not as necessary.

Vehicle Speed (Speed Sensor)

The speed sensor is a small switching device that receives a steady 5-volt signal from the logic module and breaks into eight on/off signals for each speedometer cable revolution. It is located on the transmission where the speedometer cable attaches, Figure 10–11.

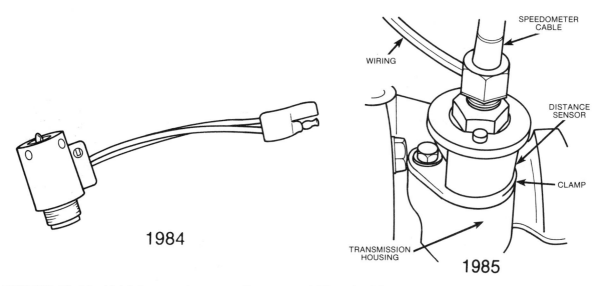

FIGURE 10–11   Vehicle speed sensor. *Courtesy of Chrysler Motors.*

Detonation (Knock) Sensor

The knock sensor is the piezoelectric type. It is screwed into the intake manifold.

Battery Temperature Sensor

A thermistor in the power module, which is near the battery, senses the battery's ambient temperature and sends the information to the logic module.

Charging Circuit Voltage

The logic module senses the voltage that the charging system is maintaining from the fuel pump power circuit.

Switch Inputs

Park/Neutral (P/N). The P/N switch tells the logic module if the transmission is in gear or not. Its input influences idle speed, and normal spark advance is not issued when the transmission is not in gear.

Electric Backlite (Heated Rear Window). If the heated rear window switch is on while at idle, the logic module will increase the throttle blade opening to compensate for the additional load that the alternator puts on the engine.

Brake. The brake switch is used as a backup in the event that the TPS fails to signal the logic module that the throttle has closed.

Air-conditioning (A/C). If the A/C is on, a signal will be provided from the A/C control circuit to the logic module so that it can increase throttle opening to compensate for the additional load on the engine during idle.

Air-conditioning Clutch. As the A/C system cycles the compressor clutch on and off during idle, the A/C clutch switch informs the logic module of the clutch's status so that the throttle blade can be adjusted accordingly to maintain a steady engine idle speed.

The two A/C system inputs were combined into one signal on later models.

Battery. One terminal of the logic module receives power directly from the battery. This input is absolutely necessary to sustain power to the keep-alive memory (KAM) when the ignition is turned off.

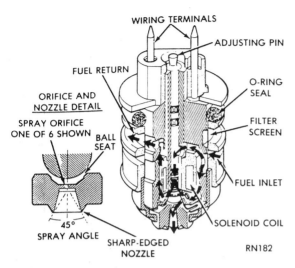

FIGURE 10–12   Low-pressure injector. *Courtesy of Chrysler Motors.*

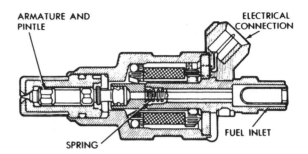

FIGURE 10–13   High-pressure injector. *Courtesy of Chrysler Motors.*

OUTPUTS

Injector/Injectors

Figures 10–12 and 10–13 show a comparison of the low-pressure and high-pressure injectors. They are each solenoid-operated injectors.

During normal operation the single-point system pulses its injector twice during each engine revolution. The multipoint system pulses two injectors simultaneously; it pulses one pair during the first revolution of the engine cycle and the other pair during the second revolution of the cycle. During cranking, this schedule is doubled as discussed earlier in the chapter. The pulse width is determined by the logic module based on input from the:

- CTS.
- MAP.
- RPM.
- TPS.
- O_2—in closed loop.
- charge temp—during cold enrichment.
- speed sensor—during deceleration.

- fuel enrichment and enleanment multipliers programmed into the logic module.

Electric Fuel Pump

The fuel pump is a DC electric motor with a permanent magnet field, Figure 10–14. It is submerged in fuel at the bottom of the fuel tank. It contains two check valves. One acts as a pressure relief valve; the other, near the outlet port, prevents fuel from flowing in either direction when the pump is not running. This helps maintain the fuel line full of fuel and reduces the occurrence of vapor in the fuel line.

When the ignition is turned on, the power module grounds the ASD relay, which powers the fuel pump. If a distributor pickup signal is not received within two seconds, the ASD ground will be removed and the pump turned off. It will remain off until the pickup signal arrives or until the ignition is turned off and on again.

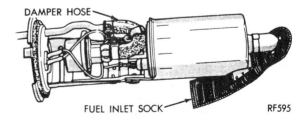

FIGURE 10–14   Fuel pump. *Courtesy of Chrysler Motors.*

Some applications such as the Shelby turbocharged units use two fuel pumps, one in the tank and one nearer to the engine.

Fuel Pressure Regulator. The fuel pressure regulator for the pre-1986 single-point system maintains a constant fuel pressure at the injector of 36 PSI; the low-pressure system maintains 14.5 PSI. The multipoint system maintains a pressure differential across the injector tip of 55 PSI, Figure 10–15. The pressure differential is maintained by connecting manifold vacuum/pressure to the pressure regulator as explained in Chapter 2. Fuel pressure as measured in the injector line varies from 42 to 62 PSI.

On the 3.0-liter engine, the fuel pressure was raised to 46 PSI to avoid starting problems during hot restarts due to fuel vaporizing in the fuel lines and injectors.

Ignition Timing

Spark advance is calculated by the logic module and using information from the:

- CTS.
- REF pickup.

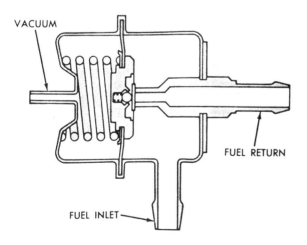

VACUUM

FUEL RETURN

FUEL INLET

FIGURE 10–15   Turbo multipoint system fuel pressure regulator. *Courtesy of Chrysler Motors.*

- MAP.
- barometric pressure (taken from MAP).

During warm idle the logic module uses spark advance as its first choice to control normal idle speed fluctuation. If a predetermined amount of spark timing change does not bring idle RPM within acceptable limits, the logic module will use the automatic idle speed motor.

Beginning in 1985 the multipoint system used a BARO-read solenoid that the logic module used to briefly vent the vacuum line to the MAP sensor. While the line is flooded with atmospheric pressure, the MAP sensor updates the BARO reading. This vent cycle occurs at least once each time the throttle closes and the RPM is below a specific value and can occur as often as every thirty seconds while conditions prevail. Using a turbocharger makes ignition timing especially critical, and barometric pressure significantly influences allowable spark advance.

Detonation Control. Beginning in 1985 the multipoint system has had the unique ability to selectively retard the timing on just the cylinder

Chrysler's Detonation Control

Chrysler has put a strong emphasis on performance in their turbocharged engine control strategy. This is reflected in their method of detonation control. There are really only two ways to stop detonation on a turbocharged engine during boost conditions, and to allow detonation to go unchecked on turbocharged engines destroys them. To reduce the boost pressure everytime a cylinder detonates hampers performance. To retard timing during boost conditions raises exhaust temperature, which is already critically high. To retard the timing on only the offending cylinder raises exhaust temperature only slightly, if at all, and still allows the other cylinders to produce full power.

that is knocking. When the knock signal occurs, the logic module checks which cylinder has just been fired. The spark advance for that cylinder is then retarded slightly. If the knock continues, the logic module will begin to lower boost pressure. In 1986 a feature was added that provides a ten-second delay from the time the throttle is moved to WOT before the logic module responds to the knock sensor.

Wastegate Control Solenoid

Prior to 1985 turbocharged units had three types of boost control. The first was the wastegate as discussed in the **Outputs** section of Chapter 3. The wastegate actuator, Figure 10–16, opens the wastegate at a boost pressure of 7.2 PSI. Secondly, boost pressure is sensed by the MAP. If the boost pressure rises above 7.2 PSI, as it will momentarily in response to snap acceleration, the logic module will recog-

nize the high pressure and will stop issuing injection pulse commands until boost pressure falls below 7.2 PSI. The third control is in response to engine RPM. If RPM goes above 6,650, the logic module will again stop fuel injection until the RPM drops to 6,100.

In 1985 a fourth feature was added. The logic module controls a boost control solenoid that when activated vents the line that pipes manifold pressure to the wastegate actuator. If operating conditions are favorable, the logic module will pulse the boost control to bleed off some of the pressure to the actuator. In this way boost pressure can actually go to 10 PSI before the actuator opens the wastegate. The inputs that the logic module considers for this decision are:

- barometric pressure.
- RPM.
- TPS.

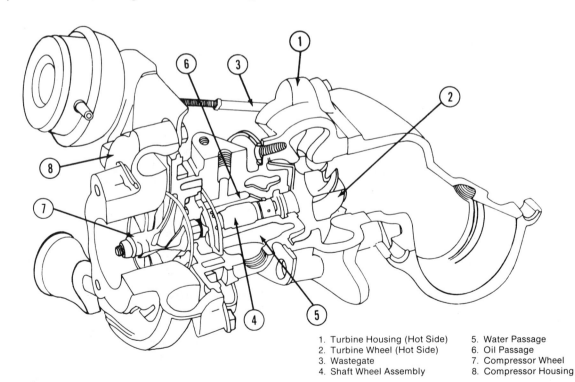

1. Turbine Housing (Hot Side)
2. Turbine Wheel (Hot Side)
3. Wastegate
4. Shaft Wheel Assembly
5. Water Passage
6. Oil Passage
7. Compressor Wheel
8. Compressor Housing

FIGURE 10–16   Turbocharger. *Courtesy of Chrysler Motors.*

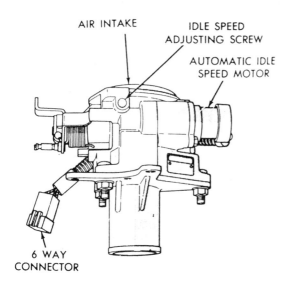

FIGURE 10-17   Multipoint throttle body with AIS motor. *Courtesy of Chrysler Motors.*

- CTS.
- charge temp.
- detonation.

Additionally the logic module considers detonation history and how long the engine has been in boost.

Revised Boost Control

Beginning in 1988, the actuator for turbo boost control on the Turbo I engine gets its pressure signal directly from the turbocharger rather than from the intake manifold. This allows the wastegate to be opened during part throttle operation and eliminates any possible boost until WOT. Without boost at part throttle:

- exhaust back pressure is reduced
- the intake charge is cooler
- tendency toward spark knock is reduced
- part throttle fuel economy is improved

Automatic Idle Speed (AIS) Motor

The AIS motor, Figure 10-17, is a reversible electric motor controlled by the logic module based on information from the:

- TPS.
- CTS.
- speed sensor.
- P/N switch.
- brake switch.

The AIS motor moves a valve to control airflow through a channel in the throttle body to bypass the throttle blade. With the valve in its most closed position, enough air passes through the throttle body to support the engine with no load at a minimum idle speed. The logic module commands the AIS motor to move the valve to provide a preprogrammed idle speed for all operating conditions. During deceleration the valve opens to prevent engine stalling and to compensate for the fuel that begins to evaporate from the intake port walls.

EGR Control Solenoid

This solenoid controls the ported vacuum to the EGR valve. When it is energized by the logic module, vacuum to the EGR valve is blocked. The solenoid is energized under the following conditions:

- coolant temperature below 21° C (70° F)
- engine speed below 1,200 RPM
- WOT

When the solenoid is de-energized, the EGR valve is additionally controlled by a back-pressure transducer, Figure 10-18. Some applications have a bleed in the vacuum line be-

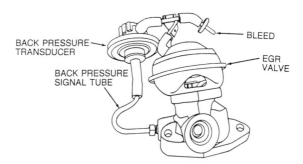

FIGURE 10-18   EGR valve and transducer. *Courtesy of Chrysler Motors.*

tween the transducer and the EGR valve. Its purpose is to prevent a pressure buildup in the EGR valve.

EGR Valve Temperature Sensor

In 1988 a thermistor-type temperature sensor was introduced on the California edition of the 1.5-liter multipoint injection engine used in the Colt. It is used to detect an EGR valve that fails to open when commanded to do so. When the EGR valve opens and exhaust gas flows through it, the temperature of valve's lower body goes up sharply. If the SMEC commands the EGR valve to be opened and does not then see a signal from the EGR valve temperature sensor indicating a temperature rise, it will store a fault code and turn on the malfunction indicator light. The EGR valve was eliminated on the turbo 2.2-liter engine in 1988.

Canister Purge Solenoid

This solenoid is in the purge line that runs from the throttle body to the charcoal canister. When coolant temperature is below 82° C (180° F), the logic module energizes the solenoid. This blocks the purge line, Figure 10–19.

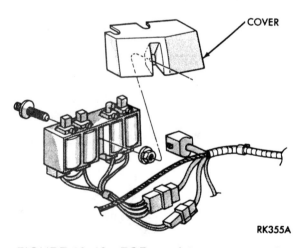

FIGURE 10–19   EGR, canister purge, wastegate control, and BARO-read solenoids. *Courtesy of Chrysler Motors.*

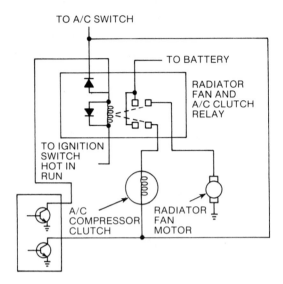

FIGURE 10–20   Radiator fan relay. *Courtesy of Chrysler Motors.*

Radiator Fan Relay

The radiator fan relay, Figure 10–20, supplies power for both the cooling fan motor and the A/C compressor clutch. The normally open relay can be activated (ground the relay coil) by the logic module or when the A/C is turned on (the A/C and blower motor are controlled by ground side switching). The logic module activates the relay when coolant temperature reaches a specified value.

Charging Circuit Control

Beginning in 1985 the logic module, working through the power module, assumed control of the alternator's output; they replaced the voltage regulator. The logic module's decisions are based on output voltage and battery temperature. When a criterion output voltage as sensed by the logic module is reached, it duty-cycles a switching transistor inside the power module. This transistor is used to ground the alternator's field, Figure 10–21. The duty cycling varies as necessary to control output voltage at the desired level. The criterion voltage varies based on battery temperature.

A/C Cut-out Relay

The normally open A/C cut-out relay is in the ground side of the A/C compressor clutch circuit and between the compressor clutch and the A/C switch. When open it breaks the A/C clutch circuit. It, too, is controlled by the logic module. If the A/C is turned on, the logic module will keep the A/C cut-out relay energized (closed) whenever the engine is running except:

- when the engine is operated at WOT.
- when engine speed is below 500 RPM.
- when the engine is cranked with the A/C on. It remains de-energized for ten to fifteen seconds after the engine is started.

Lock-up Converter

Beginning in 1988, the Torqueflite transmission was equipped with a SMEC-controlled lock-up converter for some vehicle applica-tions. The converter clutch is hydraulically applied and is controlled by a computer-actu-ated solenoid. The solenoid is mounted on the valve body transfer plate and is fed third clutch oil when the transmission goes into third gear. If the solenoid is activated (grounded by the computer) while third clutch oil pressure is present, the clutch is applied. If it is not, the non-applied solenoid allows the oil to be bled off, and the clutch is released. Its operation is similar to the systems used by General Motors and Ford. The SMEC will not apply the clutch unless:

- the coolant temperature is above 150° F
- park/neutral switch indicates the trans-mission is in gear
- brake switch indicates that the brakes are not applied
- TPS is about a minimum value

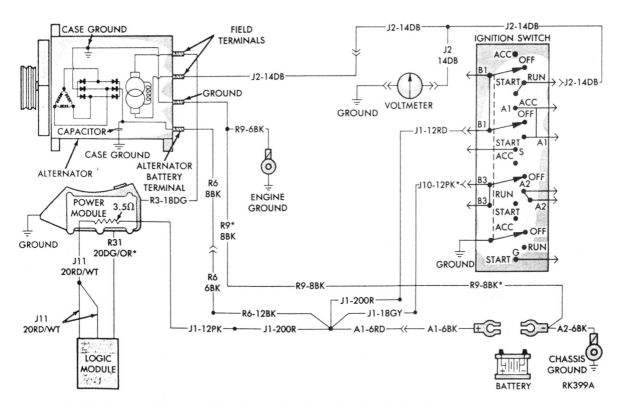

FIGURE 10–21   Charging circuit control. *Courtesy of Chrysler Motors.*

Power Loss Lamp

This is a lamp on the instrument panel and that the logic module controls. (On some early applications, it was on the logic module and could be seen through a hole in the module housing.) It is turned on as a bulb check for three seconds each time the ignition is turned on. If the logic module sees a fault in any of its most vital circuits, it will turn the lamp on and put the system in limp-in mode. If the fault is somehow cleared up, the lamp will remain on and the system will stay in the limp-in mode until the ignition is turned off. When the engine is started again, the lamp will be off and the system will no longer be in limp-in; but the code will remain in memory until the ignition has been cycled thirty times without the problem reappearing.

The Limp-in strategy was changed somewhat for 1988. On those and later applications, if a sensor fails, the computer will use a substitute value for the failed sensor and the Power Loss Lamp (referred to as a Check Engine Light on some newer applications) will come on as in the past. But if the sensor's signal comes back into proper range, the computer will go back to normal function without the ignition having to be cycled as on past models.

Top Speed Limit

Beginning in 1988, vehicles equipped with 3.0-liter or 2.2-liter Turbo I engines have a speed limit feature programmed into the SMEC. If the vehicle exceeds 118 MPH, the computer stops fuel injection to prevent speeds that exceed the speed rating of the original equipment tires.

SMEC-Controlled Cruise Control

Many SMEC-controlled systems incorporate cruise control as one of the functions that the engine control computer is responsible for. The system operates similarly to the Ford system described earlier in the **Outputs** section of Chapter 7. One feature it has in addition to those on the Ford system is a third solenoid called a dump solenoid. The ignition switch powers the dump solenoid through a fuse, the cruise control on/off switch, and the brake switch. The circuit has a fixed ground. Turning the on/off switch off or stepping on the brake will interrupt the power to the dump solenoid and it will immediately bleed off vacuum to the servo, allowing the throttle to close.

SYSTEM DIAGNOSIS AND SERVICE

A thorough diagnostic procedure is published by Chrysler Corporation for each model year and application. This *driveability test procedure* should be in the service manual you are using.

On-board Diagnostics

The logic module monitors specific input and output circuits for faults (voltage values that are too high or too low or that do not change appropriately). If a fault is detected, a two-digit code number will be stored in the module's diagnostic memory. The codes and other diagnostic information can be obtained by putting the system in one of its diagnostic modes. This can be done with the diagnostic readout box (tool C-4805) or its equivalent. The fault codes can also be presented and the input switches tested with just the power loss lamp.

Test Modes

Diagnostic Test Mode. Connect the readout box to the diagnostic test terminal under the hood. Activate the test by turning the ignition on–off–on–off–on within five seconds. The readout box displays any stored codes. After the codes are completed, a 55 is displayed.

Switch Test Mode. If the above test is completed and the 55 appears (be sure that all system input switches are turned off), turning on any of the input switches will cause the displayed 55 to change if the logic module receives the signal. Turning the switch off causes the 55 to reappear.

Circuit Actuator Test Mode (ATM). If after the diagnostic test mode is completed and the 55 appears the ATM button on the readout box is pressed and held until the desired ATM code number is displayed and then released, the ATM will be initiated. The logic module now cycles the selected actuator on and off at two-second intervals for five minutes or until the ignition is turned off. This allows the technician to verify that the actuator and circuit are operating properly. The ATM code numbers are two-digit numbers that represent each of the actuator circuits that this mode can test. They should be listed in the service manual along with the fault codes.

Sensor Test Mode. Added to system capability in 1986, this test allows the technician to select individual sensors, and the readout box will display the voltage signal that the selected sensor is sending to the logic module.

Engine Running Test Mode. This test allows you to check O_2 sensor switching, idle speed motor operation and, on turbo applications only, knock sensor performance. Connect the readout box or its equivalent to the diagnostic test connector. Start the engine and observe the display on the readout box. If the O_2 sensor is hot enough, the display will switch between 0 and 1 (0 = lean and 1 = rich). The readout box can be used to command the AIS motor to increase idle speed, and on turbo applications, striking the intake manifold near the knock sensor as directed in the service manual will cause an 8 to appear on the display along with the switching 0–1 (the 8 means that timing is being retarded).

Test Mode 10. The random access memory (RAM) of the SMEC module is less volatile than that of the Logic Module. Therefore, removing battery power from it—the procedure used to clear the memory for Logic Module-equipped vehicles—may not erase stored fault codes for several minutes. Because of this, a new sensor test mode has been added. When sensor test mode 10 is selected, fault codes are erased.

There are also some additional fault codes and some revisions in diagnostic procedures. For example, the harness connector on the SMEC is hard to probe with test instruments. As a result, diagnostic procedures may have you disconnect the harness and make open circuit tests rather than closed circuit tests.

Testing without the Readout Box. Cycling the ignition switch on and off as described earlier without the readout box connected triggers the logic module to read out the codes through the power loss lamp. It flashes on and off to indicate the code number. For example, two flashes separated by a short pause of about one-half second followed by a pause of about two seconds and then followed by three flashes separated by a short pause again indicates a code 23. Once the codes are presented, the system is ready for the switch test. If it is receiving the signals, the logic module will turn the power loss lamp on and off in response to its input switches being turned on and off. None of the other test modes can be performed without the readout box or its equivalent.

Driveability Test Procedure

The driveability test procedure is designed to use the test modes just described and to help diagnose engine performance complaints that are not caused by parts of the computer system. It should be used whenever driveability complaints are being diagnosed. The first step is to verify the complaint. This probably requires that the vehicle be operated under the same conditions in which the owner experiences the problem.

The second step is the visual inspection. The driveability test procedure provides a guide for this step. If the problem is not located during the visual inspection, the most appropriate driveability test section should be selected and used. If the engine does not start, select the no start section. If the problem occurs only with a cold engine, use driveability test cold section, and so forth. Once a driveability test section is selected, begin with the first step of that section.

PERSONAL SAFETY: Because of the high fuel pressure in many of these systems, it is very important to relieve the pressure before attempting to open the fuel system. A serious safety hazard is created if a fitting is loosened while the system is still under pressure. Only the correct part number fuel system hose and hose connections should be used when making repairs.

REVIEW QUESTIONS _____

1. Name four different operating modes of the single-point and/or multipoint system.
2. What is the logic module's power supply?
3. Name at least three things that the power module controls with instructions from the logic module.
4. What is the function of the automatic shut-down relay?
5. What purpose does the adaptive memory serve?
6. What is the purpose of the bleed hole in the MAP sensor line used on some applications?
7. What two signals does the logic module receive from the distributor on turbocharged applications?
8. Name at least four inputs that affect ignition timing.
9. Name two functions that the fan control relay controls.
10. Describe how spark knock is controlled on turbocharged applications.
11. Describe four strategies used to control boost pressure.
12. Name four inputs whose failure puts the system into limp-in.
13. Name an input that causes the engine to stop.

14. What causes the power loss lamp to come on?
15. Name four test modes used for driveability problem diagnosis.
16. What are the first two steps of the diagnostic procedure?

ASE-type Questions. (Actual ASE test questions will probably not be as product specific.)

17. Technician A says that the logic module prefers to use the AIS motor to control idle speed. Technician B says that the logic module prefers to use spark timing to control idle speed. Who is correct?
    a. A only   b. B only
    c. both A and B
    d. neither A nor B
18. Technician A says that on late-model Chrysler fuel-injected engines the logic and power modules control charging circuit voltage. Technician B says that the charging system on those vehicles no longer has its own voltage regulator. Who is correct?
    a. A only   b. B only
    c. both A and B
    d. neither A nor B
19. Technician A says that the engine running test mode of the Chrysler diagnostic sequence allows the technician to test individual actuator circuits for proper operation. Technician B says that technician A is wrong. Who is correct?
    a. A only   b. B only
    c. both A and B
    d. neither A nor B
20. A car equipped with Chrysler's EFI system is brought into the shop with a complaint of poor idle during warm-up. What driveability test section should be used to diagnose the problem?
    a. driveability test cold
    b. driveability test warm
    c. both A and B
    d. neither A nor B

11 CHRYSLER'S OXYGEN FEEDBACK SYSTEM

Objectives

After studying this chapter, you will be able to:

- identify the functions controlled by the O_2 feedback system.
- identify the system's operating modes.
- identify the conditions during which the system stays in closed loop.
- describe the three different air/fuel mixture control solenoids and their methods of control.
- describe the diagnostic procedure used to locate the cause of driveability problems on models without on-board diagnostics.
- describe the diagnostic procedure used to locate the cause of driveability problems on models with on-board diagnostics.

The oxygen feedback (O_2 feedback) system was Chrysler's first comprehensive computerized engine control system, and it has continued to be the system used for carbureted vehicles. The first system, introduced in California in 1979 on the 3.7-liter slant 6, was little more than the existing spark control computer system with the addition of an oxygen sensor and a mixture control solenoid. In 1980 the system was in limited production on four-, six-, and eight-cylinder engines. By 1981 it was in use nationwide and was beginning to take on additional responsibilities such as limited control of idle speed, EGR, and air injection switching.

As development of the system continued, some applications came to control such functions as:

- shift indicator light.
- coolant fan.
- a solenoid that controlled vacuum to the vacuum-operated secondary throttle blade for altitude engines.

Figure 11–1 presents an overview of the system.

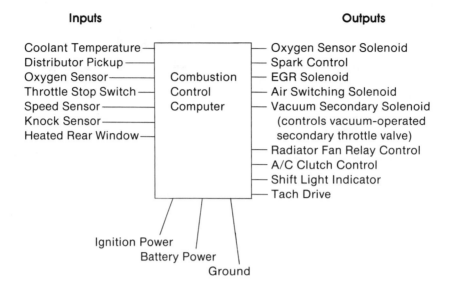

Inputs

Coolant Temperature
Distributor Pickup
Oxygen Sensor
Throttle Stop Switch
Speed Sensor
Knock Sensor
Heated Rear Window

Combustion
Control
Computer

Outputs

Oxygen Sensor Solenoid
Spark Control
EGR Solenoid
Air Switching Solenoid
Vacuum Secondary Solenoid
(controls vacuum-operated
secondary throttle valve)
Radiator Fan Relay Control
A/C Clutch Control
Shift Light Indicator
Tach Drive

Ignition Power
Battery Power
Ground

Not all inputs or outputs are used on all Oxygen Sensor Feedback applications.

FIGURE 11–1   O₂ feedback system overview

COMBUSTION CONTROL COMPUTER

On the first applications, the combustion control computer was located in the air cleaner and looked very similar to that of the lean burn system of the seventies. On later applications it can be located in other areas of the engine compartment. It is often referred to as the spark control computer. In 1985 a self-diagnostic function and memory were added to the computer to facilitate the diagnostic process.

Open Loop

In open loop the computer does not respond to the oxygen sensor. The system is in open loop during any of the following conditions:

- coolant below required operating temperature (49° C [120° F])
- oxygen sensor below its required operating temperature (315° C [600° F])

- idle
- acceleration (low manifold vacuum)
- deceleration (high manifold vacuum)
- hot engine restart
- oxygen sensor failure

Closed Loop

In closed loop the computer uses input from the oxygen sensor as well as other sensors to determine fuel-metering calibrations. The system goes into closed loop after:

- the coolant temperature reaches a criterion temperature (about 49° C [120° F]).
- the oxygen sensor starts producing voltage (must be around 315° C [600° F]).
- a predetermined amount of time has elapsed since the engine was started.

INPUTS

Coolant Temperature

Three different types of coolant temperature-sensing devices have been used on differ-

314

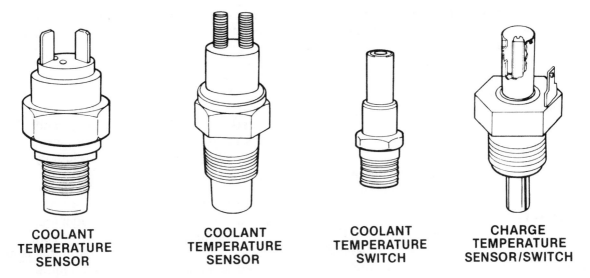

| COOLANT TEMPERATURE SENSOR | COOLANT TEMPERATURE SENSOR | COOLANT TEMPERATURE SWITCH | CHARGE TEMPERATURE SENSOR/SWITCH |

FIGURE 11-2   Temperature sensors/switches. *Courtesy of Chrysler Motors.*

ent O_2 feedback applications: a thermistor, a dual thermistor, and a switch, Figure 11–2.

Thermistors. The single-element thermistor was first used on six- and eight-cylinder engines and later used on four-cylinder engines. Still later, about 1985, some of the transverse engine applications were equipped with a dual-element thermistor. One element is

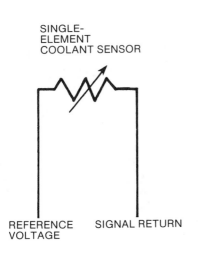

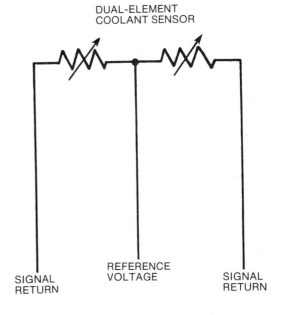

FIGURE 11-3   Coolant sensors

315

used for fuel, spark, EGR, and canister purge control; the other is used for radiator fan relay control, Figure 11-3. Some O_2 feedback systems used a separate coolant temperature sensor in the radiator receiving tank for control of the radiator fan relay.

Switch. Most four-cylinder engines through 1983 were equipped with a temperature-sensing switch instead of a thermistor. By closing at a specific temperature, the switch tells the computer whether the coolant temperature is above or below the criterion temperature.

Charge Temperature

On some six- and eight-cylinder applications, a temperature sensor is mounted in the intake manifold to sense air/fuel mixture temperature. Keep in mind that the heated air inlet system brings incoming air temperature up to normal operating temperature before the coolant reaches normal temperature. The computer's operating strategy on some O_2 feedback systems requires that it knows both air/fuel mixture temperature and coolant temperature.

Distributor Pickup

Engine speed and crankshaft position information are provided by a Hall-effect switch located in the distributor on four-cylinder engines, Figure 11-4, and by the dual pickup coils in the distributor of most six- and eight-cylinder engines, Figure 11-5.

Vacuum Transducer

Vacuum is sensed by a vacuum *transducer*. The transducer is mounted on the computer housing and contains a diaphragm that is attached to a movable metal core, Figure 11-6. Either manifold or ported vacuum is connected to the transducer. As the vacuum changes, the diaphragm and the core move. The computer passes an AC current through a coil that surrounds the core. The alternating magnetic field

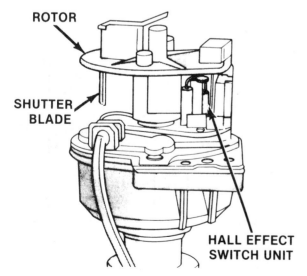

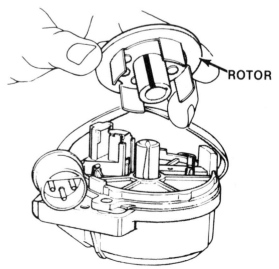

FIGURE 11-4   Hall-effect switch. *Courtesy of Chrysler Motors.*

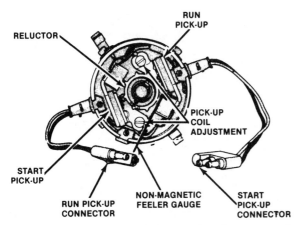

FIGURE 11-5 Dual pickup units. *Courtesy of Chrysler Motors.*

produced by the AC current produces an *inductance,* an electromotive force produced by current variation within the coil. The inductance changes as the core's position in the coil changes; during low vacuum the core is extended farther into the core and the inductance is reduced. The inductance impedes the AC current flow through the coil. This particular type of *impedance,* or resistance, is called *reactance* and is expressed in ohms. The computer measures volt drop across the coil and can therefore determine engine vacuum.

Oxygen Sensor

The O_2 feedback system uses a single-wire oxygen sensor.

Throttle Stop Switch

The O_2 feedback system does not use a throttle position sensor. It does, however, use the throttle linkage as a switch to indicate to the computer when the throttle is closed, Figure 11-7. As the throttle closes, a contact on the throttle lever contacts the throttle stop, which acts as a ground contact. The path to ground is completed, and a voltage drop occurs across the circuit's resistor inside the computer. This alerts the computer that the throttle is closed.

Note: The 1979 system used a type of throttle position sensor called a throttle position transducer.

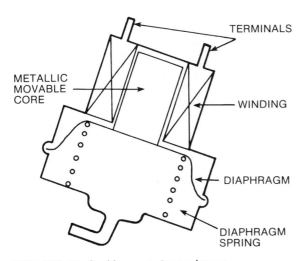

FIGURE 11-6 Vacuum transducer

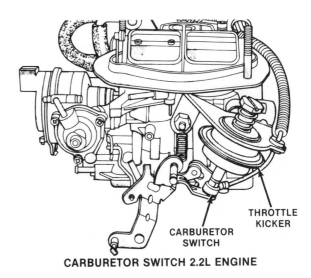

CARBURETOR SWITCH 2.2L ENGINE

FIGURE 11-7 Throttle stop switch. *Courtesy of Chrysler Motors.*

Speed Sensor

Vehicle speed information is provided on many 1985 and later systems by a pulse generator mounted on and driven by the transmission.

Detonation Sensor

A piezoelectric knock sensor is used on high-performance four-cylinder and eight-cylinder engines equipped with four-barrel carburetors.

Heated Rear Window

Some vehicles with a heated rear window provide an input signal to the computer when the window heater is turned on. This allows the computer to increase throttle opening during idle to support the additional load that the heavy current load of the heat element causes the alternator to put on the engine.

OUTPUTS

Mixture Control

The O_2 feedback system has used three different mixture control solenoids, which Chrysler calls O_2 feedback solenoids, on four different carburetors.

Holley Model 6145. This carburetor was used in 1979 and 1980 on California applications of the 3.7-liter engine, Figure 11–8. The O_2 feedback system controls idle mixture by way of the idle air bleed and the main metering mixture by way of fuel metering, similar to the operation of the solenoid used on the Holley 6149 discussed in the **Outputs** section of Chapter 7. Air for the idle and low-speed system air bleed is supplied through a fixed orifice plus a passage controlled by a metering rod. The metering rod is controlled by a vacuum diaphragm. For main metering control, a second diaphragm controls a spring-loaded rod and foot; the spring wants to drive the rod down. If the diaphragm allows the rod's foot to come down, it will force open a valve that when open allows additional fuel to enter the main metering circuit.

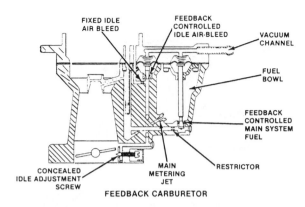

FIGURE 11–8   Holley 6145 mixture control circuit. *Courtesy of Chrysler Motors.*

Vacuum to the two diaphragms is controlled by a remote-mounted, computer-controlled solenoid, Figure 11–9. The lower portion of the unit is a simple vacuum regulator valve. It receives manifold vacuum and reduces it to 5 inches of mercury (Hg). The 5 inches of Hg is then supplied to a valve seat higher in the assembly. This valve seat is closed by a conical tip on the lower end of the solenoid armature when the solenoid is not energized by the computer. When the solenoid armature is in this position, the assembly's vacuum output port is exposed internally to the vent port, and

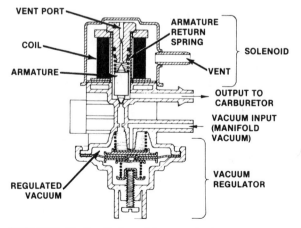

FIGURE 11–9   Solenoid-operated vacuum regulator. *Courtesy of Chrysler Motors.*

vacuum at the output port is zero. When the computer energizes the solenoid, the armature moves up and a second conical tip on the top of the armature closes off the vent exposure. The lower passage is now open, and the 5 inches of Hg is exposed to the output port. The computer duty cycles the solenoid at whatever rate is necessary to maintain the desired output vacuum (between 0 and 5 Hg) at the two diaphragms in the carburetor. High vacuum produces a leaner mixture in both idle and main metering circuits; low vacuum produces a rich mixture for either circuit.

Holley 6520 and Later Applications of the Holley 6145. In 1980 the two-barrel Holley 6520 was equipped with an O_2 feedback solenoid for four-cylinder engines, and in 1981 the same type solenoid was applied to the Holley 6145 for six-cylinder engines. This solenoid is essentially the same as the one used in the Holley carburetor used on some General Motors' CCC applications discussed in the **Outputs** section of Chapter 3.

Carter BBD and Thermo-Quad. The BBD two-barrel and Thermo-Quad carburetors were equipped with an O_2 feedback solenoid in 1980 for eight-cylinder applications. This solenoid controls the air bleed for both the idle and main metering circuits.

At idle air is brought through a passage from above the venturi area to the O_2 feedback solenoid valve, Figure 11-10. When the solenoid is energized, air is allowed into the idle circuit and the mixture is lean. When the solenoid is off, air is blocked from the idle circuit and the mixture is rich. The main metering circuit air bleed draws air from the same passage under the same O_2 feedback solenoid control, Figure 11-11. The circuit being metered depends on the circuit from which fuel is being drawn.

Ignition Timing

Ignition timing in the O_2 feedback system is still based on the strategy of the earlier lean burn system. The negative side of the ignition coil is switched to ground by the computer. During cranking, the distributor pickup signal triggers the computer to open the primary circuit. The spark plugs are fired at base timing.

Once the engine starts and if it is cold, the computer will begin to add spark advance

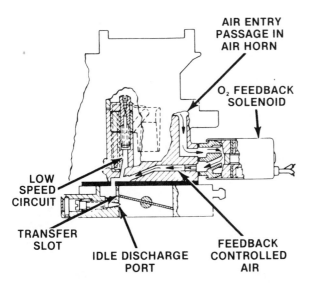
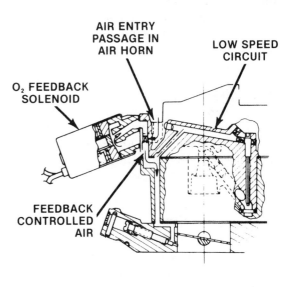

FIGURE 11-10   Carter carburetor mixture control idle circuit. *Courtesy of Chrysler Motors.*

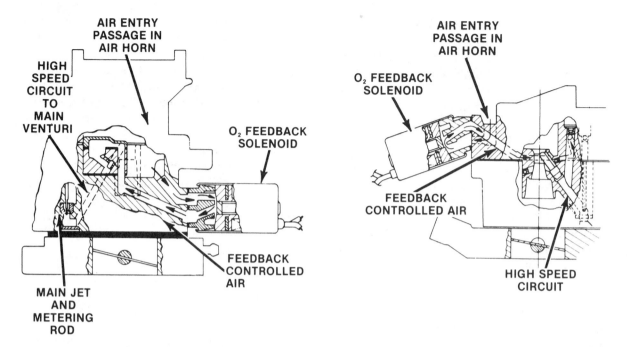

FIGURE 11-11   Carter carburetor mixture control main metering circuit. *Courtesy of Chrysler Motors.*

based on coolant temperature, engine speed, and engine vacuum. Once the coolant reaches a criterion temperature, advance is dependent on vacuum and RPM. If the throttle stop switch closes, the amount of advance issued in response to engine vacuum will be removed.

EGR Solenoid

Only some of the later O_2 feedback applications use the computer to control EGR valve operation. On these applications the computer controls a solenoid that either allows or denies vacuum to the EGR valve. Vacuum is denied during:

- cold engine operation.
- light throttle operation.
- the first minute or so following hot engine restarts.
- WOT operation.

Air Switching Solenoid

On O_2 feedback systems that control the air injection system, a computer-controlled solenoid controls vacuum to the switch relief valve, Figure 11-12. During cold engine operation, the computer commands that vacuum be applied to the valve's diaphragm and air pump air is directed upstream to the exhaust manifold. When the engine is hot enough to go into closed loop, vacuum to the valve's diaphragm is cut off and the air is directed downstream to the catalytic converter. For a brief period after each engine start-up, the computer commands vacuum to be applied so that injection air is sent upstream. In this system there is no provision for pump air to be dumped into the atmosphere except as a pressure relief function.

On systems where the computer does not control the air injection system, a similar strategy is achieved by using a coolant temperature-controlled vacuum valve in place of the computer-controlled solenoid.

Vacuum Secondary

The Holley 6520 is a progressive two-barrel carburetor with a vacuum-operated secondary throttle blade. Some later applications of this carburetor feature a solenoid-operated vacuum valve in the vacuum line that supplies venturi vacuum to the diaphragm that opens the secondary throttle blade. This computer-controlled solenoid vents the line and does not allow the secondary throttle blade to open until:

- coolant temperature reaches 60° C (140° F).
- vacuum drops below a criterion value.
- engine speed is above a specified RPM.

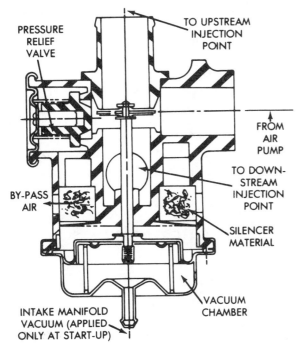

PRESSURE RELIEF VALVE

TO UPSTREAM INJECTION POINT

FROM AIR PUMP

TO DOWN-STREAM INJECTION POINT

BY-PASS AIR

SILENCER MATERIAL

VACUUM CHAMBER

INTAKE MANIFOLD VACUUM (APPLIED ONLY AT START-UP)

FIGURE 11-12 Air injection switch/relief valve. *Courtesy of Chrysler Motors.*

Radiator Fan Relay

The radiator fan relay is used on later model vehicles with transverse engines and electric cooling fans. It functions like the one discussed in the **Outputs** section of Chapter 10.

Shift Light Indicator

On 1985 and 1986 models with manual transmissions, the computer controls a light that is on the instrument panel and that indicates to the driver the optimum time to make the next upshift.

Tach Drive

On some 1986 models, the computer operates a tachometer on the instrument panel. The tach drive signal comes from the distributor side of the ignition coil.

Other features often included on engines with O_2 feedback systems but not controlled by the computer are:

- a vacuum-powered throttle kicker.
- a choke heater.
- a carburetor bowl vent solenoid.

These units are usually self-grounded and are activated by turning on the ignition switch.

SYSTEM DIAGNOSIS AND SERVICE

Diagnostic Procedure 1979-1984

Prior to 1985 the O_2 feedback system computer had no self-diagnostic capability. Diagnosis of the cause of a no-start condition or driveability complaint is performed with the aid of the driveability test procedures booklet or its equivalent. Chrysler has published such a booklet for each system application.

Diagnosis is divided into three parts: verify the complaint, do a visual inspection, and perform the appropriate test steps. The driveability test procedure outlines the visual inspection and the test steps. The test procedures are also divided into three parts: no start, driveability test cold, and driveability test warm. If the engine does not start and the visual inspection is completed, begin with the first test of the no start section. If the complaint concerns cold driveability, begin with the first test of the driveability test cold section and likewise for a driveability complaint that only occurs when the engine is warm.

If you begin with the first test of the appropriate section and then go to the next test indicated by the results of the test just completed, the test procedure will take you through all of the components and systems that could cause the driveability complaint, including systems not controlled by the computer.

Diagnostic Procedure, 1985 and Later

With the addition of self-diagnosis, the diagnostic test procedure has been modified to make use of the fault codes and other self-diagnostic features. The fault codes can be recorded for the most vital sensors and actuators.

The system is put into one of its diagnostic modes with the diagnostic readout box (tool C-4805 or its equivalent). The 1985 system has the following three different diagnostic modes:

- *diagnostic test mode*—The computer reads out any stored fault codes.
- *circuit actuation test mode (ATM test)*—The technician selects an actuator circuit and activates the test mode. The computer then turns that circuit on and off at quarter-second intervals for five minutes while allowing the technician to inspect the actuator for proper operation.
- *switch test mode*—This allows the technician to verify that input switch signals are being received by the computer.

Assuming that the no start test procedures are not required and the visual inspection is completed, the diagnostic test mode is the first test of the driveability test procedure. The latter two test modes are most useful if used when called for by one of the remaining driveability test procedures.

The 1986 system was upgraded still more by adding a sensor test mode to the other three self-diagnostic test modes. In this mode the computer reads out the sensor input voltage values through the readout box.

Reading O₂ Solenoid Calibration

A voltmeter can be connected across the O_2 solenoid to read the calibration commands being sent to the solenoid to maintain the desired air/fuel ratio. A voltage reading closer to battery voltage indicates a consistently lean command, probably in response to a rich condition. A voltage reading nearer to zero volts indicates a rich command either in response to a lean condition or an operating condition that requires a rich mixture.

Service Tips

Overcompensation. An overrich carburetor condition or excessive fuel introduced by the charcoal canister or the PCV system can cause the system to overcompensate and go too lean while in closed loop. On the other hand, a vacuum leak can cause the system to go overrich.

REVIEW QUESTIONS _____

1. What year did the O_2 feedback system come into wide-scale use?
2. List the operating modes of the O_2 feedback system.
3. List seven operating conditions that keep the system in open loop.

4. What device is used to monitor engine load?

5. List at least five engine support functions that the combustion control computer controls.

6. List the three types of mixture control solenoid used on different O_2 feedback engine applications by their carburetor application.

7. For each of the mixture control solenoids identified by carburetor application, identify the medium (air or fuel) used to control the mixture during idle and part throttle operation.

Carburetor	Idle	Off Idle
_____	_____	_____
_____	_____	_____
_____	_____	_____

8. When the engine is cold, what inputs are used to determine ignition timing?

9. What inputs are used to determine ignition timing when the engine is warm?

10. During what operating conditions is the EGR valve turned off?

11. What is the function of the vacuum secondary solenoid?

12. List the three parts of the driveability test procedure for vehicles without on-board diagnostics.

13. List the four parts of the driveability test procedure vehicles with on-board diagnostics.

ASE-type Questions. (Actual ASE test questions will probably not be as product specific.)

14. Technician A says that O_2 feedback systems do not use a throttle position sensor. Technician B says that during one model year in the early development of O_2 feedback, a type of throttle position sensor was used. Who is correct?

a. A only
b. B only
c. both A and B
d. neither A nor B

15. Technician A says that there is no easy way to determine if the O_2 feedback system is tending to run rich or lean other than by checking for oxygen sensor codes on later models. Technician B says that a voltmeter can be connected to the O_2 solenoid to determine if the system is tending to run rich or lean. Who is correct?

a. A only
b. B only
c. both A and B
d. neither A nor B

16. A 1981 O_2 feedback system is brought into the shop with a driveability problem. Technician A starts to perform the driveability test warm procedure because the customer says the problem occurs when the engine is warm. Technician B says that technician A should not start with the driveability test warm procedure. Who is correct?

a. A only
b. B only
c. both A and B
d. neither A nor B

GLOSSARY

IMPEDANCE   The total resistance to current flow within an AC circuit.

INDUCTANCE   The phenomenon by which electromotive force is induced as a result of a variation of current flow within a circuit.

REACTANCE   Resistance, expressed in ohms, within an AC circuit that is a result of inductance or capacitance.

TRANSDUCER   A device that receives energy from one system and transfers it to another system, usually in a different energy form.

12 ROBERT BOSCH CORPORATION'S MOTRONIC

Objectives

After studying this chapter, you will be able to:

- identify the operating conditions that invoke different operational modes.
- describe how the control unit determines engine load.
- identify the major input sources.
- identify the controlled functions that are standard features of Motronic.
- identify the two major steps of the diagnostic procedure.

The Robert Bosch Corporation builds engine support components and control systems that any vehicle manufacturer can purchase to use on their vehicles. The Bosch Corporation is well known for its computer-controlled fuel injection systems. Although later versions of such systems as the L-Jetronic and K-Jetronic feature an oxygen sensor, they are still single-function systems in that they only control fuel metering and therefore are not discussed in this text. The Motronic is Bosch's first multi-function engine control system. Motronic's primary concern is the control of air/fuel mixture, ignition dwell, and ignition timing. Its other function control capabilities such as idle speed control, EGR control, and turbo boost control are offered as options to vehicle manufacturers.

CONTROL UNIT

Motronic grew out of the L-Jetronic system and like its predecessor is a multipoint injection system with fuel metering controlled by pulse width. The L-Jetronic's analog computer was replaced by a digital unit with additional function control capability such as spark timing, idle speed control, and turbo boost control. An overview of the Motronic system is provided in Figure 12–1. The control unit's

Bosch and Imports

Of the imported European cars equipped with a comprehensive computerized engine control system, most use the Bosch system. It first appeared on selected BMW models. Most of the systems used on Japanese cars for export to the United States are built using Bosch patents. In fact all of the gasoline engine, multipoint fuel injection systems used on U.S.-built vehicles use some Bosch components.

mount brackets serve to dissipate heat from the transistors that drive the ignition and injectors, Figure 12–2. A thirty-five–pin connector connects to the vehicle's system harness.

Main Relay

A main relay, Figure 12–3, similar in function to the power relay used on the Ford EEC systems (Chapters 7 and 8), powers the control unit when the ignition is turned on. This relay also contains a diode to protect the control unit against accidental reversed polarity.

On-board Diagnostics

The first few model year applications of Motronic do not feature any form of self-

diagnosis except that if a fault is detected in the closed-loop operating circuit, a calculated pulse width aimed at maintaining an air/fuel mixture near stoichiometric will be used.

OPERATING MODES _____

The control unit is programmed with different operational strategies as operating conditions change. Most of the different operational strategies are conducted as part of the open-loop mode.

Cranking

Two fuel-metering programs can be employed during cranking. One is based on crank-

Inputs	Control Unit	Outputs
Airflow Meter		Fuel Injection
Engine Speed		Fuel Pump Relay
Crankshaft Position		Ignition Timing
Throttle Position		Dwell Control
Coolant Temperature		RPM Limit
Air Temperature		Peak Coil Current Cutoff
Lambda Sensor (oxygen sensor)		
Knock Sensor		The following functions are
Altitude Sensor	Main Relay	system options (available to the
Starter Signal		vehicle manufacturer):
Battery Voltage		
		Rotary Idle Adjuster
		Turbo Boost Control
		EGR Control
		Canister Purge
		Transmission Control
		Start-Stop Control

Noncomputer-Controlled Functions: Fuel Pressure Regulation, Fuel Pressure Pulsation Damper, Cold-Start Injector and Thermo-Time Switch, and Auxiliary Air Device (used in place of Rotary Idle Adjuster)

FIGURE 12–1   Overview of Motronic system

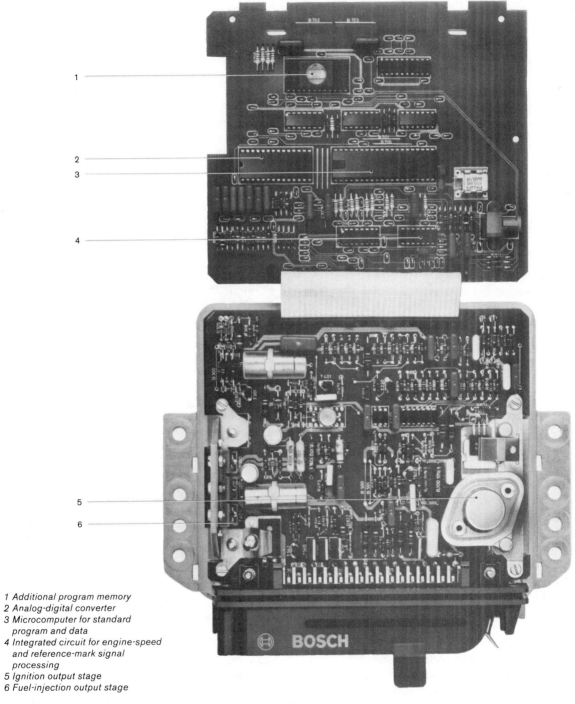

1 Additional program memory
2 Analog-digital converter
3 Microcomputer for standard
   program and data
4 Integrated circuit for engine-speed
   and reference-mark signal
   processing
5 Ignition output stage
6 Fuel-injection output stage

FIGURE 12-2   Control unit. *Courtesy of Robert Bosch.*

FIGURE 12-3   Main relay. *Courtesy of Robert Bosch.*

ing speed, the other on temperature. At lower speeds the quantity of fuel introduced is constant regardless of fluctuations in airflow (airflow sensor input is not reliable under these conditions). At higher cranking speeds, air intake diminishes slightly as a result of lower volumetric efficiency, so fuel quantity is reduced. During cold cranking the control unit superimposes an enrichment program over the speed-dependent program. Some applications provide enrichment from a cold-start injector in the intake manifold, Figure 12-4, and some use additional pulses from the port injectors. Ignition timing is also adjusted as a function of cranking speed and temperature. The cold-start injector system is the same as that described in the **Outputs** section of Chapter 5 (an example of a Bosch subsystem being used in a U.S.-built fuel injection system).

On applications that do not use a cold-start injector, the control unit pulses the port injectors several times during each crankshaft revolution as opposed to the one pulse per engine cycle used during normal operation. This is to enhance fuel evaporation and to avoid fouling the spark plugs with wet fuel. The enrichment decays to zero over a specific number of engine revolutions. The number depends on engine temperature when cranking begins. If RPM reaches a criterion value before the specified number of revolutions has occurred, the cold-cranking enrichment program is stopped. The criterion RPM also varies with the engine temperature that exists when cranking begins.

Post-start

When the engine first starts, the intake port and combustion chamber surfaces are still cool and thus cause fuel to condense. During the brief time it takes the surfaces to begin heating up, the control unit initiates a post-start enrichment program to improve idle quality and throttle response. Once initiated this enrichment decays within a brief time to zero. The actual time depends on engine temperature. In this mode the ignition timing is advanced in relation to engine temperature to provide the most efficiency with the least possible amount of fuel.

Warm-up

After the post-start period has expired, any additional warm-up enrichment necessary is provided based on engine temperature and load. An idle speed increase is also provided to improve driveability and to hasten warm-up. The idle speed increase is initiated during the post-start mode and continues into the warm-up mode. As an added means of bringing the oxygen sensor and catalytic converter up to operating temperature as quickly as possible, some applications of Motronic retard ignition timing during warm-up.

Acceleration

If throttle opening suddenly increases from a prior fixed position, a brief fuel enrichment is commanded by the control unit. This function is similar in effect to the performance of an accelerator pump in a carburetor. The base acceleration enrichment program is modified during low engine temperature.

Full Load (WOT)

During wide-open throttle operation, the control unit commands a fixed, enriched air/

fuel mixture. To avoid fuel quantity calculation errors that could occur as a result of fluctuations in the airflow sensor, fuel quantity is calculated based on engine speed. The objectives during this operational mode are to produce maximum torque and to avoid detonation.

Overrun (Deceleration)

During deceleration fuel injection is stopped but over a short time period (tapered off), and ignition timing is retarded along with the fuel cut-off. The retarded timing provides

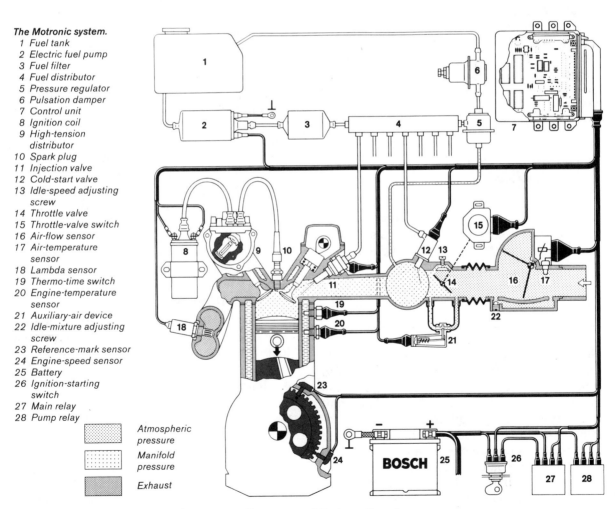

The Motronic system.

1 Fuel tank
2 Electric fuel pump
3 Fuel filter
4 Fuel distributor
5 Pressure regulator
6 Pulsation damper
7 Control unit
8 Ignition coil
9 High-tension distributor
10 Spark plug
11 Injection valve
12 Cold-start valve
13 Idle-speed adjusting screw
14 Throttle valve
15 Throttle-valve switch
16 Air-flow sensor
17 Air-temperature sensor
18 Lambda sensor
19 Thermo-time switch
20 Engine-temperature sensor
21 Auxiliary-air device
22 Idle-mixture adjusting screw
23 Reference-mark sensor
24 Engine-speed sensor
25 Battery
26 Ignition-starting switch
27 Main relay
28 Pump relay

Atmospheric pressure

Manifold pressure

Exhaust

FIGURE 12–4   The Motronic system. *Courtesy of Robert Bosch.*

better engine braking and reduces HC production in the event that fuel has been evaporated from the intake port walls. When engine speed falls below a programmed RPM or the throttle is opened again, fuel injection is restored to its appropriate level. This occurs over a programmed number of engine revolutions. Ignition advance is restored in the same way for a smooth transition. If the fuel cut-off is not used during deceleration, a fuel-enrichment program is employed to avoid engine bucking and high HC production (a result of lean misfires).

Closed Loop

The Motronic system goes into closed loop when the engine and lambda sensor reach normal operating temperature. It maintains a closed-loop mode as long as operating temperature of the lambda sensor is maintained and engine load does not require an air/fuel mixture richer than stoichiometric.

INJECTORS _____

The injectors are like those used in other U.S.-built multipoint injection systems. In fact all American manufacturers have historically used Bosch injectors for their multipoint injections.

INPUTS _____

Airflow Meter

The airflow meter is a vane type similar to that used on some of the Ford multipoint fuel injection systems (Chapter 7). Its vane shaft moves the wiper of a built-in potentiometer designed to provide accurate voltage signals to the control unit in spite of aging and rapid changes in temperature, Figure 12-5. By comparing crankshaft position input with airflow meter data, the control unit can determine the air intake for each cylinder's intake stroke. This determination is the basis for pulse width calculation.

Engine Speed Sensor

The engine speed sensor is a pulse-generating device that uses the flywheel teeth as its triggering device, Figure 12-6.

Reference Mark (Crankshaft Position) Sensor

The reference mark sensor is like the engine speed sensor except that its trigger is a reference mark on the flywheel, Figure 12-6. It provides the control unit with information about crankshaft position.

Throttle Valve Switch

The throttle valve switch contains two sets of electrical contact points: one for idle and one for WOT, Figure 12-7. Each set of contacts has its own circuit to and receives a reference voltage from the control unit. This enables the control unit to determine if each of the switches is open or closed. When the throttle blade is closed, the idle contacts are closed; during part throttle operation, both sets are open; and at WOT the full-load contacts are closed. The control unit can therefore recognize these three driving conditions. The switch is attached to the end of the throttle shaft.

Coolant Temperature Sensor

The coolant temperature sensor is a thermistor that screws into the water jacket near the thermostat housing.

Air Temperature Sensor

The air temperature sensor is a thermistor mounted in the intake opening of the airflow sensor, Figure 12-8. It measures income air temperature.

Lambda Sensor (Oxygen Sensor)

Bosch's term for oxygen sensor is *lambda sensor*.

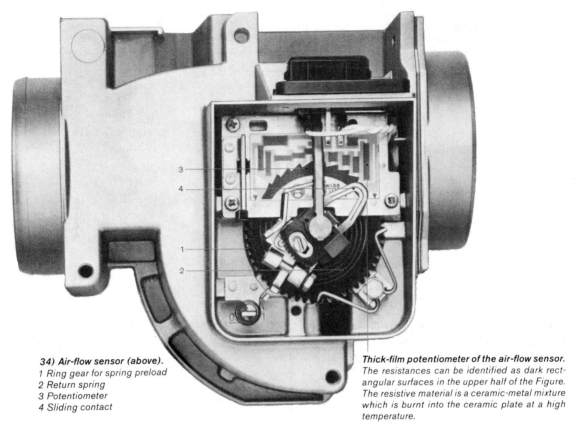

34) Air-flow sensor (above).
1 Ring gear for spring preload
2 Return spring
3 Potentiometer
4 Sliding contact

Thick-film potentiometer of the air-flow sensor.
The resistances can be identified as dark rectangular surfaces in the upper half of the Figure. The resistive material is a ceramic-metal mixture which is burnt into the ceramic plate at a high temperature.

FIGURE 12–5   Airflow meter. *Courtesy of Robert Bosch.*

1 Permanent magnet   5 Winding
2 Housing   6 Flywheel
3 Engine block   ring gear
4 Soft iron core   7 Reference mark

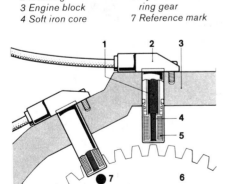

FIGURE 12–6   Engine speed and reference mark sensors. *Courtesy of Robert Bosch.*

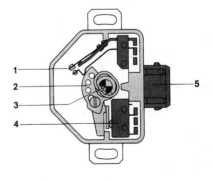

1 Full-load contact   4 Idle contact
2 Switch guide   5 Electrical
3 Throttle shaft   connection

FIGURE 12–7   Throttle valve switch. *Courtesy of Robert Bosch.*

1 Electrical connection, 2 insulation tube, 3 connector, 4 NTC resistor, 5 housing, 6 rivet pin, 7 securing flange. Arrow denotes the direction of intake air.

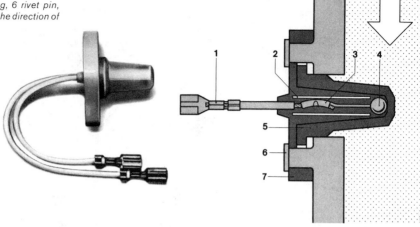

FIGURE 12–8   Air temperature sensor. *Courtesy of Robert Bosch.*

Determining Air/Fuel Ratio

Lambda (λ) is one of the Greek letters and is used here to represent a relationship that defines the air/fuel ratio. The relationship is:

Lambda = actual inducted air quantity ÷ the theoretical air quantity requirement

The theoretical air requirement is 14.7 units (by weight). If the actual air quantity used was 14.7, then lambda = 1. Using this method of expression, an air/fuel ratio richer than 14.7 is expressed, for example, as 0.9; 0.9 is equivalent to 13.23 to 1 (13.23/14.7 = 0.9). Conversely, an air/fuel ratio leaner than 14.7 is expressed, for example, as 1.03; 1.03 is equivalent to 15.14 to 1 (15.14/14.7 = 1.03).

Knock Sensor

The knock sensor is the piezoelectric type, Figure 12–9. This sensor produces a voltage signal in response to most engine vibrations, but the signal's strength is proportional to the vibration frequency. The signal produced in response to a vibration frequency of 5 to 10

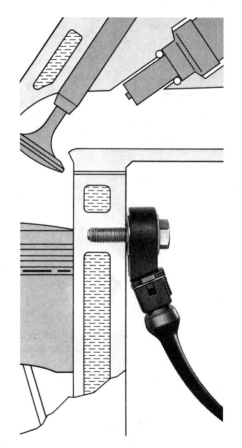

FIGURE 12–9   Knock sensor. *Courtesy of Robert Bosch.*

kilohertz is recognized as spark knock. The knock sensor is located on the block between two cylinders. On some applications two sensors are used.

Altitude Sensor

The altitude-sensing device or barometric pressure sensor is an aneroid unit attached to the wiper of a potentiometer similar to the one described in the **Inputs** section of Chapter 8. It is usually located in the air cleaner.

Starter Signal

The cranking control circuit provides a signal to the control unit when the starter is being cranked.

Battery Voltage

A system voltage input from the vehicle's electrical system is provided to the control unit so that it can monitor battery voltage.

OUTPUTS _____

Of the functions controlled by the control unit, injector operation, engine speed limit, fuel pump relay control, ignition timing, dwell control, and peak coil-current cut-off are standard features of Motronic. Other features are optional to the vehicle manufacturer.

Injectors

The injectors are solenoid-operated, pintle-style valves, as used in all of the other multipoint injection systems presented in this text. In open-loop operation, the control unit calculates pulse width based on engine load and engine speed; engine load is derived by computing airflow and engine speed. Figure 12–10 presents a three-dimensional graph or map that represents the relationship between load, speed, and air/fuel ratio for a given engine application; air/fuel ratio is expressed as lambda. This map is stored in the control unit's read-only memory and serves as the look-up table to determine pulse width. The arrows indicate increased load, speed, and richness. Each line intersection represents a relative value of air/fuel ratio.

The control unit can, however, modify the pulse width value represented by each line intersection in response to the following input information:

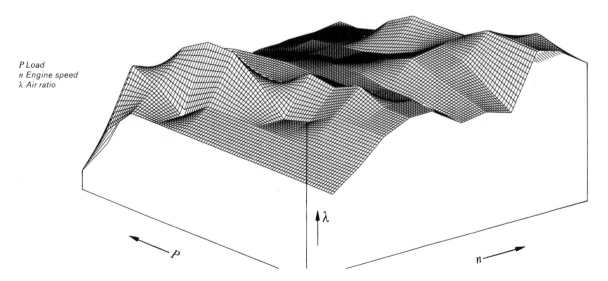

P Load
n Engine speed
λ Air ratio

FIGURE 12–10   Pulse width control map. *Courtesy of Robert Bosch.*

1 Electric fuel pump
2 Fuel filter
3 Fuel distributor tube
4 Injection valve
5 Cold-start valve
6 Pressure regulator
7 Pulsation damper

FIGURE 12–11 Fuel supply system. *Courtesy of Robert Bosch.*

- engine temperature.
- throttle position. Beyond the influences of engine temperature and throttle position already discussed under **Operating Modes,** the control unit adjusts air/fuel mixture (pulse width) at warm WOT in an attempt to avoid detonation.
- air density. The airflow sensor input needs to be corrected at higher elevations. This correction is made with input from the air temperature sensor plus, on some applications, the altitude sensor.
- battery voltage. System voltage affects the injector's response time (how fast it opens) and therefore how fuel actually flows during a pulse width. A correction factor is programmed into the control unit to modify the pulse width accordingly.

Engine Speed Limit. The control unit has a programmed engine RPM limit for each engine

application. If engine RPM exceeds that limit by more than 80 RPM, the control unit will stop fuel injection until engine speed is 80 RPM below the limit. This prevents engine damage resulting from overrevving and avoids dumping excessive HC into the exhaust system as occurs if engine speed is controlled by stopping spark delivery.

Fuel Pump Relay

The fuel pump is controlled on the Motronic system like it is on most other electronic fuel injection systems. The pump is powered by an external relay that has power available to it when the ignition is turned on. The control unit grounds the pump relay's control coil, and the pump is powered. If the engine is not turning at a criterion speed, the control unit will turn off the relay. The fuel pump is mounted inside the fuel tank, Figure 12–11.

Fuel Pressure Regulator. The fuel pressure regulator works like those on other electronic

Motronic ignition subsystem.

1 Ignition switch	5 Plug connectors
2 Ignition coil	6 Spark-plugs
3 High-tension	7 Control unit
distributor	8 Battery
4 Spark-plug leads	

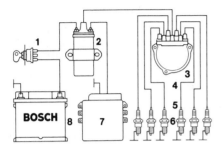

FIGURE 12-12   Ignition circuit. *Courtesy of Robert Bosch.*

fuel injection systems. It is connected to manifold vacuum so that pressure differential across the injectors is constant Figure 12–11.

Fuel Pressure Pulsation Damper. The pulsation damper, Figure 12–11, eliminates pulsation noise. It is designed much like the pressure regulator; however, it is not connected to manifold vacuum and acts as an accumulator to absorb the pressure pulses produced by the opening and closing of the injectors. It is in the fuel return line and downstream from the pressure regulator.

Ignition Timing

The control unit controls ignition by providing ground side switching for the ignition primary winding, Figure 12–12. Like pulse width, ignition timing is based on engine load and speed. Figure 12–13 presents a three-

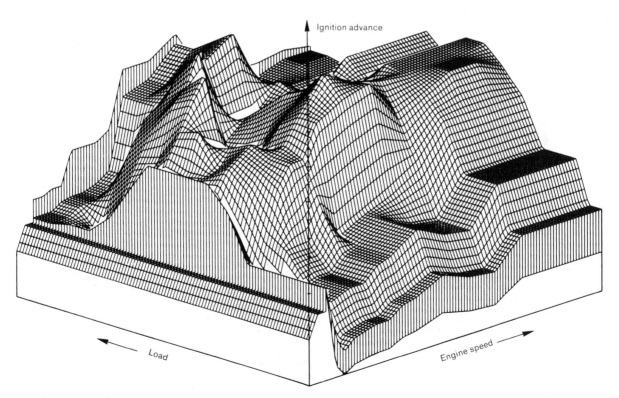

FIGURE 12-13   Ignition timing map. *Courtesy of Robert Bosch.*

dimensional map of the spark advance programming for a given engine application. The arrows point in the direction of increased load, speed, and advance. The line intersections represent the spark advance for all combinations of engine speed and load. These values are stored as the look-up table in the read-only memory of the control unit. Input data is reviewed and spark timing is recalculated between each cylinder firing. The control unit can modify the values on the map in response to inputs from:

- coolant temperature sensor.
- intake air temperature sensor.
- throttle position switch. During WOT operation timing is calculated for maximum torque with additional consideration for detonation only.
- altitude sensor (if used).

Temperature of either the coolant or incoming air most readily causes the control unit to modify the spark advance map. The modification depends on the operating condition. During cranking timing depends on cranking speed and temperature. At low cranking speeds and low temperature, timing is near TDC. At higher cranking speeds and with a warm engine, timing is still near TDC; but with high cranking speed and low temperature, timing is advanced. During a cold post-start period, timing is advanced significantly for a short time to partially reduce the need for fuel enrichment. Once the engine is warm enough to reduce the likelihood of engine stalling, timing is retarded (on some applications), at the cost of optimum torque, to hasten engine warm-up.

During warm idle, ignition timing is used to control idle speed. With no additional load, transmission in neutral, air-conditioning off, and so forth, the timing can be retarded slightly and still maintain a desired idle speed. This keeps combustion chamber surfaces warmer to reduce HC production. If a load is applied such as the transmission being pulled into gear, the idle speed will start to drop. The con-

trol unit will respond by advancing ignition timing enough to obtain more torque and bring idle speed back to the desired value. The throttle opening need not be increased, and greater fuel economy is realized. On applications with the rotary idle adjuster, idle air is increased if spark timing manipulation is not enough to maintain desired idle speed.

During WOT operation normal or higher engine and/or incoming air temperature cause the control unit to reduce spark timing at those load ranges where spark knock has the greatest probability of occurring. If a specified value of load increase occurs (acceleration), short of WOT, the control unit will reduce the existing spark advance and will then gradually begin to restore it. This tends to reduce spark knock and NO_x production.

Spark Knock Control

On earlier systems the spark knock control system was remote from the Motronic unit. On later applications it is incorporated into Motronic's control unit. Its application allows a more aggressive spark advance schedule. When the knock sensor produces a signal that the knock control module or control unit recognizes as spark knock, it reduces spark advance until the knock disappears. It then gradually readvances the timing to its original value or until the knock recurs.

Dwell Control

The control unit controls dwell in a manner similar to that of General Motors' HEI module. As engine speed increases, dwell angle is increased. In this case, however, the amount of increase is also affected by battery voltage.

Peak Coil-Current Cut-off

Allowing the ignition to remain in the on position with the engine not running can cause both the ignition coil and the coil-driving transistor in the control unit to overheat. To avoid this the control automatically turns the ignition primary circuit off anytime engine RPM drops below a programmed speed of about 30 RPM.

Idle Speed

Two different methods are used to control idle speed.

Auxiliary Air Device. On some applications warm idle speed is controlled by an adjustable screw in the idle air bypass passage, Figure 12–14. On a cold start and during warmup, idle speed is increased by an auxiliary air valve, Figure 12–14. A bimetal strip inside the valve holds the valve open at low temperature, Figure 12–15. When the valve is open, additional air is allowed to bypass the throttle blade. This air has, however, passed through the airflow sensor so that the appropriate amount of fuel is injected to mix with it. When the ignition is turned on, voltage is applied to a heat element wrapped around the bimetal strip. As the strip warms up, it gradually closes the valve

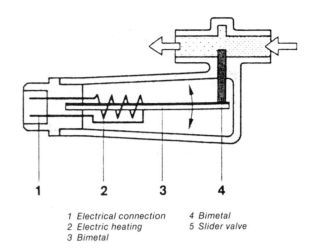

1 Electrical connection   4 Bimetal
2 Electric heating   5 Slider valve
3 Bimetal

FIGURE 12–15   Auxiliary air device schematic. *Courtesy of Robert Bosch.*

and blocks the flow of the auxiliary air. When the engine is warmed up, its heat is enough to keep the valve in the closed position.

Rotary Idle Adjuster. Other applications use the rotary idle adjuster, which is controlled by the computer, Figure 12–16. It maintains idle

FIGURE 12–14   Auxiliary air device. *Courtesy of Robert Bosch.*

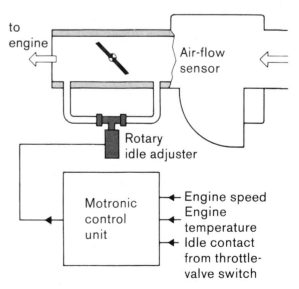

FIGURE 12–16   Rotary idle adjuster circuit. *Courtesy of Robert Bosch.*

speed on a cold or warm engine. It controls air through a throttle bypass passage by rotating a spool valve attached to the end of the armature shaft, Figure 12–17. The motor has a permanent magnet field and two armature windings. By controlling the voltage applied to the two windings, the control unit can rotate the armature and therefore position the spool valve to control idle air. The control unit has a programmed idle speed that it wants to maintain. It first uses ignition timing as a means to adjust idle speed; if that is not enough, it will use the rotary idle adjuster to bring idle speed to the desired value.

Turbo Boost Control

To control spark knock on turbocharged engine applications, Motronic uses a combination of spark retard and boost pressure reduction. Bosch feels that just reducing boost pressure puts unnecessary limits on engine performance. They also feel that just using spark retard to control spark knock is a poor choice because retarding ignition timing raises exhaust temperature. This rise in temperature can be threatening to the already high-temperature exhaust–driven turbine. Using both measures proves quite effective. When spark knock is detected, timing is retarded and boost pressure is reduced. Spark retardation is immediately effective to stop spark knock, but the drop in boost pressure takes a little longer (a few seconds because the air in the intake manifold must cool slightly) to take effect. As the drop in boost pressure takes effect, spark timing is moved back to its optimum value. Boost pressure is controlled with a wastegate control diaphragm similar to that of other turbocharged systems.

EGR Valve Control

Most European-built vehicles do not use an EGR valve to control NO_x. They rely instead on valve overlay to provide an EGR function and on the reducing function of the three-way

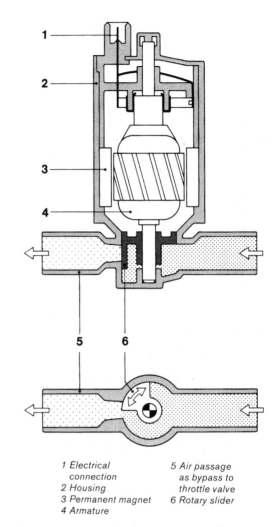

1 Electrical	5 Air passage
connection	as bypass to
2 Housing	throttle valve
3 Permanent magnet	6 Rotary slider
4 Armature	

FIGURE 12–17 Rotary idle adjuster motor. *Courtesy of Robert Bosch.*

catalytic converter. Predictions are, however, that EGR valves will become increasingly used on European-built vehicles as well. The Motronic control unit is capable of providing that function similar to the way it is done on most U.S.-built systems.

Evaporative Emissions Control

Motronic-equipped vehicles sold in the United States or Japan feature an evaporative emission control system. Vapors from the fuel tank are stored in a charcoal canister, Figure 12–18. When the system is in closed loop, the control unit activates a solenoid, which opens a valve in the purge line that connects the canister to manifold vacuum.

Electronic Transmission Control

By adding sensing devices to provide to the control unit information concerning transmission output speed (a vehicle speed sensor), gear selector position, transmission gear position, and transmission kickdown, the control unit can control the transmission's hydraulic pressure and shift points (solenoid valves replace traditional shift valves), Figure 12–19. Shift point look-up charts (one for economy, one for performance, and one for manual shifting) stored in the control unit's read-only memory provide optimum transmission performance

according to the programming (look-up chart) selected by the operator.

Smooth shifts with reduced clutch slippage and wear are achieved, especially during WOT operation, by momentarily retarding ignition timing to reduce torque output during the shift. Motronic transmission control improves fuel economy, shift quality, transmission torque capacity, and expected transmission life.

Start-Stop Control

Another Motronic feature available for manual transmission applications conserves fuel by automatically stopping the engine during idle periods and restarting it when the throttle is depressed. It requires the use of a vehicle speed sensor, a clutch pedal position–sensing switch, and a separate module.

When vehicle speed falls below about 1 MPH with the clutch pedal depressed, the mod-

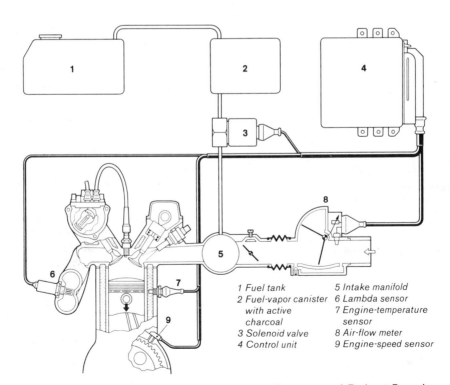

1 *Fuel tank*	5 *Intake manifold*
2 *Fuel-vapor canister*	6 *Lambda sensor*
with active	7 *Engine-temperature*
charcoal	*sensor*
3 *Solenoid valve*	8 *Air-flow meter*
4 *Control unit*	9 *Engine-speed sensor*

FIGURE 12–18   Evaporative emissions control system. *Courtesy of Robert Bosch.*

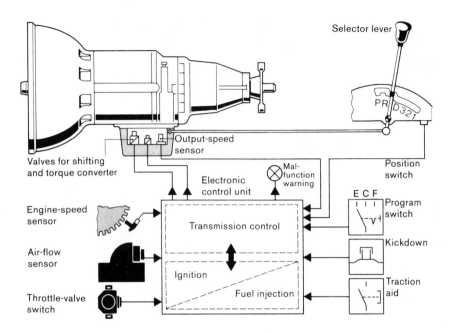

FIGURE 12–19   Electronic transmission control. *Courtesy of Robert Bosch.*

ule signals the control unit to stop fuel injection. When the driver depresses the accelerator pedal with the clutch pedal still depressed, the module activates the cranking circuit and fuel injection is restored. This function does not occur until the engine is warm.

SYSTEM DIAGNOSIS AND SERVICE

The vehicle manufacturer establishes the diagnostic procedure for each vehicle model; therefore, diagnostic procedures can vary from one make of vehicle to another. The diagnostic procedure presented here is selected as a representative sample of what should be found in the service manual.

Troubleshooting Guide

The troubleshooting guide consists of a list of driveability complaints such as: Will Not

Variable Displacement System

Another feature that Bosch has considered for the Motronic system is a variable displacement system that cancels cylinders during periods of light load and is similar to Cadillac's modulated displacement system. Plans for this system include:

- deactivating cylinders by shutting off the injectors to selected cylinders.
- having the intake and exhaust valves of the deactivated cylinders continue to function.
- providing intake and exhaust manifold valving that allows the exhaust from the working cylinders to be passed through the deactivated cylinders and thus keep them up to normal operating temperature.

Start Cold, Erratic Idle During Warm-up, Backfires, and so forth. Each complaint is accompanied by a list of possible causes. The possible causes should be checked out in the order of the easiest to check or in the order of the most likely cause and based on the technician's experience with the problem and knowledge of the specific vehicle. It is important to verify **that all of the components or systems identified in the Possible Cause list for the identified complaint that are not a part of the Motronic system are working properly before proceeding to test procedures of Motronic system components.**

Motronic Test Section

The service manual is likely to have a section that contains test procedures for each component of the Motronic system. The tests in this section should not be conducted until the engine is at normal operating temperature. In testing components and their circuits, remember that connections cause more problems than components.

REVIEW QUESTIONS _____

1. Name seven operating conditions that cause Motronic to select a specific control strategy for fuel and/or ignition timing control.

2. How does the Motronic system protect its control unit against reversed polarity?

3. What is the function of the airflow sensor?

4. What are the sources of engine speed and crankshaft position input?

5. List seven control unit-controlled functions that are standard features of Motronic.

6. Name at least six optional control unit-controlled features.

7. What does the term *lambda* mean as used in automotive application?

8. Name at least two functions affected by battery voltage.

9. What two measures are used to control spark knock on turbocharged applications?

10. What does the troubleshooting guide do?

11. What does the test section do?

ASE-type Questions. (Actual ASE test questions will probably not be as product specific.)

12. Technician A says that Motronic's pulse width during open-loop operation is determined based on data concerning engine speed and throttle position. Technician B says that Motronic's pulse width determination in open loop is based on engine speed and load. Who is correct?
    a. A only
    b. B only
    c. both A and B
    d. neither A nor B

13. Technician A says that Motronic determines engine load from engine speed and throttle position. Technician B says that Motronic determines engine load from engine speed and airflow. Who is correct?
    a. A only
    b. B only
    c. both A and B
    d. neither A nor B

14. Technician A says that Motronic controls idle speed with ignition timing and idle airflow. Technician B says that technician A is correct but there are two different methods of controlling idle airflow. Who is correct?
    a. A only
    b. B only
    c. both A and B
    d. neither A nor B

15. Technician A says that Motronic limits maximum engine speed by shutting off the injectors if the criterion speed is exceeded. Technician B says that technician A is correct but that some Motronic engine appli-

cations limit engine speed by shutting off the ignition instead of the injectors. Who is correct?

a. A only
b. B only
c. both A and B
d. neither A nor B

GLOSSARY

LAMBDA (λ)   A Greek letter used to indicate how far the actual air/fuel mixture deviates from the ideal air/fuel mixture. Lambda equals actual inducted air quantity divided by the theoretical air required (14.7).

APPENDIX: 1989 AUTOMOBILES

This appendix is included as a means of informing the reader about significant changes in electronic engine control systems and new engines and modifications that were introduced on 1989 models.

GENERAL MOTORS

Chevrolet

The 3.1L V-6 is available in 1989 as an interim option on all GMW- and A-body models. It features many of the changes made on the other engine families for 1989, including elimination of the cold start fuel injector and low restriction air filter.

On the 2.8L V-6, Chevrolet has eliminated the mass air flow sensor, relocated the manifold air temperature sensor into the air cleaner housing, and converted to a speed density engine control system. The fuel mixture is now determined by the ECM from input from the MAP sensor, MAT sensor, coolant temperature sensor, and crank reluctor.

Although more oversquare than the standard 5.7L L98 engine, the LT5 features narrow cylinder heads and duplex roller chains to drive the four overhead camshafts. As a result of this unusual design, engineers were able to mount the alternator and A/C compressor (driven by a single, serpentine belt) at the front of the en-

gine in the space between the valley and the bottom of the plenum chamber. A new Delco Remy-Nippondenso gear reduction starter fills the rear portion of the block valley. To fill the space where the starter would be, twin catalytic converters were incorporated into the tubular exhaust manifolds to gain a faster reaction time from the cats.

To help improve spark delivery and insure proper timing, the LT5 was fitted with a direct coil ignition system. In typical Chevy fashion, ignition signals are generated by a crank-mounted reluctor while spark and injector timing are controlled by an expanded ECM located directly over the battery.

The throttle body used on the LT5 is somewhat unique in that it uses three butterfly valves to control the air movement through 16 individually "tuned" intake runners (8 primary and 8 secondary). The ECM is programmed to control airflow through the secondary intake ports by directing the operation of a computer-controlled vacuum diaphragm. During part-throttle operation the secondaries remain closed and the engine breathes through the primaries. However, when the driver selects the "power mode" (by flipping a console-mounted switch), and nails the throttle, both the secondary air valves and the secondary intake valves, which are actuated by cam lobes with longer duration than the primaries, come into play.

To complete the ZR1's powertrain, Chevy designed an all new manual 6-speed gearbox, which becomes the only manual offered on the Corvette and required with the ZR1. Its main feature, in addition to the extra overdrive gear, is computer-aided gear selection, which inhibits the selection of second or third gears during light throttle driving.

Buick

Buick has made several changes to its 3.0 liter V-6. The new 3300 has been designed to operate within emission specs without the use of an EGR valve.

Also the ignition secondary coil has been altered to provide a hotter spark than in the old 3.0 liter with a resulting increase in spark plug gap from .045 to .060 inch on the 3300. The ECM carries a vehicle speed sensor buffer that is used to correct the air/fuel mix at various speeds and the intake system features an integral "hot wire" MAF sensor similar to that found on Buick 3800 engine models. Other additions to the 3300 include: steel timing chain with new link profile, new spring-loaded chain tensioner for improved chain control, redesigned starter, gerotor oil pump, single serpentine accessory belt, and tubular-welded exhaust manifolds. The 3300 is rated at 160 horsepower at 5200 rpm and 185 foot-pounds of torque at 2000 rpm.

Oldsmobile

The Toronado and Trofeo now feature a color visual information center CRT in the center of the dash, which includes the usual comfort and convenience controls along with an engine oil level display and engine systems monitor.

Changes to the HO quad four 2.3L engine in 1989 include ignition timing advanced 15° during cranking and a switch from simultaneous double-fire fuel injection pulses to an alternating double-firing discharge sequence.

Olds boast 0 to 60 times of 7.5 seconds with this optional engine and claims to be looking at both turbo and supercharging for future quad four applications.

Of note is the fact that the successful quad four will be upgraded in a special high output version in the Cutlass Calais. Available only with 5-speed manual, the HO quad four is rated at 180 horsepower at 6200 rpm and 160 foot-pounds of torque at 5200 rpm. The extra 30 horses are the result of larger valves and a compression ratio boosted from 9.5 :1 to 10:1.

Pontiac

In 1989, Pontiac expands its performance image within GM by offering a turbo-version of the 3.1L V-6 in the Grand Prix, a twentieth anniversary Trans Am with intercooled turbo 3.8L V-6, an optional 2.0L 4-cylinder in the subcompact LeMans and a continuation of the advanced 6000 STE with AWD and four-wheel antilock brakes. The antilock brake option has been expanded to all Grand Prix models and is standard on the McLaren Turbo. It can also be found on the Bonneville.

In an attempt to breath some life back into the ancient 2.5L 4-cylinder engine, subtle performance changes have been induced. The configuration of the intake and exhaust ports and valves was changed to smooth out mixture and exhaust flow and higher lift cam lobes and a larger bore TBI were added for increased performance. Available only in the Grand Am, horsepower in this version of the 2.5L is now rated at 110 at 5200 rpm and torque at 135 foot-pounds at 3200.

Cadillac

For larger models, Cadillac for 1989 still features its digitally fuel-injected 4.5L V-8 engine. The two-year-old Allanté features an upgraded version of the 4.5L that boasts 200 horsepower at 4400 rpm and 265 foot-pounds of torque at 3200 rpm.

Technical twists from Cadillac include an engine oil life indicator that reminds drivers when it is time to change oil and a new Electri-Clear heated windshield, which seems similar to Ford's Insta-Clear windshield. Cadillac's on-board diagnostics are accessed, as in previous years, by simultaneously pressing the "On" and "Warm" buttons on the Climate Control panel. The on-board system can now hold trouble codes not only for constant problems but for intermittent faults as well. Also, the on-board system can also be used for direct readings of sensor voltages and switch positions.

Ford/Lincoln Mercury

The big news from Ford in 1989 is the Probe. Viewed from the outside, the Probe's aerodynamic shape is unmistakably Ford. Structurally speaking, though, Mazda uses the same platform as a base for its 626/MX-6, so the Probe is not exactly unique.

Under the hood, the base model is equipped with Mazda's 2.2L SOHC 1-4 that features three valves per cylinder (two intake, one exhaust) and multiport fuel injection. Projected horsepower and torque ratings for this package are 110 at 4700 rpm and 130 foot-pounds at 3000 rpm respectively. More power is available when the base 2.2L is teamed with the air-to-air intercooled IHI turbocharger that is only available on GT models. With the turbo, there is a 32 percent increase in horsepower (145 at 4300 rpm), and a 46 percent increase in torque (190 foot-pounds at 3500).

The GT version also features a more sophisticated microprocessor-controlled variable assist power R&P system that varies the steering assist based on signals received from the speed and steering angle sensors. As a result of this input, assist is increased in low speeds to provide greater maneuverability and decreased at speed for improved road feel.

Despite its Ford name, the Probe's power train (including the on-board electronics) is all Mazda. For those already accustomed to working on Mazdas, especially the 626/MX-6 combination, servicing the new Probe should not pose a problem.

The big news of 1989 from T-Bird and Cougar centers is a host of changes in their power train. The 5.0L SEFI V-8 is no longer available on any T-Bird/Cougar model and the 2.3L turbocharged 1-4 has been discontinued, leaving only two engine choices: Super Coupe/XR7 models feature a standard supercharged 3.8L SEFI V-6, while all remaining models are propelled by a revised 3.8L SEFI V-6.

The base 3.8L's performance is enhanced by a new sequential fuel injection system that is similar to the SEFI design used on the corporate 5.0L V-8. On the inside, the 3.8L used in RWD models loses its counter-rotating balance shaft but gains roller tappets.

The 3.8L supercharged/intercooled V-6 that replaces the 2.3L turbocharged/intercooled 1-4 (Turbo Coupe) and the 5.0L V-8 (XR7) is an impressive piece of work. When teamed with either the standard 5-speed manual or optional 4-speed automatic, the 12 pounds of boost provided by the belt driven blower gives the 3.8 the punch it was sorely missing. After muddling around last year with 140 horsepower and 215 foot-pounds of torque, the new supercharged 3.8L, boasts 210 horsepower at 4000 rpm and 315 foot-pounds of torque at 2600 rpm.

The supercharger itself is a Roots-type positive displacement pump that's manufactured for Ford by Eaton Corp. Mechanically driven by the engine, boost pressure is controlled and regulated by engine speed (to a maximum of 12 psi at 4000 rpm). Besides the supercharger itself, other modifications have been made in this version of the 3.8L. Internally, the block, head, and crank have been strengthened to handle the increased gas loads and the compression ratio has been lowered from 9.0:1 to 8.2:1.

A conventional distributor is not used on the supercharged 3.8L. Instead there is a DIS system that is quite similar, in operation, to the Buick design introduced on the 3.8L V-6 back in 1984.

The DIS system developed for the 3.8L engine employs a combination crankshaft dampener and "chopper wheel" to time spark plug firing. Also, a synchronizer, which resides where the distributor was, is used to reference TDC of cylinder number one so that the spark goes to the right place. Other improvements worth noting on the supercharged 3.8L include the addition of a water-to-oil engine oil cooler and a new hot wire-type mass air flow meter.

At the other end of the price spectrum, the '89 Continental benefits from the same base 3.8L V-6 refinements described above. Additionally, federal models now have a EEC-IV malfunction warning light.

The other big news from Ford concerns the Taurus, which until now was hardly considered a high performance FWD sports car. The Taurus SHO series—SHO stands for super high output, and it is no exaggeration—is powered by an all-new 3.0L V-6 engine with dual overhead cams, four valves per cylinder, SEFI and a dual exhaust coupled with a 5-speed manual transaxle. With an estimated output of 220 horsepower and a projected 0-to-60 time of 7 seconds, the Taurus SHO qualifies as one of the fastest FWD "family" sedans ever produced. From a functional perspective, the Taurus SHO is equipped with four-wheel disc brakes, P215/65VR15 performance tires and a handling suspension system that features increased diameter stabilizer bars, re-valved struts, and increased rate bushings.

The new 60° 3.0L V-6 engine manufactured exclusively for the Taurus SHO by Yamaha redlines at 7300 rpm and is protected by a rev limiter that cuts out cylinders to hold a steady 7300. Incidentally, the rev limiter is not designed to protect the engine. It is there to prevent an over-revving of the accessory drive components. The engine itself has been tested to speeds greater than 8500 rpm.

To solve the problem of torque steer, numerous adjustments had to be made in the drive line, including the addition of an intermediate bearing support and equal-length half-shafts. Additionally, refinements in the suspension, engine mounts, springs, sway bars and tire compound were combined to virtually eliminate any trace of torque steer. Although the engine is built entirely in Japan, Ford supplied the DIS used on the SHO 3.0L and gave the engine total EEC-IV capabilities.

The SHO 3.0 features a unique intake system with two runners per cylinder—a long runner for low speed and a short runner for high speed. Each of the high-speed runners, or "secondaries," has an ECA-controlled, vacuum-operated throttling valve that remains closed until the engine reaches about 4000 rpm, at which point all valves simultaneously open to increase high-speed airflow. With this set-up, the engine is not overwhelmed with air at low speeds.

Ford is also introducing an all-new electronically controlled four-speed automatic overdrive transmission (E4OD) for selected light truck models with gross vehicle weight ratings over 8,500 pounds. Its electronic shift and converter clutch controls are integrated into Ford's fourth-generation electronic engine control system's (EEC-IV) on-board computer for optimized fuel economy, smoother shifting, and more accurate service diagnostics. Major E4OD features are a torque converter clutch and a 0.71:1 overdrive gear.

CHRYSLER

Single-Board Engine Computer

A new single-board engine computer appearing on New Yorker, Dynasty, Spirit, and

Acclaim models equipped with Chrysler's 3.0 liter V-6 replaces the single-module (twin board) electronic controller that was introduced in 1988. This new single-board unit is still mounted in the engine compartment to simplify wiring connections.

Chrysler has also expanded its vehicles' on-board diagnostics and enhanced the capabilities of its hand-held diagnostic tool, the Diagnostic Readout Box II (DRBII).

The DRBII is a cartridge-based scan tool incorporating a built-in "snap-shot" feature. With an updated 1989 cartridge, it will access the diagnostic programs of all on-board computer systems, including electronic height-controlled body computer, overhead console with automatic compass calibration and digital temperature display, multiplexed electronic clusters, and the new electronic automatic transaxle.

An updated cartridge/prom will allow engine electronics diagnosis with older aftermarket scan tools, but only the DRBII will handle the body computer diagnostics. Additional cartridges for the DRBII cover Jeep/Eagle and Mitsubishi vehicles.

Fully-Adaptive Electronic Automatic Transaxle

In 1989 Chrysler introduces the industry's first fully-adaptive, electronically controlled, 4-speed automatic transaxle. Code named A604, the new transmission will be teamed with the company's 3.0 liter V-6 and be offered on select 1989 models.

Among the major features that sets the A604 apart from similarly proclaimed designs are its fully adaptive electronics. The adaptive control is designed to compensate for changes in the engine or friction element torque and thereby optimize shift quality for the life of the transmission.

When a shift occurs, the A604 is programmed to sense input and output speed changes (140 times/second) within the gear train and adjust hydraulic pressure as needed. This is in contrast with traditional transaxle operation, where shifts are performed by applying hydraulic pressure through metered orifices and mechanical accumulators based on a predetermined set of assumptions about engine output and friction material characteristics.

The A604 uses clutches to change ratios so mechanical bands are not needed. Actuation and release of the clutches are controlled by ball-type solenoid valves that are designed to operate without the need for any type of intermediate element.

Further simplification within the A604 is achieved through a unique logic-controlled switching valve that permits one solenoid to control the application of two friction elements. For example, any selection of the second, third, or fourth gear elements will cause the valve to release low gear and route the solenoid valve output for lock-up control. Conversely, when a shift to low gear is appropriate, the logic-controlled switching valve follows a specific sequence of solenoid commands to shift the valve and return to low gear.

Acclaim

The Mitsubishi-built 3.0L V-6 teamed with the new electronic four-speed automatic is standard on the Acclaim LX and optional on the Spirit ES. Thanks to a redesigned intake manifold, and less restrictive exhaust, this year's V-6 gains a modest power increase over last year's model. With the aforementioned modifications, horsepower is up 5 percent (141 at 5000) and torque output increases an average of 3 percent (171 foot-pounds at 3600).

Eagle Medallion

The Medallion, like everything in the former AMC line-up, carries that company's off-board electronic engine diagnostics. The diagnostic link (located on the firewall next to the coil) is useful only with a dedicated tool (Renault MS-1700 tester), and the system does not maintain fault codes in memory. For now, the only engine offered on the Medallion is a 2.2L. It features Bendix multipoint fuel injection and an unusual-looking distributor cap—tower terminals all elbow to the left rather than outward. Timing, rpm, and FI reference are picked up at a flywheel/reluctor instead of a customary distributor-mounted pickup.

INDEX